T0276697

IET SECURITY SERIES 03

Mobile Biometrics

IET Book Series on Advances in Biometrics – Call for authors –
Book Series Editor: Michael C. Fairhurst, University of Kent, UK

This book series provides the foundation on which to build a valuable library of reference volumes on the topic of Biometrics. *Iris and Periocular Biometric Recognition, Mobile Biometrics, User-centric Privacy and Security in Biometrics*, and *Hand-based Biometric Methods and Technologies* are the first volumes in preparation, with further titles currently being commissioned. Proposals for coherently-integrated, multi-author edited contributions are welcome for consideration. Please email your proposal to the Book Series Editor, Professor Michael C. Fairhurst, at: m.c.fairhurst@kent.ac.uk, or to the IET at: author_support@theiet.org.

Forthcoming titles in this series include:

Iris and Periocular Biometric Recognition (Christian Rathgeb and Christoph Busch, Eds.): Iris recognition is already widely deployed in large-scale applications, achieving impressive performance. More recently, periocular recognition has been used to augment biometric performance of iris in unconstrained environments where only the ocular region is present in the image. This book addresses the state-of-the-art in this important emerging area.

Mobile Biometrics (Guodong Guo and Harry Wechsler, Eds.): Mobile biometrics aim to achieve conventional functionality and robustness while also supporting portability and mobility, bringing greater convenience and opportunity for deployment in a wide range of operational environments. However, achieving these aims brings new challenges, stimulating a new body of research in recent years, and this is the focus of this timely book.

User-centric Privacy and Security in Biometrics (Claus Vielhauer, Ed.): The rapid emergence of reliable biometric technologies has brought a new dimension to this area of research, allowing the development of new approaches to embedding security into systems and processes, and providing opportunities for integrating new elements into an overall typical security chain. This book provides a comprehensive overview of leading edge research in the area.

Hand-based Biometric Methods and Technologies (Martin Drahanský, Ed.): This book provides a unique integrated analysis of current issues related to a wide range of hand phenomena relevant to biometrics. Generally treated separately, this book brings together the latest insights into 2D/3D hand shape, fingerprints, palmprints, and vein patterns, offering a new perspective on these important biometric modalities.

Mobile Biometrics

Edited by
Guodong Guo and Harry Wechsler

The Institution of Engineering and Technology

Published by The Institution of Engineering and Technology, London, United Kingdom

The Institution of Engineering and Technology is registered as a Charity in England & Wales (no. 211014) and Scotland (no. SC038698).

The Institution of Engineering and Technology
Michael Faraday House
Six Hills Way, Stevenage
Herts SG1 2AY, United Kingdom

www.theiet.org

British Library Cataloguing in Publication Data
A catalogue record for this product is available from the British Library

ISBN 978-1-78561-095-0 (hardback)
ISBN 978-1-78561-096-7 (PDF)

Typeset in India by MPS Limited

Contents

Preface

This book is the second to appear in the newly established "IET Advances in Biometrics" book series. In fact, the history of this series dates back to the publication by the IET in late 2013 of "Age factors in biometric processing", which provided the impetus and set the pattern for an on-going series of books, each of which focuses on a key topic in biometrics. Each will bring together different perspectives and state-of-the-art thinking in its topic area, shedding light on academic research, industrial practice, societal concerns and so on and providing new insights to illuminate and integrate both specific and broader issues of relevance and importance.

Recent years have seen rapid expansion in the availability and power of computing and communication platforms which are convenient and user-friendly, of widespread applicability and, especially, *portable*. This has been particularly rapid in the large-scale consumer sector, creating a corresponding demand for security measures (principally concerned with personal identity) adapted to this target market, while also opening up opportunities for new applications and diversification of environments for operational deployment. The emergence of mobile biometrics has consequently accelerated, stimulating innovative technologies and techniques in the area of personal authentication and identification. While mobile biometric systems need to retain the fundamental characteristics associated with all such system implementations, there are also additional issues to address. For example, constraints in relation to power consumption, size, physical resilience, memory constraints and so on can often assume a new significance in the mobile environment. In turn, this can have implications for issues such as algorithm design and implementation, hardware design, developing standards and questions about reliability. The book demonstrates well many of these issues and, especially, deals with topics which have special relevance in the mobile world, including the range of data sources available, the importance of sensor technologies and the applicability of mobile systems across many different biometric modalities.

The contributors come from a variety of backgrounds, and the volume overall represents an integration of views from across the spectrum of stakeholders, including, of course, academia and industry. We hope that the reader will find this a stimulating and informative approach, and that this book will take its place in the emerging series as a valuable and important resource which will support the development of influential work in this area for some time to come.

Other books in the series are in production, and we look forward to adding regularly new titles to inform and guide the biometrics community as we continue to grapple with fundamental technical issues and continue to support the transfer of the best ideas from the research laboratory to practical application. It is hoped that this book series will prove to be an on-going primary reference source for researchers, for system users, for students and for anyone who has an interest in the fascinating world of biometrics where innovation is able to shine a light on topics where new work can promote better understanding and stimulate practical improvements. To achieve real progress in any field requires that we understand where we have come from, where we are now and where we are heading. This is exactly what this book and, indeed, all the volumes in this series aim to provide.

Michael C. Fairhurst
Series Editor, Advances in Biometrics Book Series

Chapter 1

Mobile biometrics

Guodong Guo[1] and Harry Wechsler[2]

1.1 Introduction

Biometrics is about using   physical and/or behavioral characteristics for personal authentication or identification, which is now a well-established discipline supporting many practical applications. Stimulated by hardware advances and a rapidly growing consumer market for increasingly powerful smart phones and similar portable devices; *mobile biometrics* has become an important trend for biometric applications. It is predicted that the global market for mobile biometrics will grow substantially in the coming years.

Mobile biometrics aim to achieve conventional and robust functionality, while also supporting portability and mobility. This is to bring greater convenience and opportunity for their deployment in a wide range of operational environments ranging from consumer applications to law enforcement and beyond. However, achieving those aims brings new challenges, such as power consumption, algorithm complexity, device memory limitations, frequent changes in operational environment, security, durability, reliability, connectivity, and so on. This has led to a significant new body of research in the past few years including a new community of researchers and other stakeholders, making this edited book on mobile biometrics a timely and important endeavor.

This book aims to bring together high quality research, while addressing the new challenges of mobile biometrics. It provides a platform both for academic researchers and industry partners to present their latest new results; stimulate discussions on current and future challenges in mobile biometrics; and report, review, and evaluate emerging solutions that will advance the state-of-the-art in this area. The book, the first substantial such survey of its kind, is expected to set benchmarks for further development in this important new field, and guide future research and new applications. The overall key feature of this book is that it brings together academic and industrial researchers to provide an integrated and informative analysis of current and future aspects of mobile biometrics.

The book is organized into six parts that cover mobile biometrics along different dimensions as detailed next.

[1] Department of Computer Science and Electrical Engineering, West Virginia University, USA
[2] Department of Computer Science, George Mason University, USA

1.2   Book organization

There are six parts in this book, with each one containing from two to four chapters that discuss some specific topics. To help readers get quick information on contents and topics, some brief overviews of the chapters are provided here.

Part 1 is about the history and challenges in developing mobile biometrics. It includes two chapters. Both chapters are written by industrial experts, who have a long-time experience in developing mobile sensors and devices. Their great experience includes rich details that may be lacking from traditional academic research publications.

Chapter 2 reviews the history of developing mobile sensors and devices, mainly focusing on fingerprint sensors. The review also covers the mobile sensors for face and iris. The mobile devices, developed by different vendors, are elaborated in this chapter. The issues, related to the development of mobile sensors and devices, address the mix of constraints such as cost, size, speed, and reliability. Finally, it presents the challenges to be encountered in development of future mobile devices.

Chapter 3 depicts more recent technological developments made to overcome particular challenges for mobile biometric sensor development, while also presenting the fundamental challenges that need to be addressed to make the mobile sensors easy to use as a credit card. The chapter focuses on the mass market for mobile biometric sensors to date, by navigating through the technology development cycle and technological approaches involving MEMS, thin films, photonics, and capitalization of the Moore's Law.

Part 2 is on mobile face biometrics, including three chapters. In traditional biometrics, face recognition is still an active research topic. In mobile biometrics, the face modality can be extended to utilize other useful and convenient cues, such as the speech signals and lip movement.

Chapter 4 presents an approach to audio–visual person recognition using mobile phone data. It has a review of speaker and face recognition systems that are developed and evaluated on a mobile biometric database. Then, it develops a mobile person recognition system using deep neural networks based on the Boltzmann Machine.

Chapter 5 describes an approach using human describable attributes for continuous authentication on mobile devices, which address both efficiency and accuracy for continuous authentication. Human attributes, such as gender, ethnicity, and age, have been shown useful for face verification, which are known to be robust for the same subject notwithstanding pose, illumination, and expression variability.

Chapter 6 explores a mobile identification system that can offer different levels of access permission for different subjects. During the use of smart phones, the lips of the person can be utilized for authentication without requesting cooperation of the subject. A Hidden-Markov-Model-based identification is developed for user authentication using lip based features.

Part 3 is about mobile behavioral and gait biometrics, consisting of four chapters. The behavioral traits, quite unique to mobile biometrics, are usually not available in traditional biometrics. The corresponding sensors, specially made for smart phones or smart watches, include embedded accelerometers and gyroscopes.

Chapter 7 shows that a smart phone user's "usage data," or the data gathered from the use of the mobile device can be utilized for verification. The usage data can be considered as behavioral traits, such as application usage or WiFi connectivity, which can be used to build a behavioral profile of the user. The behavioral profiling allows for continuous, unobtrusive identity verification based purely on behavioral measures. The chapter also provides a broad overview of the advancements made on mobile platforms with regards to behavioral biometrics via usage data.

Chapter 8 focuses on continuous user authentication during natural user-phone interaction. Two types of interaction are considered, one is touch gesture based and the other is key stroke based. For continuous authentication scenarios, it might not be appropriate to do explicit authentication. Rather, a desirable approach is to implicitly and continuously perform user identification in the background without disrupting the natural user phone interaction.

Chapter 9 introduces a smartwatch-based gait biometric system. It presents all—encompassing details ranging from hardware to how the ultimate system's fitness should be evaluated. Compared to the traditional smartphone-based systems, smart-watches have some attractive properties, for example, smartwatch screens do not orient themselves as they rotate, and a smartwatch usually has consistent orientation every time it is put on. These properties may help address the orientation and off-body problems in smartphones, without unreasonable constraints on the user.

Chapter 10 studies nonobtrusive gait biometrics by directly capturing a user's characteristic locomotion by using the embedded accelerometers and gyroscopes. A novel gait representation is proposed for accelerometer and gyroscope data, which is both sensor-orientation-invariant and highly discriminative to enable high-performance gait biometrics for real-world applications. It also addresses the issue of variability in walking speed.

Part 4 presents mobile biometrics using fingerprint and palmprint in two chapters. In traditional biometrics, there are a number of publications on fingerprint recognition and palmprint matching. Typically, the touch sensors and well-designed devices are used for fingerprint and palmprint image acquisition, whereas in mobile biometrics, the preference may be given to use the built in cameras for touchless image acquisition. The computation and storage may be limited as well in a mobile device. Thus, new challenges arise in developing mobile fingerprint and palmprint recognition systems.

Chapter 11 introduces a mobile fingerprint system, which utilizes the built-in rear camera of a smartphone to capture a photo of four fingers. This touchless fingerprint recognition system requires no extra hardware, making it different from many smart phones that use a touch sensor for fingerprint image acquisition. New challenges in developing the touchless mobile fingerprint authentication system arise and are addressed in this chapter.

Chapter 12, about palmprint recognition on mobile devices, considers issues related to complex background, large illumination variations, and uncontrolled hand locations. The chapter describes a mobile palmprint recognition system, while considering how to address the limitation of computation complexity and storage requirement in a mobile device.

Part 5 addresses attacks and spoofing in developing mobile biometrics, consisting of three chapters.

Chapter 13 provides a summary of attacks and available counter measures on the periocular biometrics with smartphones. Periocular biometrics can be used for both low security authentication, e.g. unlocking a phone, and high secure access like mobile banking. In either case, the periocular biometric systems can be attacked at the capture stage using artifact presentation (or spoofing) such as printed artifacts or electronic screen presentations.

Chapter 14 is about face antispoofing that can prevent spoofing attacks to enhance the security level of a face recognition system. A method is introduced based on exploiting the structure and texture information from a rotated face image sequence. Facial structure is salient and device-independent to distinguish a genuine subject from a spoofing photo. The combination of structure and texture cues can further improve face antispoofing performance in various environments.

Chapter 15 describes various spoofing and antispoofing techniques on mobile devices with their pros and cons discussed. It covers three modalities, i.e. fingerprint, face and iris, and presents a multibiometric antispoofing prototype, which is trait-independent and can be used for either one of face, iris or fingerprint spoofing attack detection. This approach is one important step toward the development of generalized mobile antispoofing techniques.

Part 6 brings up some additional topics related to mobile biometrics, such as biometric standards, cloud computing, and big data. Two chapters are included for this part.

Chapter 16 introduces the Biometric Open Protocol Standard (BOPS), which provides identity assertion, role gathering, multilevel access control, assurance, and auditing. The BOPS includes software running on a client device (Android, iPhone, etc.), a trusted BOPS Server, and an intrusion detection system. Standards are useful as they promote interoperability. The open standard BOPS allows maximum interoperability for mobile biometrics applications.

Chapter 17 argues that the mobile authentication plays a key role in supporting seamless transition from user owned devices carrying out their daily transactions, to secure enterprise transactions. To maximize the value of big data, users are demanding near realtime responses, more secure authentications, and prompt transactions. Mobile biometrics techniques and its solutions and architecture must embrace the change and meet the demand. The chapter also introduces cloud-based mobile biometrics identity services reference architecture and identifies emerging technologies, solutions, and standards to the architectural enablement.

In total, there are six parts in this mobile biometrics book, including sixteen specific chapters. To help readers get a quick and clear view of the organization of the book, we draw a picture to show the six parts visually, with the corresponding chapter numbers, as indicated in Figure 1.1.

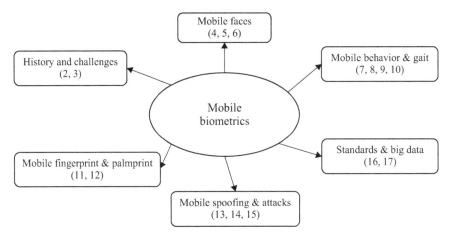

Figure 1.1 Organization for the mobile biometrics book. The chapter numbers are given for each part within a parenthesis

1.3 Acknowledgment

We express our great appreciation to all the authors who dedicated much effort and were enthusiastic to contribute their chapters to this book. It would not been possible to have this book without their contributions. We are thankful to Mike Fairhurst, who encouraged and supported this book throughout. We also appreciate the help from the IET publisher. Finally, we want to thank our families for their continuous support to bring to fruition this project.

We hope that this book can inspire future research to address both existing and emerging challenges in mobile biometrics. As everyone expects, mobile biometric products will dominate the biometric market for years to come.

Chapter 2

Mobile biometric device design: history and challenges

Peter Z. Lo[1] and Michael Rathwell[1]

2.1 Introduction

Beginning a decade ago, mobile computing devices started to become powerful enough to be honestly robust key computing devices. With the advent of viable mobile devices coupled with biometric sensors suitable for mobile use, mobile biometric processing became practical. These practical devices, coupled with the early adopter mindset that law enforcement and the military have toward biometrics, caused these devices to start to appear in "real-world" production use. Throughout this past decade, many government agencies have actively deployed increasingly capable mobile biometric devices for use worldwide. As users gained experience with them in the field, those using these devices actively request new capabilities, driving rapid advancement of mobile biometric device design.

Even with advances and growing use, until rather recently the word "biometrics" was essentially unknown outside the biometrics industry. It was *rarely* heard on "business" type news media unless the company in question had some movement in stock prices. The term was *never* heard in the popular press or the evening news by the general population. Someone in the biometrics industry trying to explain to friends and acquaintances about the biometrics often found it difficult and would usually be met with either glazed eyes or be regarded as a mad (evil) scientist when the industry was already *decades* old!

All of that started to change, recently, in a *big* way. The initial "general population" awareness of the biometrics industry started in 2012 when Apple (the most widely recognized brand name in the world) bought a little company called AuthenTec. Initially, it was a lot of buzz in the business and investing world. It *crashed* into general visibility in 2013 when the iPhone 5s [1] was released with fingerprint scanners in them. All of a sudden, the term "biometrics" was seen in *every* tech magazine rather than just the ones that might focus on biometrics. People began regularly hearing the term on the evening news and again on talk radio during their morning commute. Biometrics became mainstream and part of our daily lives. Given that it was Apple that made it happen on the "iThings," the impetus for advancement and adoption of *mobile* biometrics in particular was exponentially increased.

[1]MorphoTrak*, LLC, USA
*MorphoTrak was a subsidiary of Morpho/Safran. Morpho is merging with Oberthur Technologies.

As with most disruptive technologies released by Apple, the inclusion of a finger-print scanner on the iPhone was initially panned by reviewers as being unnecessary or even dangerous in some inchoate way. However, in Apple's usual style, they took a previously niche technology and turned it into a practical, general purpose, suc-cessful, *accepted* technology just as they had with the MP3 player, smartphone, and tablet before. In fact, when we look back at Apple's disruptive influence, we can see where the iPhone and, later, the iPad quickly drove consumer mobile computing devices ever more powerful, thus allowing mobile biometrics to equally flourish. At the time of this writing, essentially all "premium" smartphones have a fingerprint sensor of some sort and is included with increasing numbers of mid-range devices. Considering the general consumer usage forecast along with growing adoption by law enforcement and government agencies, a report [2,3] by Biometrics Research Group, Inc. finds that the inclusion of biometrics in mobile devices will generate approximately US$ 9 billion worth of revenue by 2018 for the biometrics industry. It is expected that this inclusion will be not only applied for unlocking mobile devices and security applications but also applied to multifactor (more than one biometric type) authentication services and approval of electronic payments. In fact, with the addition of Apple's iPay on the iPhone 6 using iTouch to authenticate it, biometrics is already widely used to authenticate credit/debit card purchases.

To trace how we got "here" from "there," this chapter will start with a broad overview of biometrics and related concepts, automatic fingerprint recognition and mobile biometric devices. We will review the history of automatic fingerprint recogni-tion from the early days to today, along with how these advancements were necessary for mobile biometric applications to be viable. Once that foundation is laid, we'll look at the history of mobile biometric devices and applications starting with the rather rudimentary early solutions, moving to the successful (but still niche) law enforce-ment applications and finally to the current commercial market. Finally, we'll take a glance at the crystal ball to look at what challenges might yet lay ahead.

2.2   Biometrics

What are biometrics? According to Webster, the definition of biometrics is *the mea-surement and analysis of unique physical or behavioral characteristics (as fingerprint or voice patterns) especially as a means of verifying personal identity*. That definition is … ok … and serves the general population reasonably well. However, inside the biometric industry, it is a bit more specific than that and the term "verify," as used in the dictionary definition, has a very specific meaning especially when compared to the equally important yet very different term "identify."

The biometrics industry goes rather farther than the "official" dictionary defini-tion. We find that biometric traits [4], whether physical or behavioral, can be further expanded to the following nonexhaustive list of examples:

- Physical:
  - Fingerprint, face, iris, retina, DNA, hand and vein geometry, odor, body temperature, pulse, blood pressure, ear, voice, and electrocardiogram.

- Behavioral:
  - Speech, handwriting, typing keystroke pattern, foot/shoe print, and gait.

Even given this two-level subdivision, the separation, in the context of the biometrics industry, is *all* of too blurry, coarsely, and explicitly divided at the same time. The difference between physical and behavioral is blurred, for example, whether a person's walking gait is physical or behavioral, is the gait a person exhibits purely a learned behavior or is it physically imposed due to accident/age/etc.? One could argue that the physical traits could be divided into traditional dimensional traits like fingerprint or iris and chemical traits like odor or DNA. Another subdivision might be the separation between not only hard, consistently quantifiable traits like fingerprints or DNA and soft, repeatable but also *variable* traits like speech, pulse, or gait even while these can exist in both physical and behavioral classifications. As with most things, these classifications depend on the context in which they are being applied.

When thinking about biometrics, using them for recognition has *always* been happening since prehistoric times, albeit not systematically. Both humans and most of the animal kingdom have always used biometric measures to identify other living creatures. Although we do not *naturally* use things like fingerprints or DNA in daily life, we *do* all use things like facial recognition, odor, speech/voice recognition, or even gait to identify other living beings around us. This fits the classic definition of a biometric recognition system that uses physical, chemical, or behavioral traits to recognize someone's or something's identity. Any reader of this book could instantly and uniquely identify someone they see every day in a room full of people. This kind of biometric recognition is not limited to human to human recognition only. Consider, if anyone reading this book was to walk into a room containing, for example, their orange, male, 17 pound cat along with 10 other orange, male, 17 pound cats, their cat would be identified in seconds if not split seconds using biometric traits like fur pattern and markings, gait, or other biometric identifiers. Although *automatic mobile* cat-detection devices probably are not that far along the development cycle or high on the list of priorities, it does provide a good example of just how widely varied and numerous biometric identifiers really are.

Even when we as humans are "biometrically recognizing" someone else, we can see that some of these biometric traits are more unique than the others. There is the maxim that everyone on the planet has a doppelganger. Whether true or not, consider how often in our daily lives, especially with someone we know well but perhaps see infrequently, that we mistake for someone very similar for them until we do positively identify them as being or not being our acquaintance. That process is we humans instinctively using a combination of hard biometrics (such as recognizing a face) and soft biometrics such as gait or odor (perhaps a regularly worn scent) to identify or verify an identity. Put succinctly but using the same technical terms used with systematic biometric identification, we humans do highly accurate *but not perfect* multimodal biometric identifications many times every day.

Leaving computerized solutions aside for the moment, when considering law enforcement as an example, the police have, for hundreds of years, systematically applied this natural human biometric identification to identify people and solve crimes. By using a systematically implemented hard biometric measure as a primary identifier, the investigator may add a less unique soft biometric trait, not to directly identify a person, but perhaps assist a crime investigation to narrow down the suspect list.

As noted earlier in this introduction, there are two primary types of biometric application, verification and identification [4–6], which are two very different forms of biometric matching. Put simply, identification is the unique biometric discovery of a person using a discrete set of descriptive identifiers while verification is the validation of a claimed identity by the direct comparison of a discrete set of known biometric identifiers to the biometric identifiers presented by a person. Again, leaving computers out of it for the moment, a biometric identification could be having a description of a person by height, weight, hair color, complexion, and maybe dress given to you. You would then head into a crowd of unknown people and *identify* the explicit human that is described by those identifiers. In the pattern recognition world, those identifies are often called features [4]. These features are used to verify or identify a person in a biometric recognition system. Conversely, one might have that same set of identifiers handed to them allowing the individual to come to YOU and say "I am Bob" and you verify it by the direct comparison of the known identifiers that you have. Note that the same individual could come up and say "I am Jack" and be verified as Jack. Verification is the 1:1 validation of a *claimed* identity while identification is the discrete 1:N discovery of a *unique* identity. In the biometric recognition system design, the identifiers/features to best represent the intrinsic property of the biometric trait that should be selected and extracted to achieve better accuracy. Some invariant features such as scale, translation, rotation, and time should be extracted if possible. For example, the relationship of minutiae [5] feature points of finger is a good type of feature for fingerprint recognition.

Although not specifically the focus of this chapter, there are a variety of use cases where a "verification" is a wholly suitable biometric function where a thoroughly vetted (to a particular biometric database or system of federated databases) definitively unique identity is not needed. However, there is a significant number of cases, whether for government function or commercial application, where an explicitly unique identity is mandatory for honest control over a given population base. Consider health care, where an absolutely unique identity is a definitive means to keep a person from illicitly acquiring prescription pain medication versus a solely verification based one, allowing a single individual to validly identify but with multiple identities making illicit acquisition or other medical fraud quite possible. Even with the growing general public knowledge of biometrics, it should be noted that the majority of those making general news are purely verification implementations or, at best, a one to few (very small dataset) identification.

Finally, considering that all biometric traits are directly captured from a person, some of these traits can be obtained in a less intrusive manner and may have fewer privacy concerns, whether real or perceived. Face is one of the biometrics considered to be less intrusive and with fewer privacy concerns. Conversely, fingerprint, DNA, and Iris are considered as the most unique and most reliable but significantly more intrusive to capture as they require direct contact to capture. However, fingerprints are *relatively* noninvasive of the direct contact biometrics to capture and have relatively few *real* privacy issues surrounding them. Fingerprints are not only the earliest biometric used systematically in human history but are also most mature and widely used as of today with the fewest environmental impediments to their use. As such, while we'll at least touch on other modalities, the focus of this chapter will be on mobile fingerprint devices.

2.3   Fingerprint recognition and the first AFIS system

The use of fingerprints [7–10] as a definitive biometric identifier has been in use for millennia. The Chinese began using fingerprint and handprints as evidence during burglary investigations in *200 BC*. Moreover, the inked fingerprint has been used as an invariant identifier for any important business transaction since the Han dynasty. The science of systematically using fingerprints for identification began in the middle 1800s with the practice of using fingerprints in criminal investigations beginning in the early twentieth century. Over the course of the twentieth century, fingerprints have become the most mature method of consistently and precisely identifying an individual using an invariant biometric identifier. Since 1924, the FBI holds the central US register of fingerprints collected for typically law enforcement-related business, collecting thousands of fingerprint sets every day.

Long before the advent of Automated Fingerprint Identification System (AFIS) or even computers of any description fingerprints were manually searched based on classification types classified according to the Henry classification method [11–14] developed in 1896. FBI examiners would classify their "search" print and go to storage cabinets holding hundreds or thousands of fingerprints with the same classification type to manually search for the fingerprint. Using this manual method, it would take days to compare fingerprints amongst thousands of others to hopefully find a match, making criminal investigation long and arduous. By the 1960s, the FBI's fingerprint files had grown to millions of fingerprint sets and became unmanageable using the manual search. The FBI realized that automation to compare the fingerprints would have to be the next step.

The FBI asked National Bureau of Standards, which is now NIST (National Institute of Standards and Technology), for the feasibility of automatic fingerprint searches. In 1969, the FBI formally requested bids for the automation pilot system. Two systems were created, one by Rockwell Autonetics and the other by the Cornell Aeronautical Laboratory. Following the evaluation, the FBI assigned Rockwell to build five "high speed fingerprint scanners." In 1974, the first AFIS [11–13] in the world was developed for FBI by Rockwell. Rockwell later became known as a company called Printrak that was eventually acquired by Safran Morpho who is, today, the leading provider of large-scale AFIS in the world.

The developed AFIS contained three main processing components:

- Automatic fingerprint pattern classification [15]
- Automatic minutiae detection [16]
- Automatic minutiae matching [17].

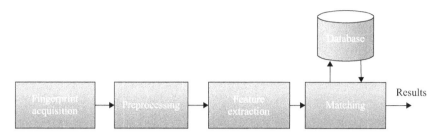

Figure 2.1   AFIS system block diagram

Since then, while AFIS technologies [5,6] have dramatically improved, these core functions, as further depicted in Figure 2.1, have not appreciably changed.

Basically, whether in 1974 or today, an AFIS does the following things in the following order:

1. Fingerprint images are captured in some way and made available to the system.
2. The invariant features such as minutiae, core/delta, classification types, quality, and direction map are extracted from the captured images.
3. The extracted features are searched against the preenrolled database.
4. A response to the search is returned.

Beyond that, there are myriad ways the workflow might be configured to meet various use cases *even in the case of mobile biometrics*. A mere handful of examples include

• The search could be either a 1:1 verification or a full 1:*N* search with or without other biometric modalities or demographics added to the search request.
• Responses might come back to the requesting device and they might be returned elsewhere. Or both.
• Searches could be "lights out" (no human intervention) or have human review for a variety of use case reasons.
• Record enrollment or update may or may not occur based on what the particular use case determines. Use case examples range from search result return only or, if a search transaction is not in the database, the search images might be enrolled as a new record to be verified against at a later time. If the search images are better quality than the enrolled images, they might replace the originally enrolled images to improve subsequent search accuracy.

The improvement of AFIS and biometric recognition technologies in general have both been driven by the need for and resulted in systems with increased ease of use, speed, accuracy, and cost. All of these factors are interrelated, each requiring the others to achieve coincidentally with each of the elements driving the others. All of them culminate in biometric systems that support viable mobile biometric identification solutions.

Early AFIS implementations (as the first "automatic" biometric systems) did their job but to be effectively used, users had to be fingerprint experts, especially in the case of latent fingerprint searching. Although a fingerprint expert may have found them to be quite functional, these early AFIS implementations were not remotely "easy to use." As the efficacy of biometric systems improved, the requirement of being a fingerprint expert lessened to where, in most use cases aside from some latent fingerprints, *any*one can productively submit, search, and receive responses from an AFIS. Improved accuracy both in biometric feature extraction and "lights out" searching automated the fingerprint expert. As the performance improved, both by virtue of better algorithms and more computing power available at lower and lower costs, nonexpert users could quickly and simply capture biometrics and have a highly accurate definitive identity returned to their workstation or other requesting device in seconds without human intervention.

As lower cost of biometric systems came coincidentally with increased accuracy and usability, progressively more agencies were able to both afford and use biometric systems effectively. The rapid proliferation of biometric systems beyond typical government agency, early adopters had biometric systems used pervasively and well beyond only capturing "important" records. With this increased usage came databases with a much greater percentage of potentially captured population in them which provided a higher percentage of identification responses to searches. By the time the law enforcement and government markets were saturated, with biometric systems becoming essentially "technology commodities," the improved accuracy, ease of use, performance, and cost had made them viable for the commercial market as well.

One of the most significant improvements to user friendliness came in the early 1990s with the proliferation of fingerprint live-scan devices; these allowed fingerprints to be quickly and easily captured and sent directly to the AFIS to be searched. Prior to that, the tedious inking and rolling followed by scanning the inked fingerprint card was the only way to get fingerprints into the system. The invention of the live scan also provided the fundamentals for scanners that came to be suitable for mobile biometric acquisition which led further, with the combination of accuracy, fully lights out automation, and essentially immediate response, to the honest viability of mobile biometric solutions. With this demonstrated real capability, came first the initial and then now, increasing demand for mobile biometric devices.

2.4 Mobile biometric devices

What is a mobile biometric device [18–29]? The authors are not even going to *try* to pin that down to a definitive set of specifics. There are too many combinations and permutations of potential mobile biometric function to discretely identify. A better approach is to describe a set of capabilities that a mobile device might encompass and fit like the following description:

> A mobile biometric device can be defined as a portable device containing a capturing sensor, a display unit, a user interface for the biometrics, a

processing unit, storage, communication capabilities, a battery and a lighting source. It should at least capable of capturing, processing if needed, matching or transmitting a biometric image and its associated information to a server. A mobile biometric device should include both the hardware and the biometric software running on these units.

If we assume that this is a reasonable description of set of capabilities for mobile biometric device, pretty much any smartphone that has been successful on the market for many years covers those capabilities. This underscores the difficulty in saying what "is" a mobile biometric device. The Apple iPhone 5s and newer iPhones, for example, have these capabilities in spades as they have the potential to be multimodal mobile biometric devices. Perhaps, if one wants a more specific definition of "mobile biometric devices," it would be better to separate mobile devices into those that are innately mobile biometric devices (such as a smartphone) and those that are specifically developed for biometric applications.

Automated fingerprint matching technology is still the most mature biometric modality in the biometrics world. It is easier to mitigate or even eliminate environmental impositions to image capture with fingerprints than it is with other biometrics. Given the strengths of fingerprints as a biometric, the first "purpose-built" mobile biometric devices were designed for fingerprint capture and processing. The first design attempt for a mobile fingerprint device was started in the early 1990s. Once again, law enforcement early adopters made the first implementations of mobile fingerprint devices for leveraging the functionality of a real-time response AFIS while in the field.

Since live-scan devices in the booking suites were starting to take hold, the first design was done by adding a live-scan-type sensor to a portable device. The devices contained a small live-scan capture device, a display unit, a simple interface, a processing unit, and a small storage unit to allow the fingerprint to be captured by law enforcement agent when they encountered a "person of interest." Due to the limited bandwidth of wireless connections at that time, real-time wireless biometric data transfer to a full AFIS server in the back end was not practical. The captured fingerprint was downloaded to a mobile computer in a nearby squad car with a relatively powerful processor where it was searched against a small, local most wanted watch list or, through that PC, submitted to a regular AFIS to search against a large enrolled database. As cumbersome as it was, the operation was still faster than taking a suspect back to the police station to capture his fingerprints through a live-scan station to search against its connected back-end AFIS.

Thus, the first purpose-built mobile fingerprint device was essentially only a small, portable live scan that could be used to capture biometric images to send to another system such as a laptop computer for processing and matching. The first prototype portable fingerprint device was developed by Digital Biometrics Incorporated (DBI) in 1993. The prototype was productized and deployed to the field in 1995 as the device known as "SQUID."

We should note, however, the "tiny, portable live scan" nature of the SQUID, when looked at retrospectively, was not a bad design idea especially given this was technologically all that could be done in this first-generation device. In fact, devices with essentially the same function, at the time of this writing, are in current production and continuing to evolve as they fill a definitive use case for a variety of customer devices. Examples of these include the very popular 3M Cogent "Bluecheck," CredenceID "CredenceOne," and the Morpho "MorphoIdent" devices. Even in a world where we find small fingerprint sensors in many widely available smartphones, these fill an important position of adding biometric capture abilities to essentially any mobile device. In addition, even on a smart phone *with* a fingerprint sensor in it, the capture area is simply not large enough to be used effectively for consistently accurate results.

By the year of 2000, wireless technologies began to advance but were still rather limited both in performance and coverage. Consider, WiFi really only started to be in limited release in the late 1990s. The term "mobile data" had not even been definitively coined even while "2G" mobile data technology officially debuted in the early 1990s albeit *very* limited in coverage and performance. Even with these limitations, the "early adopter" mentality of law enforcement with regard to biometrics coupled with increasing computing power of mobile devices, implementation of mobile devices with wireless capability for remote matching was started. The second version of the SQUID from DBI, having a remote wireless function built-in, is a good example of the initial roll out the second generation of mobile fingerprint devices.

In these early days of mobile biometrics, wireless technologies were costly and unreliable. The computation power was limited and battery life of the mobile device was short. As a biometric company with limited resources compared to giant consumer electronics firms that focused on mobile electronics, it was very hard to make advances with the mobile device technologies or even influence their development direction. The significantly higher computational needs of biometrics compared to almost every other type of data processing were a small niche case for the mobile electronics industry such that there was no significant impetus to drive them toward higher power devices. Thus, it is understandable that early designs of mobile fingerprint devices were constrained by the limitations of the mobile industry.

It was not until 2007, with the release of the first iPhone, when Apple gave birth to a mobile device category that *nobody* ever thought of: real computing power with a definitive, general purpose need for serious mobile bandwidth, in the palm of your hand. Although the RIM Blackberry had driving implementation of better data transmission, its focus was largely text data communications (email) which is significantly smaller than the needs of general computing much less biometric processing.

The iPhone and other devices that subsequently appeared after the birth of the "smartphone," as we know it today, started to require *much larger* data bandwidth than text communication mandated. By the year of 2008, wireless technologies with significant bandwidth became more mature and widespread allowing biometric images and features to be transmitted in real time. The RapID device developed and released in 2008 by Sagem, which took advantage of improving mobile data bandwidth, could be considered as an initial roll out of the third-generation specific for biometrics device.

With maturing wireless technologies and increasing demand for mobile finger-print devices, the market started to push for a true multiple biometric capability devices with true wireless capability. The L-1 Identity Solutions HIIDE 5 was intro-duced in 2010 and is a very good example in the fourth-generation design category. The device was equipped to capture, process, *and match* fingerprint, face, and iris locally. It was configured with a variety of wired and wireless and could be readily integrated into an overall biometric identity solution.

In recent years, significant focus in law enforcement solution design is to develop biometrics functionality leveraging widely available commercial off-the-shelf (COTS) hardware, software, and sensors. This approach allows easy performance improvement of the mobile biometric system without changing entire device.

The vendor can focus on their core biometric competencies by adapting the best COTS solutions on the market. Mobile biometric application design becomes one of selecting the most suitable mobile devices, biometric sensors, and software from the market and integrating their best research into this stack. By doing so, vendors can reduce their design costs and adapt the best elements from the market into their designs. The best part of this is that rapid growth of capability in the core devices is quickly driven by the vastly larger consumer market and not solely by the niche market of biometrics.

Using this approach, the design of mobile biometric solutions consists of three elements allowing the blending of the best of each into a very good overall implementation:

• Sensor selection and integration
• Mobile device platform selection and integration
• Biometric application software selection and integration.

For the consumer market, work was started to develop a biometric solution to secure a mobile computer such as laptop or notebook computer in 1997. As the mobile computers at that time had enough processing power to enable an optimized biometric algorithm, a small number of systems with a viable solution were pro-duced. However, at that time, the processor on the most powerful mobile phone was nowhere near as powerful as even the slowest mobile computer. It was not until the year 2000 that mobile-phone processors had improved enough to allow a biometric solution using fingerprints to secure a mobile phone. Leveraging newly optimized biometric algorithms and better mobile CPUs, the first mobile phone with a touch-based fingerprint sensor was introduced by SAGEM in 2000. This phone had very limited computation power and only five fingerprint images were allowed to be enrolled.

Two years later, the second commercial mobile device, a Personal Digital Assis-tant (PDA) this time, was rolled out by HP with a "swipe" fingerprint sensor. It had a more powerful processor than the SAGEM mobile phone and allowed the six best fingerprints from a finger to be enrolled. Many mobile phones were secured by a fin-gerprint since 2002 but the fingerprint enrollment and matching processes for these devices were more or less the same (and weak) until the iPhone 5s with Touch ID was introduced in 2013.

There were three issues with early biometric authentication on these consumer phones that caused them to have less than satisfying performance and, as such, they were less than successful in the market.

- The enrolled fingerprints did not cover all possible authentication fingerprints from the finger.
  - Without the largest possible authentication fingerprint features captured, even from a single user, things like finger angle could confound the matching algorithm. A finger submitted to be verified had to be carefully placed flat and in the same manner to which it was enrolled for a match to occur.
- The sensor itself: they just were not very good.
  - Early "swipe" sensors were very sensitive to the speed at which the fingerprint was swiped over it. Too slow or too fast would be difficult to enroll. Slower or faster verification swipes might not match an already enrolled finger.
- The performances of the fingerprint algorithms used or integrated were simply not as good as those available today.
  - Algorithms of the time were not tolerant of finger placement or angle and would miss the capture.

Because of these issues, a user could not reliably access his/her own phone and led to earlier phones with fingerprint authentication not being widely accepted. This changed dramatically after the iPhone 5s with Touch ID was introduced in 2013. The iPhone 5s' Touch ID sensor and software had fast, accurate, and practically fool-proof biometric authentication. This is in sharp contrast to earlier devices from other vendors which were not *quite* so foolproof. On the iPhone 5s and newer iPhones, finger enrollment is easy and well guided. The authentication operation is conve-nient and faster than typing something as simple as a simple four digit password. These iPhones can be unlocked while held in a single hand with single touch. Other smart phone vendors have been playing catch up after Touch ID was introduced. It was not until 2015, when other brand premium phones with fingerprint sen-sors were released with user experiences nearly as good as that of the iPhone 5s. Even so, these other vendors still must continue to make strides to catch up with Apple.

In the consumer market space, biometric solution design focus has been, and remains, focused on user experience: the authentication using fingerprint must be fast, accurate, and convenient. To date, to alleviate privacy concerns and to foster adoption, enrollment is always to the local device and, at most, the authentication is a 1: Few match. The iPhone, as good as it is, remains the same as its predecessors by only enrolling and matching up to five fingerprints with a sensor which is intrinsically not suitable for industrial duty 1:N search.

By looking back at the mobile fingerprint device development history, some important key milestones can be summarized in Figure 2.2.

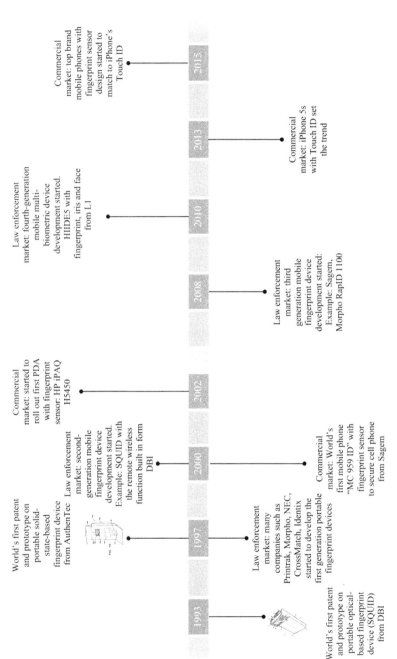

Figure 2.2 Important milestones of mobile fingerprint device developments

2.5 Features found on good mobile biometrics device design

As previously noted, the first portable fingerprint device developed for the law enforcement market was released in 1993 and the first mobile phone developed for consumer market using biometric authentication was in 2000. Since then, many versions of what can be called a "mobile fingerprint device" have been developed and released. Looking at what has been released and the evolution of them, we can see the following drivers for this evolution and what makes them successful:

- User friendly, nice styling and ergonomics, light, and rugged
- Consistently quick and easy capture of high-quality images
- Easy, seamless integration to a back-end biometric system
- Quick processing and responses
- High accuracy, security, and privacy.

Depending on the use case and target market, some of these features might be considered more important than others through the evolution of mobile biometric devices. However, all of these drivers are relevant regardless of the primary use case or target market. If we look at, for example, styling and accuracy as an "attribute pair" in two rather different markets, we can see the contextually different but semantically similar importance of those attributes. We might find that styling is an expected high-priority design criteria in the commercial market, typically for visual appeal, whereas the law enforcement market in general may attribute far higher importance to high accuracy and seamless integration to back-end systems.

Looking at this "attribute pair" in a different light finds that for the commercial market, high accuracy is important *but* if it misses on that 1:Few search once or twice to unlock the device but does consistently unlock it without consistently frustrating failure, it's not as critical as law enforcement where it *has* to be right first time, every time. Although aesthetics are typically less important to the law enforcement market, they cannot be ignored and good styling and ergonomics remain as an imperative for safe and secure operation. Considering the design driver waxing and waning driven by use case and market, we'll look at each of them discretely and cite some examples of how and why they might be important in one case but not another.

2.5.1 User friendly, nice styling and ergonomics, light, and rugged

There are several related elements here covering the overall design of a mobile biometric device. User friendliness, good physical design, and light weight are all required to help a device to be accepted by a broad audience regardless of the target market. Ruggedness has always been an important feature for law enforcement and military market as the device is almost certain to be required by the user to work first time, every time under tough conditions.

As the cost of premium smartphones has remained high, ruggedness has made its way to the consumer market to inherently protect what is typically a sizable expenditure out of an individual's budget. In fact, ruggedness can be even *more* important to the commercial market than government markets if one removes the "first time every time" requirement. Many institutional users, where "first time every time" is not necessarily a definitive requirement, have begun to eschew the need for "rugged" devices with their associated increased cost. They are finding that it is less expensive for a given program to acquire normal "nonrugged" devices and replace them if they get damaged.

Between the significantly lower purchase cost and the cost for damage replacement (which does not happen all that often), the overall program cost is *significantly* lower. Conversely, for a large majority of individual users, replacing a $700+ smartphone out of their own pocket if it is physically damaged can be a hard knock to their pocket. The potential budget impact for individuals has spawned successful market releases of a plethora of protective cases and rugged devices such as the rugged versions of the Samsung Galaxy smartphones. These variable "ruggedness" requirements underscore the fluidity of the needs for users and the importance overall in the design of the solution.

Ruggedness aside, the design of any device involves three areas: user interface, the device appearance, and the case material. Different talent might be needed for these design elements. Without careful budgeting and management coupled with the right talent, design, and industrialization cost can easily run over budget and even result in a flawed or failed design. Once released, a positive user experience is key to measure the success of the design. One need only recall the Apple "Bendgate Scandal" [30] to find an issue that made headlines even while it amounted to only *nine devices* the issue was actually attributed to. In the context of mobile biometrics, the software managing fingerprint capture process and sensor location, if poorly thought out, can significantly affect the quality of image captured, directly affecting biometric accuracy. A user-friendly graphical user interface (GUI) that guides the user to capture a good quality fingerprint will reduce or eliminate low quality fingerprint captures. A well placed fingerprint sensor can ensure quick and easy positioning of the finger for a good capture.

In the early days of mobile fingerprint systems, the capture process and associated biometric application user interface were often weakly designed. Sensor placement often might have seemed "right" to an engineer in the office but was not practical in the field. User training was lacking as, up until this time, it was booking officers and/or fingerprint examiners capturing fingerprints who had been trained how to acquire good images. To these early well-trained users and even to vendors who seemed to think "this is simple," a two-finger flat image capture did not merit training effort.

However, when these devices were released to untrained or occasional use patrol officers, either bad images were captured, or, if strong quality checks were imposed, had so many failures to capture that they gave up. A very large percentage of missed matches or failure to acquire issues could have been solved by training. However, as has been seen in more recent times, the large number of devices to be deployed makes training that many users or, worse, occasional users a training nightmare.

With this feedback, mobile fingerprint devices have matured with innovative design of friendly, onscreen guided enrollment processes, and much more attention paid to sensor location. The Morpho designed MorphoIdent and the 3M Cogent Bluecheck are examples of devices that are the result of this evolution and make earlier designs, while good in theory, seem like misguided attempts.

Furthermore, as successful as these devices have been, with thousands deployed to the field, Morpho, for example, has continued to evolve and improve the MorphoIdent. When comparing a first release MorphoIdent to a third release device, we see a device that is superficially the same but showing much more attention to sensor placement, button placement, style, and user interface.

For the consumer market, swipe or small touch sensors are typically used. The images captured by sensors used on consumer devices are smaller than those captured by the typically much larger sensor used in a law enforcement application. One reason is that consumers are not *usually* willing to pay $1,000 for a smartphone, but law enforcement agencies may sometimes pay thousands of dollars for mobile biometric devices, and are therefore willing to bear the expense of the larger and more accurate touch sensors. In the early days of biometric-secured devices, the capture software would only acquire 3–5 images with the user typically instructed to place the ball of their fingertip either carefully aligned on the sensor or not instructed to capture in other orientations to allow for variations while verifying. Although this worked for either a 1:1 or 1:Very few search, it was always "fussy" to unlock as it required careful placement of the finger to be acquired and get a valid match.

It was not until the iPhone 5s that Apple programmatically successfully addressed the issues of a small sensor area by requiring the user to enroll many finger positions during the enrollment stage to offset the weakness of the small sensor area. This overcame the earlier phone's "fussiness" by removing the need for very careful finger placement where a touch on the sensor would consistently unlock the device with a single touch. Apple's enrollment process instructed the user to capture many different positions of a finger and then intelligently selected the best combination of fingerprint captures to represent the *whole* finger so that as long as the finger is reasonably placed, the device would authenticate against it. The iPhone's Touch ID enrollment process is an excellent example for successful mobile fingerprint device design for consumer market.

When looking at government arena applications such as law enforcement, border control, or civilian population registry type functions, we are typically looking at large-scale 1:N cold search as well as potentially using mobile biometric devices for remote enrollment. This user segment tended to be rather small in the earlier days and typically able to be definitively and directly "trained" as opposed to being expected to use the functionality simply by reading a user guide or calling a help line.

However, while a typical mobile biometric application for this type of use case will be more complex than a simple verification function, user friendliness is equally important. Although a small user base can be trained, training is expensive. "User Hostile" applications can induce errors and/or require expensive retraining. Although many early mobile biometric applications were based on what were actually very good live-scan applications and well liked by booking officers who used them heavily every day. They tended to be overly complex for infrequent users to learn, remember, and use as well as usually far more complex than needed for mobile tasks.

Worse, at the time of early mobile biometric systems, these "very good" live-scan applications were often still needlessly stuck in an earlier era of biometrics. They might require operator intervention at every step, voluminous demographics entered (often redundantly), and manually shepherding the transaction from step to step. By the time mobile biometrics were coming along, even in the live-scan realm, much of this manual and/or redundant work was "because we've always done it that way" (often imposed by much earlier systems) and could have been much simpler and automated. Although perhaps only a bit painful to live with "we've always done it this way" on a system in a booking suite with a full-size screen, keyboard, and pointing device, it often became hideously difficult to use when saddled with a mobile device's small screen and keyboard.

Although these early mobile solutions worked, the complexity imposed by them, coupled with high device costs, caused their acceptance to be found only where there was a particularly large return on investment (ROI) to using them. To get beyond this relatively small subset of industrial strength users and have greater customer and user acceptance, user-friendly process management needed to be developed. This basically involved resolving the problem for the use case including simple transaction/case creation, quick and easy fingerprint image capture and search submission, and easy-to-understand responses from the central server. An overarching and continuing evolution of mobile applications is to keep weeding out manual steps and needless data entry. A patrol officer might have actually used early solutions for one of two reasons: (1) he really liked shiny new toys and would try anything new offered or (2) he was ordered to. Current "good" examples of mobile biometric solutions have very few steps required, namely: enter an identifier (*only* for definitive 1:1 verification), capture fingerprint image(s) (usually two), and get a response.

That's it.

As close as any solutions can get to needing to do one or, at most two, overt user actions with the rest happening automatically, the better the solution will be. *Maybe* the solution might require an "enroll if not found" tick box for the very few devices (even today) that are used for mobile enrollment but there should be little else. The solution offerings to government, at the time of this writing, have a few examples that are getting close to this nadir, and many that are somewhere between the best and the worst. Curiously, there are still a few really bad "because we've always done it this way" solutions that somehow get away with it often due to customers who seem incapable of moving beyond "we've always done it this way."

2.5.2 *Consistently quick and easy capture of high-quality images*

Perhaps, just as important as all the elements noted in the previous section, the ability to consistently capture high-quality images is paramount, especially in the large-scale 1:N biometric processing arena. The quality of the captured images, both in the enrolled search database and the probe images captured by the mobile device, directly affects the accuracy of the system. The key elements affecting image quality are as follows:

- The sensor
- Capture guidance
- Image quality evaluation algorithms.

2.5.2.1 The sensor

There are many types of sensors [31] that can be used for capturing fingerprint images on a mobile device. Broadly speaking, these many types of sensors boil down to two categories of sensor: optical and solid state, each with their own pros and cons.

Optical sensors were the first type of capture devices integrated into mobile fingerprint devices. These optical devices were, for all intents and purposes, live-scan sensors reduced in size allowing it to be used on a mobile device.

An optical sensor, whether integrated to a full live scan in a booking suite or mobile, uses arrays of photodiode or phototransistor detectors to convert the energy in light incident on the detector into an electrical charge. An optical scanner normally consists of four parts: prism, light source, lens, and CCD camera. To fit an optical scanner on a mobile fingerprint device was quite a challenge and expensive in the early days. Among the challenges, chief among them are physical size and power requirements. The power for the light source and detectors is quite high and early mobile devices had little hope of getting through a shift without recharging if used heavily. Although the light sources and detectors in the sensors got smaller and more power efficient, the light path lengths and focal points continue to be imposed by that pesky thing called physics and they remained rather bulky. At the time of optical sensor mobile devices, many vendors made full live-scan capture devices that needed routine calibration to account for vagaries in platen/prism/sensor alignment to keep images crisp. This is no less important in the pocket(ish)-sized live-scan sensors. This critical alignment could not be done easily (or at all) on mobile devices for all that this arrangement shared its obvious fragility to its more secure deskbound cousin.

Another significant obstacle to optical scanners used in a mobile scenario is an environmental one. Live-scan systems are almost *always* indoors in a light-controlled room. With mobile biometric devices often *not* being indoors, the illuminators and detectors are often overwhelmed by ambient light. This caused users to either curse and find somewhere shady or hold their hand over the capture area (a potentially dangerous scenario in many law enforcement situations) or for vendors to come up with all manner of ingenious yet inherently clumsy integrated shades on the device. Another weakness of optical sensors is that dirt and oil on the platen can cause the capture of low-quality images or latent fingerprints on the platen that might cause a false match. Although live-scan stations typically have algorithms in the application to negate these problems, early mobile devices just did not have the horsepower to do the same. More recent optical sensor-based devices have addressed many of these issues in various bits of technology related to them.

The cost of optical sensors is much more expensive than solid-state-based devices, largely due to critical mechanical and optical tolerances that had to be manufactured to. However, the quality of fingerprint image captured by optical sensor was superior to early solid-state sensors when the optical sensor was properly designed and maintained. Optical sensors are widely used in law enforcement applications and were the only viable choice early on for large-scale mobile biometrics.

Morpho and CrossMatch have been the top manufactures using optical sensors for mobile devices and the Morpho RapID1100 was a good example of mobile fingerprint device using optical sensor for law enforcement market at that time. Although they can still be found in the wild, mobile biometric devices with optical sensors are rapidly being supplanted by devices with solid-state sensors.

Solid-state sensors (sometimes known as silicon sensors) became commercially available in the mid-1990s at about the same time as mobile biometric devices were appearing on the market. However, the sensors that showed up during this period were decidedly "not ready for prime time." The solid-state sensor was easily damaged since it didn't have much protection during the time where the finger was applied directly to the silicon. To keep the price of the sensor reasonable, due to low volumes and the higher cost of silicon devices at the time, the sensor surface was designed to be a small size. Many of the sensors at this earlier roll out for a mobile device application were rather below 500 ppi (pixels per inch) which is typically required for consistently accurate matching and usually the minimum requirement for law enforcement customers. On top of low resolution, many of the earlier sensors were binary (1 bit gray scale, i.e. black or white) rather than 8 bit gray scale needed for good accuracy. Most consumer grade sensors were an ineffective combination of being far below this ideal minimum of 500 ppi, binary *and* very small resulting in many failed authentications and equally as many unhappy experiences. Using this small sensor without well-designed enrollment process led to low quality or limited-size enrolled fingerprints that were not able to represent the actual finger.

All silicon-based sensors consist of an array of pixels, each pixel being a tiny sensor itself. The user directly touches the surface of the silicon albeit often with some manner of protective layer applied. Based on the effects of how the physical information are converted into electrical signals that can be used by software, the sensors can divided into five types; capacitive, thermal, electric field, piezoelectric (pressure based), and a more recently emerging (and rapidly overtaking other technologies) Light Emitting Sensor. Looking at each of these types of sensors.

Capacitive sensor

A capacitive sensor is a two-dimensional array of one side of a microcapacitor plate embedded in a chip. The other plate of each "complete" microcapacitor is the finger skin itself. Small electrical charges are created between the surface of the finger and each of the silicon plates when a finger is placed on the chip. Ridges have a higher capacitance since they are closer to the detector and valleys have a lower capacitance since they are farther away from the detector. Some capacitive sensors apply a small voltage to the finger to enhance the signal to create better contrast in the image. This method is often called active capacitive sensor. Capacitive sensors can be sensitive to electrostatic discharge (shock) but they are insensitive to ambient lighting and are more resistant to contamination than some optical sensors.

Thermal sensor

Thermal sensors use the same pyro-electric material that is used in infrared cameras. When a finger touches the sensor, the ridge temperature is measured where the finger ridges contact the sensor surface but not the valleys of the ridges. A fingerprint image is created by the skin-temperature ridges and the ambient temperature valleys.

The biggest drawback to this type of sensor is that it only takes about a tenth of a second for the sensor surface between the ridges and valleys to reach the same temperature which then creates a solid black finger-shaped blob.

Pressure (piezoelectric) sensor

Pressure sensors can be made very thin and are often used on electronic devices. Earlier pressure sensors had to make a trade-off between durability and quality because any protective layer on the sensor surface would diminish the contrast of the impression. There are two types of pressure-sensing detectors available, conductive film detectors and micro electro-mechanical devices (MEMS). Conductive film sensors use a double-layer electrode on flexible films. MEMS is a newer technology that uses extremely tiny silicon switches on a silicon chip. When a fingerprint ridge touches a switch, it closes and is detected electronically. Although not technically limited in practical size as other silicon sensors are, this technology has the rather large drawback of being binary. The "micro switches" (however they are executed) are on or off with no gradation between the two.

Electric field sensor

Electric field sensing employs small, embedded metal electrodes connected to oscillators that produce high-frequency AC (alternating-current) electric fields. When anything that conducts electricity reasonably well, such as a person's finger in this case, comes within a certain distance of the electrodes, the electric field fluctuates. The nature and extent of this change is translated into the image of the fingerprint.

Light emitting sensor

The light emitting sensor (LES) film developed and patented by Integrated Biometrics is a multilayer, polymer composite. Dispersed within the film, at the nanoscale level, are particles that luminesce (give off light) in the presence of an electric field. When a finger is placed on the film, the live skin of the ridges of the fingerprint completes the low-level electric circuit which causes the particles in the film to luminesce narrow wavelength light, producing a highly accurate, high-resolution analog image of the fingerprint. The resolution of the fingerprint luminescence is between 1,200 and 1,500 ppi. These sensors are becoming increasingly popular due to their low power requirements and available large sizes. In addition, in spite of the FBI originally stating unequivocally that solid-state sensors would never be given "Appendix F" certification, they now have been and remain the only solid-state sensor to achieve this level of certification.

Since solid-state type of sensors can be cheaply and consistently produced coupled with their inherent ruggedness and low power requirements, they are more readily adaptable to mobile biometric applications. Apple's AuthenTec, Fingerprint Cards AB, Synaptics, and, increasingly, Integrated Biometrics are among the top manufactures of solid-state sensors.

There are so many types of sensors from so many vendors, that the question arises of which one to select and how to evaluate these sensors. In the early days, sensor selection (of any type of sensor) was not an easy task. There was not a definitive "apples-to-apples" method of comparison or even definitive technical ranking. When looking at some of the choices at a deeper level, "apples-to-apples" comparisons were more like "apples-to-elephants" comparisons. Selection of a good sensor is a crucial step of any biometric capture solution and no less important for mobile fingerprint device design, especially in accuracy critical government or other large-scale biometric systems.

2.5.2.2   Sensor certification

To minimize the risk to accuracy and interoperability imposed by "questionable" devices and software, the FBI working with NIST jointly developed compliance specifications [32–38] to define and regulate the quality and format of fingerprint images and certify offline/live-scan scanners. NIST has subsequently worked with other US government entities to develop biometric standards relevant to their needs as well. These very detailed and definitive standards ensure a consistent quality of fingerprints captured and stored. By doing this, all vendors and customers have a solid basis on which to design toward and/or evaluate biometric solutions including mobile biometric solutions. Rigorous testing of these standards is managed by the US government's contractor, Mitre Corporation.

Typically, all Requests for Proposal from government entities require the fingerprint sensors to be certified. If the sensor is one of the types that cannot explicitly be certified to a certain standard, RFPs will request that images produced by sensors must at least be of a particular standard's quality. Although the standards defined by NIST were born in the United States, they have been adopted and/or requested by essentially any entity in the world given the robust and consistent nature of these standards.

The two standards currently in use for fingerprint quality that wholly apply to mobile biometrics as well as other systems: Appendix F and PIV-071006:

- Appendix F
  - Appendix F has stringent image-quality conditions, focusing on human fingerprint comparison and facilitating large-scale machine many-to-many matching operation.
  - Appendix F certified devices are not only capable of 1:1 matching but also can handle 1:N identification.
- PIV-071006
  - PIV-071006 is a lower level standard focused on support for one-to-one fingerprint verification.
  - Certification is available for devices intended for use in the Federal Information Processing Standard Publication 201 PIV (Personal Identify verification) program.
  - Includes the capability to collect a single finger flat impression with a minimum size limitation.

The FBI's fingerprint certification also includes certification of fingerprint printers, fingerprint card scanners, fingerprint image compression algorithms, and live-scan devices of multiple types based upon the appropriate standards. In all cases, a certified unit is a configuration of specific hardware and driver/support software optimized for

Table 2.1 Sensor certification

Sensor certification	Specification	Capture image size (in. × in.)
FAP 10/PIV single finger	PIV-071006	0.5 × 0.65
FAP 20	PIV-071006	0.6 × 0.8
FAP 30	PIV-071006	0.8 × 1.0
FAP 40	PIV-071006	1.6 × 1.5
FAP 45	Appendix F	1.6 × 1.5
FAP 50	Appendix F	2.5 × 1.5
FAP 60	Appendix F	3.2 × 3.0

usage with fingerprints. For large-scale mobile fingerprint capture and processing, all of these standards implicitly apply even if not explicitly "required" depending on the customer and their application.

One element of certification is the size of the sensor. The category is subdivided into several levels by fingerprint acquisition profile (FAP) number, based upon device capture dimensions, the image quality specification applied, the number of simultaneous fingers that can be captured, and the ability to capture rolled fingerprint images. FAP 30, 45, and 60 are larger size sensor which are typically needed for $1:N$ large database search. In 2014, there was a push by the FBI to *require* a minimum of FAP 30 for any device submitting to IAFIS. However, when the images of "Appendix F quality" on sensors as small as FAP 20 produced were found no measurable reduction in accuracy, this pending requirement quietly faded away. Typically, for mobile biometric solutions, flat images only are required where verification or identification is the goal. However, with the advent of Appendix F certified FAP 45 solid-state sensors, an honest "mobile booking system" is feasible and being marketed by a handful of vendors.

Table 2.1 details the currently defined sensor certifications. It is currently considered that FAP 45 is the smallest suitable for rolled fingerprint impressions.

A device of FAP 45 and above is capable of capturing optional rolled as well as plain fingerprints. For example, a device of FAP 60 allows four fingers to be captured simultaneously. Typically, a scanner/sensor with a specified level of certification is required by most government customers. A certified sensor standardizes a crucial parameter when considering sensor performance by providing consistent image capture with consistent quality. Even for nongovernmental selection, such as a commercial large-scale $1:N$ environment, might leverage these standards. As they are defined with proven results, this is a perfectly good requirement for any customer to insist on. When looking at sensors, the following parameters that directly impact fingerprint quality become checklist items for sensor selection and evaluation:

- Resolution: This indicates the number of dots or pixels per inch (DPI or PPI).
  - o 500 ppi is the minimum resolution for FBI-compliant fingerprint scanners and is met by many commercial devices.

- o 250–300 ppi as a minimum resolution overall (when combined with other parameters) that allows extraction algorithms to locate the minutiae in fingerprint patterns, thus providing good overall biometric performance.
- Area: The size of the area sensed by the captured object is a fundamental parameter.
  - o For a fingerprint, a larger sensor area will allow more of the fingerprint to be captured making it more distinctive and allowing more captures for identification or verification to successfully match. An area greater than or equal to 1×1 square inches (as recommended by FBI specifications) permits a traditional full-plain fingerprint impression to be acquired and is, in truth, driven by the box size on inked fingerprint cards that have not changed for decades.
  - o Given the proven lack of performance loss by using a smaller sensor along with hugely improved algorithms never thought of in the days of inked fingerprint cards, many vendors offer smaller sensors. These smaller sensors both keep the physical size of a mobile biometric device smaller as well as often reducing power requirements and device cost.
  - o Small-area scanners may allow a whole fingerprint to be captured and users "blessed" with a "less-than-optimal" biometric solution may encounter difficulties in representing the same portion of the finger leading to false nonmatch errors. This was a big issue in the early days which led to failure to make the smaller fingerprint scanner being accepted by users. Modern algorithms and careful physical design of sensor integration with a device which "guides" the finger to the correct placement for match greatly reduces false nonmatch errors. An "enrollment" station, whether desktop of mobile, will have a larger sensor so the fingerprint gallery matched against has full images obviating even more false nonmatch errors.
- Number of pixels: The total number of pixels on the sensed area.
  - o It can be derived by the resolution and the area: a scanner with resolution PPI R over an area of height (H in.) $\times$ width (W in.) has $R * H \times R * W$ total pixels on its surface.
- Dynamic range (or depth): This denotes the number of bits used to encode the intensity value of each pixel.
  - o The FBI standard for pixel bit depth is 8 bits, which yields 256 levels of gray.
  - o This proven pixel depth is now a de facto standard for any deployed large-scale system.
- Geometric accuracy: This is usually specified by the maximum geometric distortion introduced by contact between the finger and the acquisition device.
  - o Expressed as a percentage with respect to x and y directions.
  - o As close to 1:1 as possible is required to reduce match errors.

Regardless of the sensor selection, design, development, and integration of a good, user-friendly fingerprint capture and image quality evaluation tool is key. The program should automatically guide to the user on finger placement and then accurately evaluate the image quality to ensure a good quality fingerprint image to be enrolled for matching. For example, the tool should provide instruction for the user to place the finger in different positions to form a "larger view" of the finger than the sensor can get in a single capture if the sensor is a small one or guidance for pressure/position on a larger sensor. While some of these functions are newer software developments, many of the current crops of capture and evaluation tools were not possible in the early days of mobile biometrics since the computation power was significantly limited on mobile devices.

2.5.3 Easy, seamless integration to a back-end biometric system

Now that a good image has been captured, what are we going to do with it? When one is looking at local enrollment/comparison on a device, regardless of the target market, the ability to integrate to a back-end biometric system is not so important. However, as soon as we look at any government application, whether law enforcement, military, or civilian government, this becomes very important. The processing and matching power on a mobile biometric device, even today, is limited. Normally, only a most wanted list or other limited list is stored on the mobile biometric device for a local search. For example, even a mobile device with big storage and a powerful processor would be significantly taxed to search against a database of more than 100,000 records on the device. When looking at any government solution and increasingly prevalent commercial implementations, one is usually talking about databases in the multiple hundreds of thousands to many millions. An extreme example is the Indian Unique Identification Authority of India (UIDAI) project is a massively heavy user of mobile biometric devices where, as of January 1, 2016 over 950 *million* unique identities have been enrolled. That is not going to fit on a mobile biometric device. As such, being able to readily integrate to the back end from *whatever* device is in use, including a myriad of mobile devices, is paramount.

For extreme cases like UIDAI ranging down to the small police AFIS with 250,000 records, interoperability is crucial. A mobile biometric device is increasingly considered as "just another workstation" connected to a large already deployed back-end biometric system. The remote server might be deployed by a different vendor than the mobile biometric device vendor, and in the case of very large systems, it's almost certain. There are two elements to interoperability; interface and content payload.

2.5.3.1 Biometric system interface

To connect and use the server, the device must follow a standard interface and security protocol. In law enforcement, the captured image and processed features typically follow a standard format defined by NIST *regardless of who developed the solution*. This NIST standardization, while often thought to be "Lowest Common Denominator," has proven successful enough that these standards have been adopted not only outside law enforcement, but also outside of the United States. Increasingly, however, the NIST standards, especially *outside* law enforcement, are beginning to be just one of a multitude of ways to interface to a back-end biometric system. In the early days of biometrics, a biometric matching system was an oddball system that the local IT department did not know anything about and did not want to; they just provided a network connection and let the department that owned it self-support with the help of the vendor. In recent times, IT departments the world over are working harder at employing enterprise standards and best practices which biometric systems now must live in which often includes how they are interfaced with and in a lot of cases…it's not NIST. While the ICD (Interface Control Document) *might* continue to be NIST based, it does not need to be and the transport is now no longer driven by the biometric provider.

2.5.3.2 Biometric payload

Once again looking back to the early days of biometrics, each of the many vendors had "their way" of doing things which usually had not one single chance of every working at all with another vendor's "way" of doing things. While this is still largely true, the payload also must be able to, at some level, deal and work with other systems. The interchangeability of the payload might be at image level or at biometric feature level.

Early on in mobile biometrics, mobile bandwidth was at an extreme premium. As such, to keep the size of the payload to a minimum, a fingerprint image might be captured on a mobile device but never left it. The mobile device extracted the biometric features and sent this rather smaller data when compared to the relatively large fingerprint image. That was all well and good but a mobile device from one vendor would not be able to extract and send a feature set to a back-end system from another vendor. To get around this problem, NIST created a feature template standard called "MINEX format." NIST also provides certification tests for coder and matching to achieve better compatibility between the vendors at feature level. A MINEX certification test [37–39] was started in 2004 which requires the coder/feature extractor and matcher to meet a certain requirement [39] in order to be certified.

In 2015, NIST launched a more strict certification test "MINEX III" for coder and matcher. Thus, the design not only includes the protocol and format implementation but also a better coder (template creation) design. While this work continues for the government, many systems in other use cases run in geographic regions where mobile bandwidth just is not available so these efforts remain relevant.

However, in more developed areas, the restriction to sending a very small feature set is rapidly becoming unnecessary as mobile bandwidth is so wide that a fingerprint image can be sent in less time than just a feature set could be in the early days. Even at that, while a fingerprint image is suitable to be sent, at present, the two "accepted" formats for fingerprint images are the FBI's WSQ image format as the leader with JPEG gaining increasing acceptance. Now that more and more systems directly support submitted JPEG images, we might surmise that the days of WSQ are perhaps limited.

Even while the preceding two key elements of integration have relatively few considerations, the combinations and permutations of them result in a very large number of potential solutions. As such, it is incumbent on mobile biometric device vendors to either make their solution eminently configurable or, to guard against bloat, potentially having multiple offerings with different market segment foci.

2.5.4 Quick processing and fast responses

A responsive solution, *end-to-end*, is key to providing a positive user experience. Slow processing on any computing device or network transmission is one of the most frustrating things that can be experienced especially now that we've become accustomed to snappy performance in our technology. To be readily accepted and adopted, performance of the entire solution needs to have matching accomplished in seconds and verification preferably in subseconds.

While user experience is important, when looking at a mobile biometric device used in law enforcement, rapid responses become critical in terms of safety; it's always best to know *fast* if it's a bad guy in front of you or not. A key design element, where local processing is done, is to optimize the biometric programs making the biometric coder and matcher faster. A reasonably powerful mobile processor with enough memory should be considered in the code optimization to tune for best performance. While most back-end biometric systems typically return responses in short order, some slower ones still exist and, considering the interface discussion in the previous section, if a transaction needs to be processed very quickly, a priority tag should be considered so that more important transactions can be returned quickly.

2.5.5 High accuracy, security and privacy

All of the previous design elements are all for naught if the overall solution is not accurate, secure, and provides good privacy *regardless of the application*. Put simply, a biometric system/device is useless if a person cannot be verified or identified.

Technically speaking, the accuracy [5,6] of biometrics recognition can be measured by false rejection rate (FRR) and false acceptance rate (FAR). FRR is the probability that the system incorrectly rejects an actual match biometric input with a preenrolled template. FAR is the probability that the system incorrectly matches the biometric input with a preenrolled template that is *not* a match.

Current technology is never 100% accurate, but top performance of a fingerprint system can achieve 0.6% FRR at 0.01% FAR using the best proprietary templates according to NIST PFT II benchmark test results [40] and can achieve 0.88% FRR at 0.01% FAR using a interchangeable template according to NIST MINEX III benchmark test results [41]. The accuracy requirement for identification is much higher than that of verification, especially for a large database. In the case of mobile biometric identification application, the identification is a lights out operation; i.e. no human review of the match results before returning to the requesting device. The matched case from the identification need to be not only ranked the first and be the most deviated from the rest of the responses in order for the system to confidently to say this result is the "one."

What affects mobile biometric accuracy? Simply put, basically everything in the previous sections on biometric system design.

One element of solution accuracy not really touched on in depth so far is that the accuracy of a single biometric [40,42–44] might not be enough to achieve desired accuracy under some circumstances. For example, potential environmental impacts on captured images sometimes cannot be controlled by the designer or the user. Considerations, when looking at the three most popular and mature biometric modalities (finger, face, and iris), that could induce a need to design for using more than one biometric are fingerprints that may have no visible ridges on the finger due to worn ridges, not enough iris area or blind, or strong facial make up. Each one of these scenarios could cause any of the modalities in a unimodal use case to fail.

However, a carefully designed multimodal biometric system [45–47] can overcome this by using each modality in concert to the point where, even if each modality in use is weak, the individual results fused into a single score can greatly increase overall system accuracy.

It does not matter how accurate or fast a biometric device/system is if fake/spoofed biometrics can be used to fraudulently match against a stored biometric gallery or real images can be obtained or produced from a device or back-end system and used fraudulently. The vulnerabilities of biometric systems have been a concern since they rolled out and, with increasing general knowledge of biometric systems weighed against the scourge of identity theft, this fear is growing. The fear is exacerbated by sensationalist news articles overstating biometric system risk and fictional "defeats" portrayed in cinema and is taken as "gospel" by the general population. That is not to say that there are no risks.

There is no such thing, today, as a usable *and* perfectly secure system. Biometric images could be stolen from a back-end system and subverted. The face is ridiculously easy to acquire; we show them to the naked eye, and coincidentally, to cameras every day. Given these examples of images that could be fraudulently used and acquired, fake/spoofed biometrics detection and biometrics data protection should always be considered and striven for in the design of any system, especially for mobile biometric device since they are often not physically protected and easily accessed.

Some technologies, such as liveness detection [48], can be used to reduce spoofed biometric attacks and another approach is to employ continuous authentication since the attacker may not be easily able to access and use the spoofed biometrics or practically simulate live motion for continuous matching [49]. Multibiometrics can be used for accuracy improvement but can also be used to reduce potential attacks since the multibiometrics can be harder to use simultaneously to spoof. Multiple layers of security processes can be added to enhance security. These technologies come at a cost, but the cost is usually far outweighed by potential losses, especially considering that one can change a password, but it is rather difficult to change a fingerprint.

Privacy has been a concern ever since biometric systems were first implemented. This is the case even in law enforcement criminal biometric systems. Not only do criminals still have rights, if they are exonerated, released, or any number of statutory situations that might occur causing them to be considered "upstanding citizens," a leak from a criminal system could cause undue harm. In the consumer market, rightly or wrongly, users do not want biometric data such as fingerprints to be stored on a server. This is not possible in many large government or commercial systems that are used to *explicitly* ensure uniqueness of identities since they *must* be enrolled to match against. However, this can be reduced significantly for simple verification processing.

Thanks to the FIDO Alliance [50], a set of standards, tools, and devices have been created to address the issue by having biometric authentication only done locally at the device in a standards based and secure manner. Biometric data is encrypted and stored locally without leaving the device when conducting online transaction verification.

The Mission of the FIDO Alliance, which does include some major, large-scale biometrics providers, is to change the nature of online authentication by

- Developing technical specifications that define an open, scalable, interoperable set of mechanisms that reduce the reliance on passwords to authenticate users.
- Operating industry programs to help ensure successful worldwide adoption of the Specifications.
- Submitting mature technical Specification(s) to recognized standards development organization(s) for formal standardization.

FIDO provides two user experiences to address a wide range of use cases and deployment scenarios. FIDO protocols are based on public key cryptography and are strongly resistant to phishing.

The passwordless FIDO experience is supported by the Universal Authentication Framework (UAF) protocol. In this experience, the user registers their device to the online service by selecting a local authentication mechanism such as swiping a finger, looking at the camera, speaking into the mic, entering a PIN, etc. The UAF protocol allows the service to select which mechanisms are presented to the user.

Once registered, the user simply repeats the local authentication action whenever they need to authenticate to the service. The user no longer needs to enter their password when authenticating from that device. UAF also allows experiences that combine multiple authentication mechanisms such as fingerprint + PIN.

2.6 History of mobile biometric devices

Now that we've discussed what mobile biometric systems and devices *are,* let's have a look at how we've gotten to where we are. Although some of these designs may seem almost comical in light of today's technology, we invite you to look to Google for *other* technology of the day. It is certain that the reader will also find that desktop computers, mobile phones, and game consoles seem equally silly. However, in their day, they were the cutting edge and often predicted the road map of how we've gotten where we are today. Considering the pace of change and that mobile biometric devices got really started within the last two decades, in two decades hence, it is likely that the best of today will seem…comical.

2.6.1 Law enforcement market devices

The first patented portable biometric device titled "Portable fingerprint scanning apparatus for identification verification" was filed to US patent office on 11/19/1991 by Glenn M Fishbine and Robert J. Withoff from DBI (Digital Biometrics Inc.), and it was awarded on 6/22/1993. DBI is the one of earliest companies which developed

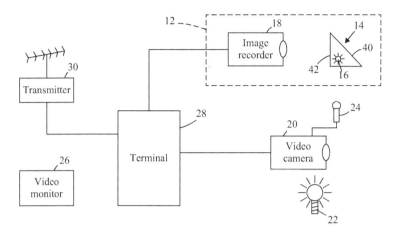

Figure 2.3   Invention idea block diagram from US Patent 5222152

biometric capturing devices of any kind, mobile devices being just one of them. DBI was acquired by Visionics in 2001 and later became part of Morpho after subsequent corporate acquisitions.

The DBI device might be considered as the first documented design attempt for mobile biometric device. The following figure shows the block diagram for the patented device and consists of a camera, an optical fingerprint scanner and a terminal which controls the two sensors to capture, select and store fingerprint and mugshot images. The terminal also contains a monitor to allow the images to be displayed and a radio-frequency transmitter allowing transmittal of the captured images to a mobile unit for identification. The radio frequency transmitter was standard RF wireless carrying data to communicate between the law enforcement agents. The radio frequency communication bandwidth was very limited. In these early days, to use RF data to transmit images was a big challenge and would have presented difficulties if the invention idea was implemented. This early patent did not depict an enclosure design of the portable device and was not disclosed in the patent. Figure 2.3 illustrates the invention idea of US patent 5222152.

DBI issued a press release about a device designed based on the idea in the patent application on 2/19/1992. In the news release, it stated the device was developed for the specifications of the FBI's National Crime Information Center (NCIC) in connection with its NCIC 2000 project [51]. The second patent application on same title was filed by Glenn M Fishbine, Robert J. Withoff, and Theodore D. Klein on 3/31/1993. The patent was awarded on 11/14/1995. The patent application was a continuation and extension of the earlier filed patent. In this patent, it further disclosed a lightweight fingerprint scanner and gave a detailed disclosure of the device housing. The outward design of the device is shown in Figure 2.4. There were four claims added in the patent. In 1995, a version of the first mobile device called (SQUID) [52] was developed by DBI according to news [53] released on 11/2/1995.

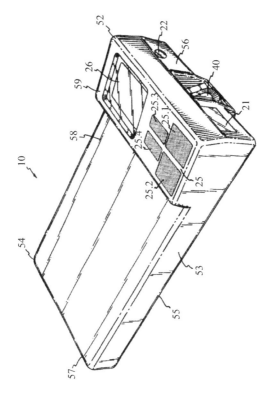

Figure 2.4   A sketch drawing for a mobile biometric device from US Patent 5467403

As such, this new release might be only a prototype. According to the disclosed information from the new release [54] of 1999, SQUID was to be used in police cars in connection with a pilot program in conjunction of NCIC 2000 project. SQUID was not a standalone device and was only a portable, handheld, device capable of capturing and downloading the fingerprint image to a laptop in the squad car. Even so, this device could be considered as the result of the first design attempt for the mobile fingerprint device design with optical sensor. Figure 2.4 shows a representation of one embodiment of a package used to house the portable image collection unit for the invention of US patent 5467403.

From 1995 to 1999, several AFIS companies such as Printrak, Morpho, NEC, Identix, and CrossMatch also developed similar types of portable fingerprint devices. These devices were still bulky and lacked wireless communication capability. As with the SQUID, these are other examples of the first generation of mobile fingerprint devices that, in the strict sense, were only portable fingerprint capture devices with wired communication capability. The best way to categorize the early devices is to think of them as limited, special purpose live-scan devices. The following sections show some examples of these early biometric devices.

Figure 2.5 A mobile fingerprint device "SFS 2000" from Printrak

Figure 2.6 A mobile fingerprint device "MV5" from CrossMatch

2.6.1.1 Examples of early mobile biometric devices

Printrak developed a portable device called SFS 2000 as shown in Figure 2.5 which provided one-to-one fingerprint matching capability according to the news release [55] on 12/10/1996. The device contained advanced features of the time including optical sensor to capture fingerprint and also sensors to distinguish left hand to right and thumb from index finger. The other key features also included real-time quality control, real-time display of scanned prints, prompts for image centering and recapture, and self-calibration.

The device contained an on-board 50-MIPS computer processor, touch screen interface, 500-dpi resolution optical scanning device and Printrak's specialized image processing software. Weighing less than 2.5 pounds, the SFS 2000 was approximately $13 \times 7.25 \times 4.75$ in. in size. The captured images might be sent directly to a host computer via network transmission. The SFS 2000 was "internet ready" and supported 10Base-T, RS-232, and PCMCIA interfaces.

In 1999, CrossMatch Technologies Inc. started to develop a mobile fingerprint capture device which was called MV5 [52,56,57] as shown in Figure 2.6. The device was released in middle of 2000. The key feature of the device was very small and light with rechargeable battery which made it very attractive for law enforcement at that time.

In 1999, DBI started to develop a second generation of SQUID that would be delivered together with a new developed AFIS server named IBIS which could accept remote submission from the new SQUID. The product consisted of a remote capture device (SQUID), multiple communications interfaces, a central server, and a real-time relational database [54]. This device could be considered as an example of second generation design for a mobile fingerprint device with remote identification using wireless capability. The device was still bulky and wireless capability was limited due to constraints of the wireless technology at that time.

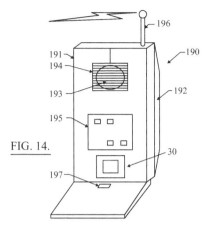

Figure 2.7   A sketch drawing of a mobile fingerprint device from US Patent 6088585 by Authentic

AuthenTec built a handheld wireless device prototype with capacitive fingerprint sensor in 1997. A few functioning prototypes were built and demonstrated using a 2-Mbits wireless Ethernet connection. Two fingerprints could be captured, transmitted wirelessly, searched against a database and the red light/green light result returned to the device to notify the operator if there was a match or no match. The design idea was filed as a patent application [58] and was awarded on 7/11/2000. Figure 2.7 shows the diagram of patented mobile device.

Subsequent to AuthenTec's release, other biometrics companies such as Sagem Morpho, Cogent, Identix, and CrossMatch also released similar devices with wireless capability. Examples of these include the Morpho RapID™ 100 and Cogent Blue Check.

In 2008, Morpho released Morpho RapID 1100 [59] as shown in Figure 2.8 which could be considered as an example of the third generation designs of mobile biometric devices. The feature characteristics different from the second generation are as follows:

- From bulky to small and good looking
- Better user interface: the communication and information is well integrated
- From single biometric fingerprint to two biometrics: fingerprint and face
- Ability for seamless connectivity and mobility
- Robust & high performance mobile platform
- On board and remote capability
- Much better accuracy.

The Morpho RapID 1100 was an early example of biometric companies developing mobile biometric devices by leveraging their core competencies coupled with mobile device manufacturer core competencies. The RapID 1100 was built on the Psion Teklogix iKon™, a new rugged handheld PDA highly resistant to dust and rain.

Figure 2.8   A mobile biometric device "Morpho RapID 1100" from Sagem Morpho

Increased wireless communication options WiFi, cellular and BlueTooth with GPS localization for rapid data transfer from any location. As a PDA, this was a very advanced device from the time and beyond what essentially any company not explicitly in the mobile device business to create. A 500-ppi forensic-quality ruggedized scanner captures digital images of a subject's fingerprints. Using highly accurate Morpho Automated Biometric Identification System (ABIS) technology, these prints could instantly compared against an on-board watch list of up to 180,000 records, or transmitted wirelessly for automated identification against a centralized database. Mug shots can also be captured with the integrated 2-mega-pixel camera for facial recognition.

Search results are returned for review on the Morpho RapID 3.7-in. screen with customer defined data, such as mug shots, rap sheets, wants/warrants, etc. Fast positive identification with full reporting means an officer can assess risk on-site very quickly, without having to return to the station or even the police cruiser. The entire automated process can range from near real-time to less than 3 min, depending on such factors as information requested, communication, and database size. In 2009, Cogent also released a similar device called Cogent MI2 [60] with similar features.

In 2010, L-1 Identity Solutions, released a device named HIIDE 5 [61]. HIIDE 5 as shown in Figure 2.9 is a Handheld Interagency Identity Detection Equipment as shown. The HIIDE series [62,63] of devices could be considered as an example of the fourth generation of the mobile biometric device. Considering all the design criteria for a successful solution cited in earlier sections, the HIIDE must have gotten something right. While a *highly* specialized, target market specific and *expensive* device, nearly 100,000 of them have been deployed and are in active use not only in the US Armed Forces but with other governments around the world.

The advances implemented beyond third generation devices include:

• From two biometrics to three biometrics: fingerprint, face, and iris
• Better seamless connectivity and mobility
• Better mobile platform
• Having much better on board identification capability
• Improved accuracy.

Figure 2.9 A mobile biometric device "HIIDE 5" from L-1

The system was customizable to fit a wide range of military missions, homeland security applications, law enforcement, and transportation security programs.

The HIIDE 5 was the latest release of the HIIDE series of devices designed for "hasty enrollment." The unit weighs 3.1 pounds and fits into a cargo pocket for optimal portability. HIIDE 5 has a full-quality optical sensor for single- or two-finger flat-and-roll fingerprint capture. Simultaneous dual iris capture, a full-function camera for face capture, and built-in image quality checks come standard. GPS and wireless transmission capabilities supporting 802.11, Bluetooth, 3-G/4-G cellular, WiFi, Tactical Radio and WiMax are also included.

The device could capture multibiometrics that can search and match onboard the device, or remotely over a wireless connection. The HIIDE simultaneously captures dual-irises, slap or roll fingerprints using an optical sensor, and captures facial images. The key features include:

- Ruggedized, Windows based, Hand-Held, Mobile computer including Touch-Screen, daylight readable display
- Capable of running Windows applications, record digital photos, video, biometrics, GPS audio, and text data
- Biometric capability includes iris, fingerprint, and facial recognition, with on-board real time matching up to 1,500,000 records
- Multiple, Flexible network connectivity options include Ethernet, WiFi, 3G plus interfaces to Military Networks (such as Bowman).

Given the clear success of a specialized device like the HIIDE, several companies such as CrossMatch, 3M, Northrop Grumman, and Credence ID subsequently developed similar types of mobile biometrics devices as shown in Figure 2.10 for fingerprint, face, and iris. The names of the devices are CrossMatch SEEK Avenger [64], 3M Fusion, Northrop Grumman BioSled [65] Credence Trident [66]. The figures of these devices are from the companies' product brochures. Some of the companies that developed these devices have been merged into other entities, but some new companies are borne. As of the today, there are many companies that develop mobile biometrics aimed at government and large-scale commercial applications. The need for industrial duty mobile biometric devices continues to the point where new players

*Figure 2.10   Examples of mobile devices similar to "HIIDE 5" from the
companies: CrossMatch, 3M, Northrop Grumman and Credence ID*

pop up; some with vapor and some with excellent products. Amrel is an excellent
example of a new company with an excellent product. This rugged device maker
now markets a multimodal mobile biometric device with many of the same functional
properties of the HIIDE 5 and embodies the key elements necessary for a successful
device cited above.

2.6.1.2   Newer devices on the market but prior to current generation

To provide a "snapshot" of what a biometrics company participating in the large-
scale biometrics business offers, we'll look at Safran Morpho as an example. Along
with companies like NEC Biometrics Division, and 3M Cogent Systems, each offering a similar range of capabilities, Morpho's range of mobile biometric devices and
solutions will give us an understanding of today's state of art offerings for law enforcement applications. Each of the three cited companies remain among the *very* few in
this large-scale biometrics market and each have similar histories of organic product
development and release coupled with technology acquisitions.

Safran Morpho has more than 40 years' experience in biometrics and was a pioneer in biometric matching and a leader in providing large-scale biometric solutions.
Morpho continues to grow its business and technology portfolio both organically
and also by external acquisitions. Morpho is also a very good "snapshot" of recent
consolidation in the large-system biometrics providers where both large and small
companies are rolled up into a broad-spectrum biometrics conglomerate. Morpho
acquired both L-1 Identity Solutions and Motorola's Biometrics Division (formerly
Printrak) to form two subsidiaries in the United States named MorphoTrust and MorphoTrak. L-1 Identity Solutions contained a rich portfolio of biometric technologies
both through initial acquisitions and then subsequent ongoing development.

The origin of L-1 Identity Solutions can be traced back to Identix (pioneer of
biometric scanners), DBI (who created the first mobile fingerprint capturing device
in 1992), Visionics and Visage (originally competitive pioneers in face recognition),
Iridian (developed the first iris recognition system), and Bioscrypt (a pioneer in access
control using biometrics). Printrak was a market leader in law enforcement AFIS.

As market leader in biometrics, Morpho and its subsidiaries have participated and
made great contribution to the mobile biometric community. Morpho, as an aggregate
amalgamated corporate entity, has been providing mobile identification solutions for

Figure 2.11   Examples of mobile devices developed by Morpho in recent history

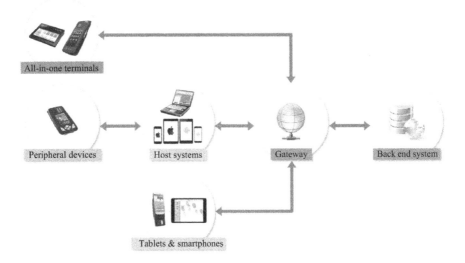

Figure 2.12   Example of mobile devices integrated with Backend AFIS system

more than 20 years beginning with that first device from DBI. Morpho's leading technology, being the end result of ongoing research and development built on internal and acquired technology provides this very good example of mobile technology as it mirrors also the progression of the handful of other top tier biometrics companies in the world.

Many different versions of mobile devices have been designed and released by Morpho and their acquired entities in the last 20 years. Figure 2.11 depicts a few of the mobile biometrics devices from recent history.

Today, Morpho offers mobile biometrics devices for many different markets, including criminal justice, public security, border control, civil ID, and corporate solutions. The mobile devices cover the following three types:

*   All-in-one terminals
*   Peripheral devices
*   Data capture attachments and applications for COTS/Consumer devices.

As seen in Figure 2.12, whether from Morpho or any vendor, these devices can be readily be integrated with back-end technology from a variety of biometric vendors due to the NIST standard and its variants.

*Figure 2.13   An example of all in one mobile biometric device "MorphoTablet"
by Morpho*

2.6.1.3   Examples of Morpho newly released mobile biometric devices

The MorphoTablet [67], as shown in Figure 2.13, is an in-house designed all-in-one unit that can be provisioned with user-friendly Morpho biometric applications for a wide variety of solutions. The device can perform identification and verification (on the device or remotely) in multiple situations and environments. MorphoTablet is an enterprise-class, touch-sensitive device that adds the security of biometrics (fingerprint and face) and cryptographic functions to mobile operations. Beyond the law enforcement market, MorphoTablet has equal potential in commercial solutions and has the facilities to ensure secure mobile data operations, including mobile identification controls. Key features include the following:

- Android-based tablet with incorporated FBI Personal Identity Verification (PIV) IQS, and Standardization Testing and Quality-Certification-certified optical fingerprint sensor
- Embedded secure element, cryptographic functions, and secure device management
- Integrated NFC and contact/contactless smartcard readers
- Cellular (voice and data), WiFi, and Bluetooth communication
- Ready to host customer specific developments (SDK available) or Morpho proprietary applications.

Morpho RapID 2 [68] as shown in Figure 2.14 is another Morpho-designed mobile all-in-one device targeted at government and law enforcement markets with a primary focus of biometric identification rather than more general application like the tablet. It is a ruggedized device that is highly resistant to dust and rain with very

Figure 2.14   An example of all in one mobile biometric device "Morpho RapID 2" by Morpho

robust communication. It can perform fingerprint or face identification on device or remotely. The key features include the following:

- Biometric capture: FAP 30 and FBI PIV IQS certified fingerprint sensor, 8MP camera with autofocus and flash
- Contact and contactless card readers (optional)
- 2D barcode reader (optional)
- MRZ reading capability (using device rear camera)
- Communication: 4G/LTE (3G for US market), WiFi, Bluetooth
- Rugged design: elastomer protections, IP54, -20 to $60°$C operating range.

Morpho's peripheral device is essentially a discrete fingerprint capture device that is designed to be paired with a smartphone, tablet, or mobile PC. Continued existence of this device is a key indicator that some of the "early" designs that were largely the sensor part of the equation did not have it all wrong. MorphoIdent [69] as shown in Figure 2.15 is an example of Morpho's peripheral devices. Morpho sells thousands of these devices around the world, as does 3M Cogent with their "BlueCheck" and NEC with their recently released "NeoScan 45," to a variety of customers and requirements. MorphoIdent's key features include:

- Light and palm sized
- FBI PIV IQS optical scanner
- Vibration alert, color screen with intuitive GUI

Figure 2.15   An example of mobile peripheral biometric device "MorphoIdent" by Morpho

Figure 2.16   Examples of mobile attachment for standard smart phones by Morpho

- On-device (up to 1:5,000) or remote biometric system identification
- Multicase management, multiple requests processed simultaneously
- More than 8 h of continuous use from a single battery charge.

To take advantage of the best commercial off the shelf solutions and smart phones, Morpho developed data capture attachments as shown in Figure 2.16 and both application software and developer kits to provide enriched solutions for all markets and different customer demands. The top brands of commercial smart phones can be easily integrated using tools from Morpho's mobile biometric offerings.

Morpho's data capture attachments unit contains the following capture features:

- Optical fingerprint scanner
- Contact and contactless card reader
- MRZ swipe reader

Figure 2.17   Examples of mobile attachment for standard smart phones by Morpho

- Magnetic stripe reader
- 2D barcode reader
- Integrated battery.

MiMs (Morpho iMobile Suite) [70] contains mobile biometric solutions enabling real-time operations in the field. MiMs can perform the following applications:

- Enroll persons of interest
- Verify secure identification documents
- Identify individuals on a local watchlist or back-end authoritative database
- Perform direct capture of latent prints at crime sites.

Some of application screen shots of MiMs are shown in Figure 2.17. Morpho's mobile application software can be used both on Morpho designed mobile devices and commercial smart phones or tablets running Android or iOS.

MiMs can be used for contactless capture of biometrics and biographic information in a wide range of scenarios and can locally match the biometrics against a database on the device. In addition, it can also manage transactions to submit search requests and display results on mobile platforms. It is compliant with FBI American National Standard for Information Systems (ANSI)-Information Technology Laboratory (ITL) standards.

MiVerify is an application component of MiMs specifically developed for remote fingerprint and face identification searches. MiMs also supports iris enrollment through integration with third-party Software Development Kits (SDKs). Using a mobile device, fingerprints, faces, and demographics are captured, sent to a local watchlist or authoritative database for search.

Another component of MiMs is MiCrimeScene which can capture latent fingerprint, live facial and SMT images. It can be used to capture prints to identify deceased soldiers. MiCrimeScene works with cameras on phones and tablets, and enables captured digital images to be transmitted instantly from remote locations to an ABIS.

2.6.2   Commercial/consumer market devices with biometric capabilities

In the consumer market, the main application currently employed by a mobile biometric device is to use mobile biometric to secure a device or make a transaction based on biometric *verification* which is rather different from law enforcement applications where the primary function is *identification*. Many different biometrics, such as voice, face, and handwriting, along with fingerprints have been tried for the verification in consumer market. With all these variations tried on mobile or even desktop devices, it seems that the fingerprint remains the best and consistently reliable single mode biometric as of today. Low cost (both technologically and device build cost) of capturing biometrics and reliable small sensors have been top priorities in the mobile phone design consideration. As such, the first attempt at commercial release for a mobile phone with fingerprint sensor was a few years after the solid-state sensor was introduced.

One of the earliest attempts [71] to roll out a commercial mobile phone secured using fingerprint was in 1998 by Siemens. The prototype as shown in Figure 2.18 was developed in 1998 and shown off at CeBIT in 1999. Unfortunately, the prototype was never commercialized due to unexpected cost, space size and sensor issues during the commercialization stage.

Between 1998 and 2000, computer companies such as HP, Acer, Compaq, and NEC-integrated fingerprint sensors on "mobile device" notebook computers to secure them. While these were reasonably successful, there were two challenges to design a mobile telephone secured by fingerprint. Compared to the laptop computers just noted, there was simply just not that much room to place the sensor and related boards of the day inside the then shrinking mobile phones and the processor was really not very powerful. To make it work on a mobile phone these two challenges had to be overcome. In addition, the cost of sensor and a powerful enough processor made the biometrically secured mobile phones significantly more expensive than their PIN-secured siblings. Sagem overcame this well enough during design in 1999 that in 2000, Sagem unveiled the world's first commercially released dual-band GSM mobile

Figure 2.18 A prototype cell phone with fingerprint sensor developed by Siemens

Figure 2.19 MC 959 ID with fingerprint sensor developed by Sagem

phone MC 959 ID [72] that included a fingerprint sensor on the back for enhanced user security. The MC 959 ID mobile phone with fingerprint sensor weighed under 4 oz with 150 h standby and 3.5 h talking.

The Sagem MC 959 ID as shown in Figure 2.19 was the first commercial mobile phone that used a fingerprint sensor to replace PIN codes where the user could store a maximum of five fingerprints. The sensor used to capture the fingerprint image is a ST/Upek touch fingerprint sensor placed at the back of the phone. Sagem had overcome the issues faced by Siemens and gotten a viable device to market.

Figure 2.20   iPAQ h5450 PDA with fingerprint sensor developed by HP

Two years later, in 2002, the second commercially released mobile device secured with a fingerprint was rolled out. The HP iPAQ h5450 PDA [73] as shown in Figure 2.20 had much more powerful processor than the Sagem MC 959 ID and had a great looking design for the time which was also a harbinger of smartphones to come. Many early adopters bought the device when it rolled out. The PDA came with the Atmel AT77C101 thermal swipe sensor and was equipped with a 400 MHz Intel Xscale and 64 MB memory with a weight of 7.6 oz.

The swipe sensor was placed at bottom center right under Navigation Button. To unlock the device, one needed to use two hands; i.e. one hand to hold device and the other hand to swipe the sensor. It was not as convenient as touch based sensor which can be used singlehanded if correctly designed. To unlock the device the finger had to be swiped across the sensor in certain way to get a good quality fingerprint and did have its "issues" with consistent performance usually due to the need to be *very* consistent in swiping the sensor which just was not that easy to do all the time. The nice aspect of the swipe sensor at the time was that they cost much less than touch based sensors and took less physical space on the case. However, along with the difficulty to use consistently, the sensor was not well protected against scratching and wear, leading to other reliability issues.

All that said, the biometric function of PDA worked *well enough* that it was the standard to meet at that time. Before 2013 (with the release of the iPhone), the enrollment process of most *usable* devices was, or became, more or less similar to the iPAQ and is a good example of an early implementation that worked reasonably well and was quite user friendly. Looking at it directly as presented by Hewlett Packard:

1. From the Start menu, tap Settings, and then Password.
2. Select the option from the down arrow list that you choose for your security setting.

3. Enter either your PIN or password when prompted.
4. On the Fingerprint screen, tap the finger on the display you want to enroll.
5. Swipe the selected finger. It is recommended that you gently swipe your finger downward across the sensor. You should begin the swipe at the first joint of your finger and continue downward with a smooth slow motion.
6. Monitor the status to see if your fingerprint was correctly swiped. If OK, the perimeter of the oval turns green and a message displays indicating the quality was good. If the captured image is poor quality, the perimeter of the oval turns red and a message displays.
7. Repeat steps 5 and 6 to acquire a sufficient number of quality swipes (notice the progress bar) to enroll the print.
8. Tap OK.
9. Tap Yes.

While this worked and was the best of the day, it still had issues. During the enrollment process, the device stopped enrolling after 6 of the 8 slots for good quality fingerprints were captured. Since HP was *not* a biometrics company and did not understand many of the ramifications of biometric matching, the assumption was that 6 good ones in a row were all that was needed. However, the 6 out of 8 good quality fingerprints enrolled did not deviate much from each other such that a slightly different speed or position swipe during authentication caused a failed match. Hewlett Packard did not provide instruction, much less the capability, to the user to enroll different positions of finger to account for this. The designers assumed that "good enrollments" *also* meant that they should be all the same. Where this falls down is that during the enrollment, not only good quality fingerprints need to be enrolled, but also all possible angle and position swipes of good quality fingerprints need to be enrolled so there were not failed matches of *other* good captures but, for example, just at a different angle.

Since 2002, many major mobile device vendors [74–76] have tried to roll out fingerprint sensor based devices each year, but they just did not gain much traction. The majority of these devices had technology that failed to live up to user expectations. Sensor scratch and wear protection was gradually solved over the years, but poor overall biometric application user experience was not solved until the Touch ID was introduced on Apple's iPhone 5s in 2013. The excellent user experience of Touch ID changed the earlier bad perceptions of biometrics and made biometrics a household word. iPhone 5s soundly demonstrated that a device secured by a fingerprint could be convenient and secure when correctly designed and used.

We might surmise that this was truly successful due to a couple of key and coincident factors:

1. Apple is a company obsessed with positive user experience. If they have a corporate group tasked with unboxing new Apple products to ensure that the unboxing is a pleasant experience, they'll make sure all elements are as pleasantly designed as possible.
2. Apple bought a *real* biometrics company, AuthenTec, including their biometric domain expertise; not just a sensor from a component provider as had been

Figure 2.21   iPhone 5s with touch ID sensor developed by Apple

done up until this time. By bringing significant biometric expertise in-house, Apple was the first commercial product company that definitively understood the ramifications of biometric processing. Prior to this, other manufacturers without this domain expertise tried to make fingerprints as much the *same* as possible like really long *identical* PIN numbers. Apple, with AuthenTec's expertise, accounted for that fussy human variability factor.

Even 3 years on (a lifetime in the consumer electronics market) and with many competitors releasing biometric enabled devices, iPhone's Touch ID can be considered state-of-the-art in the consumer market using a fingerprint for authentication. The iPhone's Touch ID capacitive, flat image sensor is built under the traditional iPhone "home button" as a scratch resistant button made from sapphire crystal—one of the clearest, hardest materials available. The button is usually the first physical control the user touches so the biometric capture happens automatically in the process users are already accustomed to. The steel ring surrounding the button, looking like the bezel on previous devices, detects a finger and tells Touch ID to start reading your fingerprint. Figure 2.21 shows the Touch ID sensor.

iPhone 5s with touch ID led the hot wave for fingerprint based smartphones. The total number of smartphone models with fingerprint sensors rolled out in the three years since iPhone 5S released in 2013 is greater than the number of *all* rolled out devices with fingerprint capabilities before 2013.

The sensor captures the fingerprint image when a finger touches on the button. Touch ID compares the captured fingerprint with the preenrolled fingerprints to verify whether the new captured fingerprint is from the same person who enrolled fingerprints on the device or not. If the match is successful, the phone is unlocked. Otherwise, 5 attempts are allowed before a passcode is needed to unlock the phone.

The sensor on the phone is *very* small, especially when compared to the typical size found on biometric function specific devices. The size is less than 0.7 cm on each side and, based on 500 DPI resolution; the image is less than 140×140 pixels. Apple overcame this by capturing different fingerprint images from different parts of the first joint of a finger by enrolling many images from different portion/position of the finger to model the whole first joint fingerprint. The enrolled templates enable a user to be able to use almost any area of first joint finger to unlock the device. With the rotation invariant matching used, the user can touch the sensor from any fashion (any orientation) to unlock the device. This made the iPhone 5s as an outstanding device which has excited the whole world about biometrics.

iPhone 6s used more powerful processor and better-optimized algorithms. The phone can be instantly unlocked when the user touches the sensor. The 6s also incrementally adds new portions of the person's fingerprint to the enrolled fingerprint/template to improve matching accuracy over time. The Touch ID feature allows the user to not only unlock the device but also can be used to authorize credit card payments (another Apple innovation) and iTunes store purchases. The iPhone 7, and we can be sure any subsequent devices, continue to build on these capabilities. In addition, Apple has recently purchased another definitive biometrics company, Real-Face, and we can be sure that multimodality embodying rigorous biometrics best practices are not far off.

Apple filed two patents to protect Touch ID. One is a about a matching method "Finger biometric sensor providing coarse matching of ridge flow data using histograms and related methods" on 3/15/2013. The matching method was awarded on 8/25/2015 and US patent number is 9117145. The other one is "efficient texture comparison." The application was filed on 3/12/2013. The publication number is US20130308838. The second one focuses on protecting the algorithms and the other focuses on protect the device and the algorithms in situ.

From the claims and block diagrams of the patents, we can see the matching algorithm consists of the following process: it extracts the feature map and a low resolution pattern which represents the captured fingerprint. It uses the low resolution pattern of the input image to match against with the enrolled low resolution enrolled pattern associated with enrolled images and return few candidate pattern's feature maps for detail matching.

Since Touch ID was introduced on iPhone 5s in 2013, all major smartphone vendors have tried to catch up to the functionality and wholly positive user experience of the Apple devices. The definitive user benefits, not to mention customer acceptance and demand for fingerprint authentication caught their eyes. Competitors may be able to catch up to this level of function in 6 to 18 months, but we can assume that Apple will have gone further by that time, setting the bar even higher. In 2015, the top smartphone models [77–79] with fingerprint sensors include the latest Google Nexus devices, Samsung Galaxy S6, and Huawei's P8 and Ascend Mate 7. Some users claim the user experience of these phones is better than that of iPhone 5s. However, they may either be enjoying the fast speed of the authentication of these new phones or it could simply be manifestation of the Andrioid anti-iOS ethos. The fast authentication speed on these phones is not surprising since the processors on these phones are faster than that of iPhone 5s which has now also been eclipsed by subsequent iPhone models. Some of the non-Apple phones put the fingerprint sensors on the back, whereas some place on the front in a similar location to the Apple iTouch. Although not all devices have the same silky-smooth user experience of the well-executed iOS devices, all devices have now progressed to where the typical user experience is a good one with simple to use, reliable, and fast one touch biometric-secured access to their devices. While the Apple devices use their now in-house AuthenTec sensors, the majority of the fingerprint sensors on other devices are from Fingerprint Cards AB (FPC). FPC carries a variety of fingerprint sensors [80] for a variety of applications.

2.7   Future and challenges

What does the future bring for mobile biometrics and what are the challenges? Looking in a crystal ball has always been an endeavor fraught with peril. One needs only to look as recently as the movie "Back to the Future 2" from 1989 to see just how wrong the predictions depicted for "Back to the Future Day," October 21, 2015, really were. Your authors currently do not have a flying DeLorean or a flying *any*thing in our garage at home. Do not get us started on how wrong economists can be about even the next quarter.

That said, we have decades of experience in this "new" biometrics industry working for companies that have even more decades delivering biometric systems. While we *may* eventually be proven spectacularly wrong, we at least have extensive direct experience with a significant research and engineering foundation on which to base our prognostications. As such, we'll probably end up being only *somewhat* wrong.

If you pay any attention to the biometrics industry, business, or the biometrics industry *and* business, we are sure you've seen the now commonplace "*<insert biometric modality here> biometric market expected to grow <insert astronomical growth figure here> in the next <insert timeframe here> years*" statements. It is our opinion that these statements should be taken with, perhaps, a shovelful of salt for a few reasons. We are not saying that biometrics, including mobile biometrics, will not have significant growth in the coming years. However, most of such statements are based simply on a formulaic arithmetic progression built on historical evidence. Where they are not purely formulaic and some actual research on the biometric market was done by an economist, most of the recent business activities have been based on biometric *verification* projects which are rather different from *identification*.

Further, although some of the modalities are projected to have some astronomical growth, there are significant technical aspects to be overcome for them to be as consistently reliable as fingerprints. Although biometric verification implementations do an excellent job of validating a person against a claimed identity, we can be sure that, soon enough, it will become imperative to also definitively ensure a discretely unique identity for many use cases. If one does an internet search simply for "biometric companies," dozens, if not hundreds, of discrete companies will pop up.

However, there remain only a handful of companies *in the world* who are definitively in the large-scale biometric identification market. When these factors eventually do get factored into market forecast statements, we can be sure they will look rather different with more definitive foci. If we add in that adding biometrics to many market segments will take a definitive investment and infrastructure change by the business world as a whole, this upsurge may end up a bit flatter in practice.

Throughout this chapter, we've commented on the upsurge in interest and acceptance of biometrics initiated by the hugely successful release of the Apple iPhone 5s and improved upon by subsequent iPhone models and will continue to be with new models not yet seen. Fortunately, for ongoing consumer acceptance of biometrics, releases from other vendors have not yet been comparatively substandard enough to sour the consumer's taste for biometrics. As a result, the days of biometrics only appearance in "popular" media being when a publicly traded biometric company did something noteworthy, prompting 30 s on MSNBC's "Mad Money" with Jim Cramer are long gone. With consumer acceptance firmly in place and evolving to become consumer *demand* we can be sure that R&D money will continue to flow making biometrics on mobile devices constantly more reliable and easy to use. We can already see signs of this where only a few short years ago biometric capabilities in the consumer market were only on a *handful* of top tier or specialized devices while at the time of this writing are making their way down to mid-range devices.

As demand grows, beyond merely being a commonly standard feature on any decent mobile device, biometrics capabilities may well be a requirement to be even remotely successful in the marketplace. Already we see biometrics being used to physically secure and authenticate a mobile device. The release of the iPhone 6 with Apple's iPay using iTouch biometrics to authenticate made credit/debit card payments as secure as the credit/debit card companies should have years ago while still maintaining the convenience of old school plastic. We can expect to see continued growth for POS purchases in this manner along with soon to be expected application to online shopping driven by consumer demand for payment processing security. Already, many "apps" in the iOS realm utilize iTouch for their authentication including for purchases such as with Amazon.

Looking a little farther out and considering what is being done by the FIDO alliance [50], it is likely only a matter of time until mobile biometric enabled devices are ubiquitously used to definitively authenticate a claimed identity for almost any purpose where proof of who you say you are is important.

With the rapid roll-out of smartphones using biometrics today, authentication using fingerprints for many transaction types is becoming ever more popular. However, while many of these transactions such as contactless biometrically enabled credit card payments are becoming more common, there remains multiple alternative ways to complete the same task *without* a biometric enabled smartphone. As adoption moves onward to where this is not an alternative but a requirement, devices will need to be available to all different groups of people and the device will need to work under many different environments.

Today, only the top models of the top brands of smartphones are equipped with fingerprint sensors and only a few use fingerprints for anything other than unlocking the device. Also, the majority of the device owners and users are in higher income brackets, especially considering Apple products. The condition of their finger is generally much better than those who engage in manual labor. The fingers with labor intensive work are more easily damaged and the ridges of the fingers may be rather worn down.

This means the accuracy and reliability of the algorithms that match these fingerprints will be much lower than the current crop of more affluent early adopters. Even beyond those engaged in manual labor, so often the affluent engage in "weekend warrior" type activities that can just as readily damage fingers. This will require improving accuracy of fingerprint matching along with potentially solving some of the environmental issues that can impact the accuracy of other modalities so that they may be used in concert with or as a backup to fingerprint matching.

Consistent accuracy and reliability *under any circumstances* will be paramount for biometric devices to become the norm rather than the exception when employed for whatever types of transactions the world may dream up.

For large-scale government applications, the main function is identification, which is *much* harder than simple verification since it needs to accurately compare millions of prints to decide whether a captured print definitively matches one inside the database or not. As with commercial applications, the imperative will be to continue to improve capture and match capabilities to account for many different environmental issues. Already we are seeing mobile biometric devices being deployed to be used during crime scene investigation. When they become used more and more for the direct capture of evidence, they will need to have biometric algorithms strengthened to support potentially nonfingerprint-expert users in the field.

Devices with the currently (for a mobile device) very large FAP 45 sensor will increasingly be used as mobile booking/enrollment devices and will need to evolve to honestly support this and we may see the traditional live-scan station go the way of the cassette tape. Real time latent image capture and identification will certainly be improved, albeit with some challenges, with improvements to sensors, cameras, assistive lighting, biometrics processing and matching algorithms. Many of these improvements are actually already on the way.

Biometric research communities and industries are working hard to improve the accuracy of biometric recognition. The main areas of research have been (including both capture and processing):

* Single type feature algorithm improvement,
* Multifeature from single modal and fusion,
* Multibiometric to multimodal biometrics.

Many great improvements on accuracy [81] have been seen in recent years. Many of these improved technologies can be integrated into new devices or systems now and in the future to help improve accuracy.

Ever since biometrics was first electronically processed, the privacy issue of using biometric has been a hot topic and concern. This is amplified in recent time with all the data breaches that have made the news. The information captured in recent well known data breaches is data that is directly human usable while biometrics are not without special effort. However, while the existence of biometrics as a data processing element is well known by the general population now, they are still little understood by the general population.

As a result, coupled with how news outlets might *possibly* over-amplify items to make ratings, extreme and unfounded fears abound. That said, protection of biometric data is and will become extremely important. While not *readily* usable outside their native environment, biometrics are not *impossible* to use outside their native environment. While a password can be changed, one's fingers are not so easily changed.

Key players such as the old guard from government project context National Institute of Standards, Technology and the National Security Agency and industries now starting to use biometrics outside their traditional government setting must and inevitably will work together to create processes and policies that allow for more rapid innovation while still embracing the necessary security guidelines. The policies from government driven down to industry need to be in place and understood by the individuals involved. During design, encryption and security protection for data handling from transferring to storing need to be implemented. The FIDO Alliance [50] can and should provide a good framework for all the players to address the issue.

In commercial biometric applications, the vulnerability of a biometric system is always a concern. With physical consumer devices, it is particularly a concern since the device could be lost and fall to the wrong hands potentially allowing the biometrics of a person could also be stolen. With new and better algorithms, new concerns will emerge as "real" rather than hysteria we see often now.

For example, fingerprint extraction algorithms that can capture fingerprints from an image rather than needing contact mean fingerprints left some places such as glass or even high resolution photos of a person displayed on a public website such as Facebook or LinkedIn could be subverted. This is just as real of a concern, perhaps even more so, as someone hacking a database or the ever famous "gummy bear" fear, where a "fake" fingerprint is made from gummy bear material and used on a fingerprint sensor. One easy way to address the issue is to combine the biometric authentication with password protection which helps in a simple use case but does not put biometrics in primary control *or* address issues where biometrics are the requirement to uniquely *identify* you in a huge government or corporate database.

For the longer term, biometric companies will need to and will invest in better liveness detection [82,83], which checks the biometric samples used for authentication as to whether a biometric is from a live person or not. Much research [48] has been done on liveness detection in recent years. For fingerprint liveness detection, there are two types of approaches: hardware based and software based. The hardware-based approach incorporates different types of sensors beyond the fingerprint image sensors to measure life sign features such as finger temperature, pulse, blood pressure, electric resistance, and odor. The software approaches try to detect perspiration [84], skin deformation [85], uniformity [86], and pores [87], among others [88,89].

For face liveness detection, many of the methods [90–94] have been proposed based on such as Fourier spectra, facial texture change and landmark tracking. Face recognition devices such as the Samsung S4 face unlock, simple liveness detection has been incorporated to try to prevent the attempt of using a face photo/picture of a person.

A lot of progress has been made, but the liveness test accuracy is still not very high and still some "nonlive" faces get through. Liveness detection techniques are often different for each biometrics. The detection is normally optimized and done for the each type sensor. A liveness detection technique for one sensor might not be work on other sensors even if for the same biometric.

Other technologies [49,95–98] have also been proposed to solve the issues for some applications. Normally, spoofed biometrics are not placed on the sensor but injected or, if on a sensor are utterly motionless. By using continuous authentication [95] to look for and identify those little movements that any live human being makes this vulnerability is greatly reduced. Multibiometrics [45–47] can be used to improve authentication accuracy, but also can be used to reduce vulnerability since it is hard to create fake multibiometrics and place them on the device simultaneously.

A picture of a person's face might be easily stolen but a fingerprint or iris of a person, without skill, might be harder to produce. The vascular/vein of finger or palm of a person would be very hard to get. These technologies could be combined with liveness and continual authentication technologies to make biometrics authentication on the mobile device almost impossible to attack without, quite possibly, more effort than it is worth to attack.

To integrate these technologies, costs will increase, and some of the technologies are not yet mature enough to be placed on production devices. Protection will need to be implemented based on a level of security requirements to achieve a balance between the accuracy, security, convenience, and cost. With the technologies improved and getting cheaper as time goes on, some of these technologies can, and will be, integrated. With education, instruction from biometrics providers, and users understanding the implications and responsibilities of device security, coupled with technology, will lead these devices to being more secure with the fewest possible opportunities for attack.

The biometrics business is booming especially with more biometric devices rolling out every day. The strong demand for biometrics sensors will drive more research funding to sensor and algorithm research. While many of such forecasts are taken with *much* salt by the authors, one is probably pretty accurate. According to a new market research report [99] for fingerprint sensor market, the market is expected to reach $14.5 billion by 2020 from $5.5 billion in 2014 which is an average of 17.1% growth rate each year. Figure 2.22 is a financial graph that shows the revenue growth from the main fingerprint manufacture FPC last few quarters. From its growth rate, we can see that the sensor business is really booming. The investment to the sensor research also proportionally increased. This can be seen from the recent financial growth as shown in Figure 2.22 which was extracted from FPC's Q3/2015 financial report presentation [100] page 18.

Even if many of the market research reports are overly optimistic, we know that biometric processing will continue to be improved and advanced. Liveness detection might be part of sensor since most of the techniques are designed for certain sensor type. Pulse sensors might be incorporated together with fingerprint sensors. Capturing on move and at distance technologies [101,102] for face and iris have advanced in

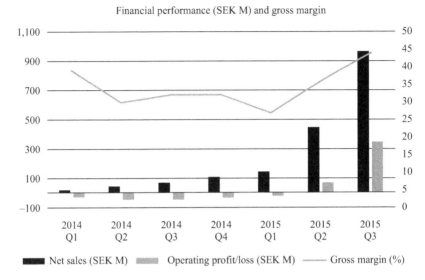

Financial performance (SEK M) and gross margin

Net sales (SEK M)   Operating profit/loss (SEK M) ——— Gross margin (%)

Figure 2.22   Financial growth of FPC from Q3/2015 financial report

recent years. Touch-less capturing of fingerprint, palmprint, and latent fingerprints [103] are also on the way.

However, all technologies have to overcome the "noncontrolled environments" problem since the device must be able to be used at anytime, anywhere. With the technologies improving, it would not surprise us if some type of mobile device using touch-less capturing of fingerprint, palmprint, and latent print for law enforcement application will be released in the near future. The research [104,105] toward using a display as fingerprint sensor is also under away. Apple was awarded a patent for fingerprint sensing underneath the touchscreen on 9/8/2015.

Another recent development in fingerprint sensor technology, the flexible optical sensor [106], if proven in practice, could perhaps be an enormous impetus to biometric adoption. All current fingerprint sensors are typically flat and square. This new technology, being flexible, will allow fingerprint sensors to be placed where they could never be practically be placed before such as, for example, a handgun grip or the inside grasping surface of a door handle.

Furthermore, the flexible nature of the sensor means that any curved surface designed as much for aesthetics as function could become a surface capable of capturing fingerprint images. Where an "optional" biometric capability might have been eschewed because an "ugly" flat sensor would harm the look of a device, when capturing on a curved surface is possible…the sky is the limit. If we consider that this new technology not only works in the visible spectrum of light but *also* infrared which allows sensing *below* the surface of the skin, the potential for adding significant liveness detection presents itself. This approach along with other finger and palm vascular/vain capture sensor might be improved one day to allow integration with fingerprint sensors on mobile devices.

There might be a lot challenges ahead in the design, but they will be solved. From the history of fingerprint device development, it is clear that the mobile biometric device will be much more widely used in the future. So, while, as we said in the introduction to this section, we might be somewhat wrong on a few of these things, one prediction we can make for sure; it's going to get big. Really big. And it will be an interesting ride in the process.

References

[1]   T. Brewer, and N. Kerris. Apple Announces iPhone 5s—The Most Forward-Thinking Smartphone in the World. Cupertino, CA. 2013. Available from http://www.apple.com/pr/library/2013/09/10Apple-Announces-iPhone-5s-The-Most-Forward-Thinking-Smartphone-in-the-World.html [Accessed 26 Jan 2016].

[2]   Biometrics Research Group Inc New Releases for Mobile Biometric Report. Oct. 2015. Available from http://www.prweb.com/releases/2015/10/prweb13048469.htm [Accessed 26 Jan 2016].

[3]   Rawlson King. Special Report: Mobile Biometrics Market Analysis. Oct. 2015. Available from http://www.biometricupdate.com/201510/special-report-mobile-biometrics-market-analysis [Accessed 26 Jan 2016].

[4]   J. Wayman, A. Jain, D. Maltoni, and D. Maio, Eds. Biometric Systems, Springer-Verlag: New York, 2005.

[5]   S. Prabhakar, A. Jain, D. Maltoni, and D. Maio, Handbook of Fingerprint Recognition; Springer-Verlag: New York, 2003.

[6]   N. Ratha, and R. Bolle, Eds. Automated Fingerprint Recognition Systems; Springer-Verlag: New York, 2007.

[7]   History of Fingerprints Time Line. Available from http://www.fingerprintamerica.com/fingerprinthistory.asp [Accessed 26 Jan 2016].

[8]   Fingerprint History. Available from http://onin.com/fp/fphistory.html [Accessed 26 Jan 2016].

[9]   U.S. Marshals Service for Students. Available from http://www.usmarshals.gov/usmsforkids/fingerprint_history.htm [Accessed 26 Jan 2016].

[10] J.G. Barnes. The Fingerprint Sourcebook. Chapter 1: History. 2010. Available from https://www.ncjrs.gov/pdffiles1/nij/225321.pdf [Accessed 26 Jan 2016].

[11] Rob Horton, 40 years of AFIS, Safran News Magazine, page 17–19, issue 6, Jan. 2015.

[12] Kenneth R. Moses, Peter Higgins, Michael McCabe, Salil Prabhakar and Scott Swann. *The Fingerprint Sourcebook*. Chapter 6: Automated Fingerprint Identification System(AFIS). 2010. Available from https://www.ncjrs.gov/pdffiles1/nij/225326.pdf [Accessed 26 Jan 2016].

[13] Fingerprint Recognition. https://www.fbi.gov/about-us/cjis/fingerprints_biometrics/biometric-center-of-excellence/files/fingerprint-recognition.pdf [Accessed 26 Jan 2016].

[14] The Henry Classification System. New York, NY. International Bio-
      metric Group. 2003. Available from http://www3.pittsfield.net/sandbox/
      groups/lauraschneider/wiki/d37f3/attachments/9353c/Henry%20Fingerprint
      %20Classification.pdf?sessionID=7b21a52d8e141778cda9806effefcf8b5cdd
      1620 [Accessed 26 Jan 2016].

[15] Automatic Pattern Processing System, Filed: September 10, 1976, Rockwell
      International Corporation, US Patent Number: 4,151,512.

[16] Binary Image Minutiae Detector, Riganati; John P., Vitols; Visvaldis
      A. Rockwell International Corporation, September 10, 1976, US Patent:
      4,083,035.

[17] Minutiae Pattern Matcher, Riganati; John P., Vitols; Visvaldis A. Rockwell
      International Corporation, September 10, 1976, US Patent: 4,135,147.

[18] Shuo Wang and Jing Liu. Biometrics on Mobile Phone. Yang, D.J. (Ed.),
      2011, Recent Application in Biometrics, pp. 1–22. Available from http://cdn.
      intechweb.org/pdfs/17035.pdf [Accessed 26 Jan 2016].

[19] Secure Phone. European Commission, 2004. Available from http://www.
      secure-phone.info/ [Accessed 26 Jan 2016].

[20] Welcome to MOBIO (Mobile Biometry). Available from http://www.
      mobioproject.org/ [Accessed 26 Jan 2016].

[21] R. Ricci, G. Chollet, M.V. Crispino, *et al.* (2006). "The SecurePhone: a
      mobile phone with biometric authentication and e-signature support for deal-
      ing secure transactions on the fly," Proc. of SECRYPT 2006, Int. Conf. on
      Security and Cryptography, pp. 9–16, Setúbal, Portugal.

[22] Jassim, S.A., and Sellahewa, H. "Face verification schemes for mobile per-
      sonal devices," Zanin, Kurdish Scientific and Medical eJournal, issue 1,
      vol. 1, pp. 21–30, 2005.

[23] Guillaume Dave, Xing Chao, and Kishore Sriadibhatla. Face Recognition in
      Mobile Phones. Department of Electrical Engineering Stanford University,
      USA, 2010.

[24] S. Marcel, C. McCool, C. Atanasoaei, *et al.* MOBIO: Mobile biometric face
      and speaker authentication, in: Proc. IEEE Conference on Computer Vision
      and Pattern Recognition, 2010.

[25] S. Marcel, C. McCool, P. Matejka, *et al.* On the results of the first mobile
      biometry (MOBIO) face and speaker verification evaluation, in: Proceedings
      of the ICPR 2010 Contests, 2010.

[26] Sung-Uk Jung, Yun-Su Chung, Jang-Hee Yoo, and Ki-young Moon.
      Real-Time Face Verification for Mobile Platforms. In Advances in visual
      computing, pages 823–832. Springer Berlin Heidelberg, 2008.

[27] S. Kurkovsky, T. Carpenter, and C. MacDonald. Experiments with simple
      iris recognition for mobile phones. In Information Technology: New Genera-
      tions (ITNG), 2010 Seventh International Conference on, pages 1293–1294,
      2010.

[28] Jennifer R. Kwapisz, Gary M. Weiss, and Samuel A. Moore. Cell phone-based
      biometric identification. In 2010 Fourth IEEE International Conference on
      Biometrics: Theory, Applications and Systems (BTAS), pages 1–7. IEEE,

September 2010. (This paper describes a system that uses phone-based acceleration sensors, called accelerometers, to identify and authenticate cell phone users).

[29]   Julio Angulo and W Erik. Karlstad, Sweden. Exploring Touch-screen Biometrics for User Identification on Smart Phones. 2011. (Draw pattern based). Available from https://www.kau.se/sites/default/files/Dokument/subpage/2011/11/exploring_touch_screen_biometrics_for_user_identif_16753.pdf [Accessed 26 Jan 2016].

[30]   Joe Rossignol. 2015. Apple Addresses 'Bendgate' By Strengthening Weak Points of 'iPhone 6s' Shell. Available from http://www.macrumors.com/2015/08/10/iphone-6s-addresses-bendgate-video/ [Accessed 26 Jan 2016].

[31]   Jean-Francois Mainguet on Fingerprint Sensors: 2004–2015. Available from http://biometrics.mainguet.org/types/fingerprint/fingerprint_sensors_productsi.htm [Accessed 26 Jan 2016].

[32]   American National Standards for Information Systems—Data Format for the Interchange of Fingerprint Information; ANSI/NIST-CSL 1-1993; National Institute of Standards and Technology, U.S. Government Printing Office: Washington, DC, 1993.

[33]   American National Standard for Information Systems—Data Format for the Interchange of Fingerprint, Facial & SMT (Scar, Mark, and Tattoo) Information; ANSI/NIST-ITL 1a-1997; National Institute of Standards and Technology, U.S. Government Printing Office: Washington, DC, 1997.

[34]   "Criminal Justice Information Services (CJIS) WSQ Gray-scale Fingerprint Image Compression Specification," Federal Bureau of Investigation, Document No. IAFIS-IC-0110 (V3), 19 December 1997.

[35]   American National Standard for Information Systems—Data Format for the Interchange of Fingerprint, Facial, & Scar Mark & Tattoo (SMT); ANSI/NIST-ITL 1-2000, NIST Special Publication #500-245; National Institute of Standards and Technology, U.S. Government Printing Office: Washington, DC, 2000.

[36]   American National Standards for Information Systems—Data Format for the Interchange of Fingerprint, Facial, & Other Biometric Information; ANSI/NIST-ITL 1-2011, NIST Special Publication #500-290; National Institute of Standards and Technology, U.S. Government Printing Office: Washington, DC, 2011.

[37]   National Institute for Standards and Technology. MINEX: Performance and Interoperability of INCITS 378 Fingerprint Template (NISTIR 7296); March 6, 2005.

[38]   P. Grother, M. McCabe, C. Watson *et al.* MINEX: Performance and Interoperability of INCITS 378 Fingerprint Template; NISTIR 7296; National Institute of Standards and Technology, March 21, 2006.

[39]   Ongoing MINEX: 2010–2015. Available from http://www.nist.gov/itl/iad/ig/ominex.cfm, created 13 May 2010, updated 21 September 2016 [Accessed 5 July 2017].

[40] Proprietary Fingerprint Template Evaluation II: 2010–2014. Available from http://www.nist.gov/itl/iad/ig/pftii.cfm, created 2 April 2010, updated 22 December 2016 [Accessed 5 July 2017].

[41] MINEX III: 2015–2016. Available from http://www.nist.gov/itl/iad/ig/minexiii.cfm, created 15 April 2015, updated 12 June 12 [Accessed 5 July 2017].

[42] E. Khoury, B. Vesnicer, J. Franco-Pedroso *et al*. The 2013 speaker recognition evaluation in mobile environment. In Biometrics (ICB), 2013 Inter-national Conference on, pages 1–8, 2013.

[43] Face Challenges: 2011–2015. Available from http://www.nist.gov/itl/iad/ig/face.cfm [Accessed 26 Jan 2016].

[44] Iris Recognition Homepage: 2011–2014. Available from http://www.nist.gov/itl/iad/ig/iris.cfm [Accessed 26 Jan 2016].

[45] J. Kittler, and N. Poh, Multibiometrics for Identity Authentication: Issues, Benefits and Challenges, in: IEEE Conference on Biometrics: Theory, Applications and Systems, pages 1–6, 2009.

[46] Arun Ross and Anil K. Jain, "MULTIMODAL BIOMETRICS: AN OVERVIEW," Proc. of 12th European Signal Processing Conference (EUSIPCO), (Vienna, Austria), pp. 1221–1224, September 2004.

[47] J. Daugman. Cambridge University. Combining Multiple Biometrics. Available from http://www.cl.cam.ac.uk/~jgd1000/combine [Accessed 26 Jan 2016].

[48] Amani Al-Ajlan. Survey on fingerprint liveness detection. In Biometrics and Forensics (IWBF), 2013 International Workshop on, pages 1–5. IEEE, 2013.

[49] Heather Crawford, Karen Renaud, and Tim Storer. A framework for continuous, transparent mobile device authentication. Computers & Security, pages 1–10, May 2013.

[50] Fido Alliance. Available from https://fidoalliance.org/ [Accessed 26 Jan 2016].

[51] Digital Biometrics Announces Portable Fingerprint and Mugshot Unit. PR Newswire Association LLC. 1992. Available from http://www.thefreelibrary.com/DIGITAL+BIOMETRICS+ANNOUNCES+PORTABLE+FINGERPRINT+AND+MUGSHOT+UNIT-a011927847 [Accessed 26 Jan 2016].

[52] Darrell Geusz. In the Beginning: Fingerprint Biometrics in Mobile ID. 2013. Available from http://mobileidworld.com/in-the-beginning-fingerprint-biometrics-in-mobile-id/ [Accessed 26 Jan 2016].

[53] Digital Biometrics Announces Portable Fingerprint and Mugshot Unit. PR Newswire Association LLC. 1992. Available from http://www.thefreelibrary.com/DIGITAL+BIOMETRICS+ANNOUNCES+PORTABLE+FINGERPRINT+AND+MUGSHOT+UNIT-a011927847 [Accessed 26 Jan 2016].

[54] Digital Biometrics Previews the Identification Based Information System (IBIS) for Law Enforcement Investigators. MINNETONKA, Minn.

PRNewswire. 1999. Available from http://www.prnewswire.com/news-releases/digital-biometrics-previews-the-identification-based-information-system-ibis-for-law-enforcement-investigators-76958512.html [Accessed 26 Jan 2016].

[55] Paula B. Bordigon. Printrak announces new portable fingerprint ID solution. Anaheim, CA. 1996. Business Wire. Available from http://www.thefree library.com/Printrak+announces+new+portable+fingerprint+ID+solution. -a018927682 [Accessed 26 Jan 2016].

[56] CrossMatch MV 5. Available from http://commlogik.com.br/products/26/395 [Accessed 26 Jan 2016].

[57] Cross Match Technologies Inc. Announces the Release of MV5 Mobile Biometric Fingerprint Capture Device. Available from http://www. mobic.com/oldnews/2001/05/cross_match_technologies_inc.htm [Accessed 26 Jan 2016].

[58] The Patent Application Titled "Portable Telecommunication Device Including a Fingerprint Sensor and Related Methods" by John C. Schmitt and Dale R. Setlak. US Patent Number US 6088585.

[59] N. Jullien, and C. Coudert. Sagem Sécurité introduces Morpho RapID 1100 for the first time in Qatar. Paris, France. 2008. Available from http://www. morpho.com/en/media/20081117_sagem-securite-introduces-morpho-rapid-1100-first-time-qatar [Accessed 26 Jan 2016].

[60] Cogent MI2. Available from http://multimedia.3m.com/mws/media/773452O/ 3m-mi2-mobile-identification-handheld-device.pdf?fn=MI2.pdf [Accessed 26 Jan 2016].

[61] Morpho HIIDE 5. Available from http://www.morpho.com/en/public-security/check-id/mobile- devices/hiide5#sthash.iVm31Lw1.dpuf [Accessed 26 Jan 2016].

[62] "Biometric Automated Toolset (BAT) and Handheld Interagency Identity Detection Equipment (HIIDE)," NIST XML & Mobile workshop, Biometric Task Force. 9/19/2007.

[63] User guide of HIIDE S4, L-1 identity solutions.

[64] Cross match Seek Avenger. Available from http://findbiometrics.com/cross-match-introduces-new-seek-avenger-handheld/ [Accessed 26 Jan 2016].

[65] Northropgrumman BioSled. Available from http://www.northropgrumman. com/Capabilities/IdentityManagement/Documents/BioSled_datasht.pdf [Accessed 26 Jan 2016].

[66] Credence Trident. Available from http://mobileidworld.com/credence-id-launches-trident-android-based-tri-biometric-enrollment-and-verification-device-02131412/ [Accessed 26 Jan 2016].

[67] MorphoTablet. Available from http://www.morpho.com/en/public-security/ check-id/mobile-devices/morphotablet [Accessed 26 Jan 2016].

[68] Morpho RapID 2. Available from http://www.morpho.com/en/public-security/check-id/mobile-devices/morphorapid-2 [Accessed 26 Jan 2016].

[69] MorphoIdent. Available from http://www.morpho.com/en/public-security/ check-id/mobile-devices/morphoident [Accessed 26 Jan 2016].

[70] Mopho MiMs. Available from http://www.morpho.com/en/public-security/ check-id/software-suite/mims [Accessed 26 Jan 2016].

[71] The First Mobile Phone Prototype with Fingerprint Sensor from Siemens. Available from http://www.bromba.com/protoe.htm [Accessed 26 Jan 2016].

[72] Sagem News Release for MC959 Mobile Phone with Fingerprint Sensor. Available from http://www.thefreelibrary.com/Sagem+adds+fingerprint+ recognition+to+GSM+mobile+phones.-a059670994 [Accessed 26 Jan 2016].

[73] HP News Release for iPAQ h5450 PDA with Fingerprint Sensor. Available from http://m.hp.com/us/en/news/details. do?id=170696&articletype= news_release [Accessed 26 Jan 2016].

[74] The History of Mobile Device with Fingerprint Sensor. Available from http://biometrics.mainguet.org/types/fingerprint/fingerprint_history.htm [Accessed 26 Jan 2016].

[75] The History of Smart Phone with Fingerprint Sensor. Available from http://biometrics.mainguet.org/appli/appli_smartphones_history.htm [Accessed 26 Jan 2016].

[76] The History of Selective Mobile Fingerprint Devices for Law Enforcement Applications. Available from http://biometrics.mainguet.org/appli/ appli_gov_portID.htm [Accessed 26 Jan 2016].

[77] Top 10 Android Smartphones with Fingerprint Sensor. Available from http://thedroidguy.com/2015/09/top-10-android-smartphones-capable-of-fingerprint-recognition-1049781 [Accessed 26 Jan 2016].

[78] Top Android Smartphones with Fingerprint Sensor. Available from http://www.phonearena.com/news/Top-Android-phones-with-fingerprint-scanner-protection-Late-2015-edition_id72976 [Accessed 26 Jan 2016].

[79] Top 2015 smartphones with fingerprint sensor. Available from http://gadget note.com/top-smartphones-fingerprint-sensor [Accessed 26 Jan 2016].

[80] FPC Fingerprint Sensors. Available from http://www.fingerprints.com/ products/ [Accessed 26 Jan 2016].

[81] NIST Biometric Accuracy Tests. Available from http://www.nist.gov/ itl/iad/ig/biometric_evaluations.cfm [Accessed 26 Jan 2016].

[82] Stephanie Schuckers, Larry Hornak, Tim Norman, Reza Derakhshani, and Sujan Parthasaradhi, Issues for Liveness Detection in Biometrics, CITeR, West Virginia University, Presentation, 2003.

[83] Hyosup Kang, Bongku Lee, Hakil Kim, Daecheol Shin, and Jaesung Kim. A study on performance evaluation of the liveness detection for various fingerprint sensor modules. In Knowledge-Based Intelligent Information and Engineering Systems, pages 1245–1253. Springer, 2003.

[84] Sujan TV Parthasaradhi, Reza Derakhshani, Larry A Hornak, and Stephanie AC Schuckers. Time-series detection of perspiration as a liveness test in fingerprint devices. Systems, Man, and Cybernetics, Part C: Applications and Reviews, IEEE Transactions on, 35(3):335–343, 2005.

[85] Jia Jia, Lianhong Cai, Kaifu Zhang, and Dawei Chen. A new approach to fake finger detection based on skin elasticity analysis. In Advances in Biometrics, pages 309–318. Springer, 2007.

[86]   Shankar Bhausaheb Nikam and Suneeta Agarwal. Curvelet-based fingerprint anti-spoofing. Signal, Image and Video Processing, 4(1):75–87, 2010.

[87]   M. Espinoza and C. Champod. Using the number of pores on fingerprint images to detect spoofing attacks. In Hand-Based Biometrics (ICHB), 2011 International Conference on, pages 1–5, 2011.

[88]   Jia Jia and Lianhong Cai. Fake finger detection based on time-series fingerprint image analysis. In Advanced intelligent computing theories and applications. with aspects of theoretical and methodological issues, pages 1140–1150. Springer, 2007.

[89]   Jiamin Bai, Tian-Tsong Ng, Xinting Gao, and Yun-Qing Shi. Is physics-based liveness detection truly possible with a single image? Proceedings of 2010 IEEE International Symposium on Circuits and Systems, pages 3425–3428, May 2010.

[90]   Face Anti Spoofing Evaluation, 2013. Available from https://www.tabularasa-euproject.org/evaluations/icb-2013-face-anti- spoofing [Accessed 26 Jan 2016].

[91]   Face Anti Spoofing Evaluation, 2013. Available from https://www.tabularasa-euproject.org/evaluations/ijcb-2011-competition-on-counter-measures-to-2d-facial-spoofing-attacks [Accessed 26 Jan 2016].

[92]   J. Li, Y. Wang, T. Tan, and A.K. Jain. Live face detection based on the analysis of Fourier spectra. Defense and Security. International Society for Optics and Photonics, pages 296–303, 2004.

[93]   Jukka Komulainen and Abdenour Hadid. Face Spoong Detection Using Dynamic Texture. Computer Vision-ACCV 2012 Workshops, pages 146–157, 2013.

[94]   H Fronthaler and K Kollreider. Assuring Liveness in Biometric Identity Authentication by Real-time Face Tracking. Proceedings of the IEEE International Conference on Computational Intelligence for Homeland Security and Personal Safety, (July):21–22, 2004.

[95]   Koichiro Niinumaa, and Anil K. Jain. Continuous User Authentication Using Temporal Information, Proc. SPIE 7667, Biometric Technology for Human Identification VII, 76670L (April 14, 2010); doi:10.1117/12.847886.

[96]   P. Tome, R. Raghavendra, C. Busch *et al.* The 1st competition on counter measures to finger vein spoofing attacks, In The 8th IAPR International Conference on Biometrics (ICB), May 2015.

[97]   Qin Bin, Pan Jian-fei, Cao Guang-zhong, and Du Ge-guo. The anti-spoofing study of vein identification system. In Computational Intelligence and Security, 2009. CIS '09. International Conference on, volume 2, pages 357–360, 2009.

[98]   J. Galbally, J. Fierrez-Aguilar, J. Ortega-Garcia and R. Cappelli, "Fingerprint Anti-spoofing in Biometric Systems," in Marcel, Sébastien, Nixon, Mark S., Li, and Stan Z. (Eds), Handbook of Biometric Anti-Spoofing, Springer, 2014.

[99]   The Source of the Biometric Sensor Market Report. Available from http://www.marketsandmarkets.com/PressReleases/fingerprint-sensors.asp [Accessed 26 Jan 2016].

[100]   http://www.fingerprints.com/wp-content/uploads/sites/2/2013/09/FPC-Q3-15-Earnings-Presentation.pdf [Accessed 26 Jan 2016].

[101]   SRI Iris on Move and at Distance. Available from https://www.sri.com/engage/products-solutions/iris-move-biometric-identification-systems [Accessed 26 Jan 2016].

[102]   Mopho Iris at Distance and on Move. Available from http://www.morpho.com/en/media/20140311_iris-distance-power-behind-iris [Accessed 26 Jan 2016].

[103]   Available   from   http://cherup.yonsei.ac.kr/files/article/CCBR2015.pdf [Accessed 26 Jan 2016].

[104]   Embedded Authentication Systems in an Electronic Device US Patent Number: 9128601.

[105]   Display Screen with Integrated user Biometric Sensing and Verification System. US Patent Publication Number: 20140133715 A1.

[106]   Technology Site Description of Flexible Optical Sensors. Available from http://optics.org/news/7/1/17 [Accessed 26 Jan 2016].

Chapter 3

Challenges in developing mass-market mobile biometric sensors

Richard K. Fenrich[1]

WARNING: This chapter presents concepts that are not commonly synthesized and discussed in open forums. With tongue in cheek, perhaps the reason is that nobody is interested in what will be said. But it is far more likely the reason is that much of the information relates to companies that make at least a portion of their living by leveraging biometrics into solutions. Company-business practices and intellectual property are the keys to unlock these revenue streams and to protect competitive advantage. Yet, these items are rarely discussed. Much of what is presented is derived from the author's experiences and conclusions after having been involved in the biometric community for many years. Every effort has been made to give the reader an appreciation of the challenges in developing mass-market biometric sensors from a holistic perspective.

The journey begins with an admission that the title assumes the reader is at least familiar with the concepts mentioned in the title. For example, the term mass market implies that the sales opportunities meet certain business criteria like 'there is a large volume needed'. Consequently, there must be an identified need or pull for these products. In addition, the reader may be tempted to assume that the challenges presented will be all technical in nature. Nothing could be farther from the truth! In fact, depending on the company, perhaps, the biggest challenge of all is to ensure they do not run out of money on their path to commercialization.

To place a framework around the challenges discourse, the first section provides background information that places the discussion into the correct context. After briefly refining the meaning of the words in the chapter title, use cases are discussed so that we can bridge the gap between marketing and engineering. Some technical background information is then discussed. After defining the term 'biometric sensor,' the three sensor types considered herein are defined. Finally, a simple New Product Development (NPD) framework is introduced as the vehicle through which the challenges will be presented.

The following section then considers the primary challenges in developing mass-market biometric sensors. Business and technical issues will be covered within the

[1] Identification International, Inc., USA

NPD framework. The completeness of this presentation is not guaranteed; there are many other challenges that can arise in the development of mass-market biometric sensors.

At the end of the primary challenges, the business discussion will have ended and the reader will have a taste of how business and technical challenges interplay with one another. But to hopefully provide some more satisfaction, the final section briefly discusses some interesting technical directions that could prove to enable significant mobile-biometric capabilities in the future.

With this short introduction, let's start with the background first.

3.1 Background discussion

What is meant by the words 'developing,' 'mass market' and 'mobile' within the context of this chapter? It will be assumed that the reader is familiar with the term biometric as a physiological trait that can be used to uniquely identify individuals. Certain characteristics of biometrics will be more fully developed later in the chapter.

While interpretations of the singular word 'developing' could cause dramatic contention at a pub bordering a university and a technology park, the words that follow 'developing' in the title place it in a specific commercial context. Within this context, developing should be interpreted with the intent of pursuing the nearest term goal of generating a sellable biometric sensor that has the following features:

- the cost to manufacture the sensor is as low as possible,
- the failure rate of the sensor is minimal,
- the sensor is easy to use for any consumer who uses it and
- the sensor does its job, always.

Moving on, let us make the simple definition that a mass market is any market in which more than 100,000,000 sales opportunities exist per year. This number could be different, of course, but for our purposes, this market size incorporates smart phones, tablets, smart watches, health monitors and various classes of devices that have been discussed in the scope of the Internet of Things (IoT).

Not so long ago, systems were sold into biometric markets that were mobile because one could put them on a trolley to move them from one place to another! The view of what a mobile biometric sensor is has changed over the last 10 years with the understanding now being guided primarily by the mobility of electronics such as tablets and smart phones. Therefore, for the purposes of this chapter, let's consider mobile biometric sensors as sensors that can be embedded into devices one carries in a pocket, purse or satchel in an inconspicuous manner. If the opportunity arises to stretch the size of the purse to meet the goals of this chapter, the stretching will be called out so as to not offend the reader. Implicit in the discussion of mobile sensors is the concept that the sensor will be embedded or tightly integrated into a host system. This means that the sensor needs to be as small as possible while requiring as little power as possible. In today's product landscape, biometrics in these environments tend to fall into one of two areas: smart phones and tablets.

These simple definitions give us a glimpse of the tip of an iceberg. From a business perspective, they cajole us into deepening our knowledge by getting a firm grasp on the use cases for the sensor, the types of conditions under which the sensor will be used, the environment into which the sensor is integrated, the manufacturability of the sensor, the marketing and sales of the sensor and so on. Let's dig into a high-level definition of mass-market use cases on which the remainder of the chapter will be based.

3.1.1 Use cases

Use cases describe the particular function the biometric will be used for. Specifications defining the sensor and host system are extracted from the use cases. The use cases from that point on are understood more deeply through the specifications. As an example, the primary use case for mass-market biometric sensors has been authentication for convenience – the process of guarding access to a device using the biometric instead of a password or a Personal Identification Number (PIN). This use simply means matching the stored version of a biometric against the biometric captured on the sensor.[1] Device access is granted when the stored biometric matches the input biometric, otherwise access is not granted. While generating the specifications associated with this use case, the following items should be considered:

1. the type of biometric sensor to be used, e.g. fingerprint, iris, facial,
2. the size, weight and power consumption of the sensor,
3. the error rates that are acceptable,
4. the speed at which the authentication occurs,
5. the environmental conditions the sensor must work in,
6. the ease of use for the user,
7. the sensor's failure rate and maintenance requirements,
8. the sensor's cost and
9. the extendibility of the sensor to other potential use cases.

Many other factors should also be considered while enumerating the specifications and taken together the factors serve to define a very usable, low cost and reliable approach to completely address the use case.

While authentication for convenience has been the primary use case for sensors, there are use cases that extend this usage. The largest amongst these is authentication for financial transactions. If your smart phone stores your electronic wallet that contains your bank account, credit card and electronic commerce logins, you should be interested in paying bills and purchasing goods using this information. The primary difference between the authentication for convenience and authentication for financial transactions is that the consumer should demand a much higher level of security to ensure their sensitive information is protected to a much higher level. Authentication

---

[1]Authentication typically has the characteristic of matching one input biometric against a small number of stored biometrics tied to identities. Identification, on the other hand, typically matches that input against a much larger set of stored biometric/identity pairs.

for convenience allows a consumer onto their device with little trouble and therefore may give the consumer the belief that the device is secure enough even though only a small number of people access the device. Moving forward, this perception trap needs to be carefully addressed by the specifications derived from the use cases.

This example brings up the point that a single platform can be used for multiple use cases if the embedded biometrics provide the functionality to support those use cases. For example, the United States Department of Defence had a programme for many years called Biometric Access Tools (BAT) – the larger than a purse BAT system collected and managed several biometrics with the intent on applying these biometrics to a specific set of about 20 use cases. This functionality extension to new use cases will come around again later in the discussion about the business continuity strategy.

Day identifies 12 primary biometric applications [1], and within these 12 applications, the two authentication use cases just discussed are the largest mobile mass-market use cases. Many of the other applications cited commonly require collection of a person's biometrics to enable identification of a person against databases that have many entries. For example, criminal background checks with fingerprints identify people against a database with 104 million entries in the United States of America [2]. In these non-mobile scenarios, standardization and interoperability features of the sensors take on a much larger role. A mass-market use case for identification has not been realized yet.

For the purposes of the remainder of the chapter, the focus will be placed on authentication for convenience and for financial transactions.

3.1.2   Biometric sensors

The role of a biometric sensor is to capture a person's specific physiological characteristic in such a manner so as to minimize that attribute's variance considering a wide variety of factors. In other words, the goal is to generate a sample, so it is identical to all other samples from the same source. Again, according to the Encyclopaedia of Biometrics [3], some sources of variability include the following:

1. intrinsic biological variability – broad variability exists between people;
2. environmental variability – lighting, humidity and temperature conditions affect capture;
3. sample presentation variability – pressure varies for touch fingerprint sensor and the angle a face or iris is captured at can change;
4. biological target contamination – a person might have dirty fingers or be wearing glasses or a hat;
5. acquisition losses, errors and noise – the sensor peculiarities may result in a poorly captured sample.

The goal of a good biometric sensor is to consider these variables and capture the signature of the given physiological trait with as small of a variance over time as possible. If the sensor is placed into an environment where the biometric being captured is almost always from the same person, then there is a better chance of capturing consistent samples for comparison because the intrinsic biological variability factor is reduced. Fortunately, this characteristic works to our advantage in the two authentication use cases and is partly the reason why existing mass-market sensors are being well adopted today. On the other side of the coin, as more secure devices are desired, sensors need to consider the variability between people much more carefully.

In the most basic form, a biometric sensor converts an energy impulse into a digital representation of the physiological trait. The energy could be visible or infrared (IR) light, pressure, electricity, acoustical or thermal. Today's sensors universally output digital data which is then consumed by various software and hardware processes that implement the biometric extraction and matching engine. In the remainder of this text, what we mean by a biometric sensor is that component or set of components that converts the incident energy into a digital representation of the biometric. Therefore, a camera that is formed from an imaging chip, associated electronics, lens and lens adjustment mechanism is a sensor. Likewise, a swipe scanner for fingerprint capture is a sensor.

A sensor in a mobile environment cannot in and of itself do the work the use case demands. No, the sensor is much like a member of an orchestra or a sports team wherein one part intimately relies on another to deliver the best overall performance. In mobile systems in particular, the level of integration with the electronics and the software has to be very tight in order to deliver such a highly compact and functional solution. Thus, in some places within this text, challenges properly outside of the sensor itself will be described simply because features of the sensor relate to the challenges described.

The Encyclopaedia of Biometrics [4] identifies at least 20 different biometrics. For the purposes of this chapter, fingerprints, iris and facial will be the three biometrics that are focused on only because they have had the most work done on them in the mobile environment. It is indeed true that the camera of a smart phone can capture video (for gait analysis) or images for palm vein or hand geometry, but the ubiquity of these first three defines them as the likely candidates for near to mid-term mass-market adoption. The other biometric candidate that could readily be viewed as a candidate for consideration is voice analysis due to the fact that there are microphones on smart phones and tablets. But the massively noisy environments that would interfere with the use of voice as a viable mass market alternative leads to a lower prioritization of this biometric herein.

3.1.3 New product development

As tempting as it may be to focus on technical issues as the primary challenges to developing sensors, it would likely completely miss the primary point of this chapter. Rather, starting from the premise that mass-market adoption of biometrics will continue to accelerate, this chapter seeks to create an appreciation for the challenges in all phases of successfully bringing mass-market biometric sensors to the consumer. From this higher level viewpoint, the technology of mobile biometric sensors certainly needs to be considered, but more importantly, the business environment that the sensor development is enveloped in plays a much larger role as to whether or not such a biometric sensor will be commercially successful. For this very reason,

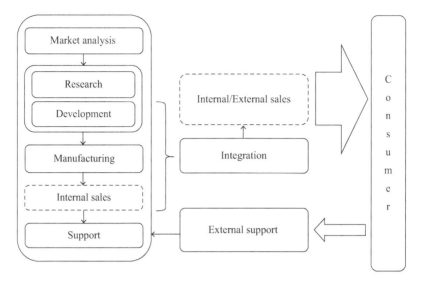

Figure 3.1   New product development process

this chapter wraps the discussion of the primary challenges into a NPD business process. In reviewing each step within the NPD process, challenges arise at every twist and turn. Inability to meet these challenges can relegate a company to the corporate graveyard but surmounting the challenges can lead to incredible success.

Quite a few NPD approaches exist in the literature and in practice. Companies tend to have their own unique implementation of a NPD approach that has been refined over time. Often these processes define a series of decision-making points, or gates, which must be passed in order for a new product development to proceed to the next step. These gates are designed to ensure good decision-making processes so the overall NPD optimally uses human and capital resources, i.e. they are designed to avoid wasting money. A starting point that contains various NPD approaches can be found on Wikipedia [5].

For the purposes of this chapter, let's settle on a hypothetical NPD process that clearly illustrates the overall elements involved in bringing a product to market. It is not an optimal process, and it is not recommended for your company's approach as it misses many details and nuances, but it delivers the message about the inherent challenges in developing mobile mass-market biometric sensors. This process is shown graphically in Figure 3.1.

The NPD process starts with a market analysis in which the use cases, requirements, market size, price sensitivity and competitors are identified. A firm grasp on these constituents is built into a business plan incorporating an analysis of strengths, weaknesses, threats and opportunities. The business development team presents this business plan and associated budget to the decision-making staff that decides to move forward in the process or not. Remember that a business plan is ever changing, and you will discover the need to adapt this plan as the NPD process moves forward.

After your programme has been approved, research efforts commence to solve the issues that do not have a readily apparent solution. The resulting intellectual property is protected, and once the technology reaches a high enough maturity level, it is passed off to the developers who build low volume repeatable and reliable products. Extensive testing is integrated into the development process. Later, the research-and-development steps are discussed in the same section, but they are often considered as two different steps.

At this point, the development team, who should have been working with the manufacturing team, turns the responsibility for building the sensors over to the manufacturing team. If a good job was done up to this point, the manufacturing team can hit the ground running and a reasonable process for repeatedly making thousands upon thousands of units can begin, right? This rarely, if ever, happens and there is often a lot of tweaking and process adjustment to get the manufacturing to the point where it is acceptable.

The internal sales group in Figure 3.1 is in a dashed box because they will not be talked about a lot in this chapter. Even though they are likely the most important part of reigning in commercial success, their role in helping develop the sensor is more to collect customer feedback for product enhancements and new product ideas.

During the development, trial manufacture or early sales processes, some special customers might have been given the opportunity to see early sensor versions, and they may be working with the development team to complete integration. Hopefully, other customers are found through the sales force, and the integration of the sensors with these companies may be done with less interaction with the development team. In either of the cases, the company integrating the sensor interacts with the internal sales group, and when the integration and product development are complete at the device manufacturer, the external sales department with possible support from the internal sales department starts selling the device to consumers.

Feedback and support issues are channelled through the device manufacturer's support staff and feedback from those interactions is relayed to the sensor manufacturer support staff. The support function ensures customers have a good experience. They often work hand in glove with the sales force, and between the two, product issues and enhancements can be identified and communicated back to the marketing team so that future enhancements can be worked on.

In practice, boundaries between these different steps are fuzzy because a lot of simultaneous activities may be occurring. While from the top level, the entire process may look chaotic and appear to be like herding a large set of house cats, and as long as efficient and effective communication is used between team members, the progress toward a success commercial launch can be steady and fast. The stage is set. Let's now start looking at the primary challenges in developing a mass-market biometric sensor.

3.2 The primary challenges

A biometric sensor history from which an industry appreciation can be fostered could start at this website [6]. This site has been maintained for many years by an industry expert, and it contains a significant amount of information about historical biometric sensors. Challenge yourself to see how many of these sensors are on the market today. How many sensor companies have been bought, sold or even made it to market with a product? How many have failed? To further reinforce this point, two examples come to mind for mobile systems that incorporate biometric sensors: one a failure and one a success.

A fine example of challenges that were not managed properly was a company named Raptor Identification Systems, Inc. This company had developed an exciting mobile multi-biometric-capture platform that met with a very warm reception from many potential customers, but overwhelming challenges to their business approach resulted in the company failing in a flash. Unfortunately, the lessons learned from this experience cannot be relayed in this forum other than to say the failure was not due to a technological impediment.

On the other hand, properly managing the business plan and risks can result in successes like Apple Inc.'s iPhone 5S – the first phone with a well-adopted fingerprint sensor. This product set the bar for mobile biometric sensors as arguably the first truly mass-market launch. Apple's path to success was not assured but their methodical approach set the odds of success in their favour. Two of the key reasons for this success were the absolute simplification of the use case for the fingerprint scanner and Apple's control over the entire ecosystem. As a result, we now see these seeds of the original fingerprint sensor concept extending functionality with Apple Pay.

It is time to now consider the initial marketing challenges that are faced during the NPD process.

3.2.1   Market relevance

The quality of a house is reflected by the quality of its foundation. The same principle applies here. By completely understanding the use cases of the proposed biometric, the company that integrates with that sensor then will know the advantages and disadvantages that could be carried into the market with that sensor. Because of this, sensor manufacturers need to be synchronized with the potential sensor customers, and the sensor manufacturer needs to understand the ease of use, security provisions, standards, interoperability requirements, cost, size, colour and many other requirements associated with the sensor. With this in depth knowledge, the marketing team can seek to understand the size and demographics of the possible customer base.

There *must* be an identified market need for a mobile biometric sensor prior to a company engaging in sensor development. Clearly, no market need implies no revenue which in turn means one could only hope at best to lose money. Therefore, armed with the use cases, the marketing team ensures that a market for the product exists. During their investigation, they uncover who the competitors are and how large the market size is. Although these data points are not necessarily difficult to obtain, the better the data that is collected by the marketing division, the better foundation the sensor development programme has to build upon.

As an example that we keep going back to, it is now clear that Apple demonstrated a significant market exists for fingerprint scanners built into smart phones.

The use case was simple: to unlock the phone. By simplifying this usage of the fingerprint sensor (while having a much broader strategy in mind), Apple unlocked the fingerprint sensor market for smart phones even though it was not the first to sell a fingerprint enabled smart phone. Apple did their market research, generated an assessment of the potential for success, enumerated the downside risks, came up with risk mitigation plans, wrapped that analysis in their strategic vision for usage of fingerprint sensors and made the decision to move forward. They had both the advantage and disadvantage of being the first one to successfully mass market a fingerprint sensor embedded into the smart phone but they had to weigh that possibility against the risks.[2]

Another example of a potential for mass-market biometrics started playing out in 2015. Fujitsu introduced the ARROWS NX F-04G with an iris scanner built in. Very recently, Microsoft announced iris scanning in two of their Lumia phones and just as recently both Samsung and LG announced that they will release at least some smart phone models with built-in iris capture and recognition capability in 2016. With the prevalence of selfie capture proficiency, will this technology translate into a second golden opportunity?

This question leads us into an attempt to understand more about what makes biometrics work for a consumer. In [7], Zhang, Guo and Gong describe seven factors that affect the suitability of a physiological characteristic as a biometric. The seven factors are as follows:

1. Universality – does everyone have this physiological characteristic?
2. Uniqueness – is this physiological characteristic unique between individuals?
3. Permanence – is the physiological characteristic permanent?
4. Collectability – can the physiological characteristic be collected, or sampled, easily?
5. Performance – how accurate are classification engines at using this biometric?
6. Acceptability – how widely is this biometric accepted?
7. Circumvention – is it easy to fool or work around this biometric?

Notice that, the use cases defined by the marketing team will be affected by the answers to these seven points. When needed, refinements to the use cases might be able to be made according to what the technological limits are for these sensors. This is one way to help optimize a plan for success.

Collectability and acceptability are the most likely barriers for iris biometrics whereas acceptability and performance are the likely barriers for facial recognition. From the acceptability perspective, battle lines exist today with issues being raised by groups such as the American Civil Liberty Union. Primary amongst the issues are: how does an individual protect their privacy in an environment where biometrics can be captured without a person's knowledge? For example, claims have been made that iris recognition can be fooled using only a high quality photograph of a person and such a photograph can easily be taken of a person without their knowledge. To be even handed, this same issue applies to face, iris and fingerprint biometrics because

---

[2]Motorola actually sold the first smart phone with a fingerprint sensor but the Atrix phone was not nearly the success that the iPhone has been.

now each one of these modalities can be captured in a stand-off manner, i.e. at a bit of distance from the person.

It is almost certain that even if consumers' embrace the technology, there will need to be legal frameworks created over time to address the privacy concerns. Related to this sticky point is the apparent mutual exclusivity of privacy and security. Should one give up some amount of privacy for higher security? The security aspects will likely end up having a legal framework built around them as well. Time will inform us about these very high level challenges to biometrics, not just sensors, as we see the dance play out with biometric sensors on mobile platforms.

With this background, several factors must be carefully addressed in the marketing analysis including:

1. What is the secondary approach to take when the sensor does not work?
2. Are there privacy versus security versus convenience versus cost trade-offs that have to be made and how are these trade-offs affected by issues outside of consumer acceptance?
3. Should the sensor support the Fast IDentity Online (FIDO) Alliance [8] specification and/or a compliant Trusted Execution Environment (TEE) [9] platform?
4. Do the use cases have to support any other standards? Each set of standards implies more resources to build and test the sensor. Are these items really needed to penetrate the mass market?
5. Does the sensor need to produce data that will be used to interoperate between different systems? For example, will the sensor need to capture Federal Bureau of Investigation Image Quality Standard compliant fingerprint images so the images can be used to match against much larger databases[3]?

3.2.2 Research and development

By the very nature of research and development, unknown factors exist that can impact the entire business case pretty quickly. Because of this direct impact, approaches that identify and mitigate the riskiest areas first can be employed with continuous feedback to refine the entire team's fluctuating confidence in whether or not such a sensor can be brought to market. Often, the research and development can eat up 2 years, or even longer, when developing a new biometric sensor.

Each sensor type has a different set of primary challenges associated with it while all of the sensor types share the common challenge of supporting enhanced system security. For fingerprint sensors, development challenges relate to improving what will shortly be called biosecurity – the error rates associated with matching on images that do not contain enough features. With iris imaging the primary challenges include biosecurity but in the form of iris image quality and public acceptance. Finally, facial imaging has the same biosecurity challenges due to image quality concerns

---

[3]Use cases exist for mobile biometric sensors that require images compliant to these standards but most, if not all, mobile biometric sensors do not generate these compliant images today.

and the inherent matchability of facial images. Facial imaging is also challenged by public acceptance. In both the iris and facial case of public acceptance the issue is the possibility of unexpected capture without a person's knowledge.

Let us take a closer look now at the fundamental challenges that must be addressed or continue to be addressed during the research and development phase. Some of these challenges are well-known and improvements to these challenges have been made over the years but opportunities still exist for more improvement. The primary technical challenges to be addressed for mass-market biometric sensors are as follows. The observant reader will note that the first two of these challenges relate back to the seven factors that make a physiological attribute a good biometric.

1. Security – is the biometric data kept secure enough and does the biometric provide a low enough error rate to support the use case risk profile?
2. Performance – is the use case implemented well enough to prevent people from becoming frustrated enough to stop using the sensor?
3. Integration – with the ever increasing demands on functionality and move towards more highly integrated environments, can the sensors continue to improve their support for such mobile platforms?
4. Image quality – on camera-based biometric capture, how do we minimize motion effects, focus quality and speed, and image processing algorithms to enable better matchability?

3.2.2.1   Security

There are at least two different levels of security that need to be considered in depth on mobile platforms: the security that can be offered by the biometric itself, let's call this biosecurity, and system level security. Biosecurity is concerned with the biometrics ability to uniquely identify an individual and as such it relates directly to performance as discussed in the next section. System security, on the other hand, relates to the environment that uses the biometrics. Because a version of the biometric is likely stored and processed on the system, system security concerns itself with areas such as protecting the biometric samples, the processing of unknown samples against the known samples and spoofing of authentication results that result from an application of matching.

Both security challenges are present and real in today's devices and both present challenges to the sensor. At the risk of making generalization errors, it is found that iris imaging has the best error rates followed by fingerprint then followed by facial. Many caveats to these results can be given but fingerprints have likely led the way on mobile platforms due to the public's longer exposure to this biometric as well as its ability to meet current consumer needs. From a system level security perspective, sensors need to be concerned about security since once a biometric is compromised there is no easy way to make the biometric uncompromised: consumers cannot change their biometric like they can change a password. Thus, platforms like Synaptics fingerprint sensor [10] that provides on board matching and Iritech's iris sensor that provides encryption capabilities [11] have promise in aiding the system security challenges. This section considers both types of security.

Biosecurity can be improved by improving the errors of matching as defined in the next section of this chapter. NEXT Biometrics does a fine job explaining limitations of biosecurity on their website as it relates to fingerprints [12]. Increasing the sensor size, sensor resolution and matching algorithm, all improve fingerprint biosecurity but there are cost and real estate trade-offs. For iris-matching improvement, biosecurity means improving the image quality and perhaps incorporating the irises of both eyes and for facial recognition this means improving image capture quality and matching algorithms. In general, matching more factors for the authentication process will make the overall process more secure as well. Perhaps, multi-factor authentication will apply to improving biosecurity.

Extending use cases into the secure authentication realm requires improvements in the host environment security infrastructure. Industry recognizes this: they have cast a wide net by forming the FIDO Alliance which promulgates specifications and provides certification for a secure authentication framework. Industry also developed the TEE concept now maintained by Global Trust and microprocessor companies have been supporting these specifications in their hardware. As more and more financial transactions move to the mobile platform, these biometric sensor enabled trusted architectures will become more prevalent and the systems they are placed within will become dramatically more scrutinized.

FireEye, a computer security firm, recently published an industry note about the security risks they observed in mobile platforms [13]. They then followed up this note with a BlackHat conference presentation that pointed out other specific issues with fingerprint implementations [14]. As pointed out in the research note, these platforms are rich environments for collection of sensitive information. FIDO and TEE are pushing forward very quickly to help stem the inevitable attacks on these platforms and encryption and biometrics are quickly becoming engrained in much more secure authentication processes. But as smart devices become more connected to one another through web services and other technologies and as more apps become available, the threat vectors increase simultaneously with the increased attention to improving security. Sensors that have built in security and on-board matching like those of Synaptics and IriTech mentioned above could support a more secure user experience.

A quiet shift from authentication for convenience to authentication with security is underway and maintaining device usability during this shift will be rather important for mass-market adoption. Will the FIDO and TEE standards form complete security underpinnings for this shift or will new more universal security approaches be developed and adopted in a wider manner? The industry will also need to find itself through the maze of privacy versus security concerns. Society is on the forefront of understanding that these issues need to be confronted and it is quite likely that as the market matures there will be impacts on the development of biometric sensors.

3.2.2.2   Performance

Mobile applications will be used anytime and anywhere by anybody. At the same time, a primary objective of a sensor is to capture intra-class samples (what varies for a specific person) with the highest uniformity possible and inter-class samples (what varies between people) with the highest variance possible. The sensors in use today have tended to focus on the intra-class problem since the use case has been primarily authentication for convenience. When considering inter-class challenges in more depth, the sensor, matching algorithms and security must be considered in a much broader scope.

At a minimum, mass-market sensors need to be low cost and high enough performance to meet consumers' needs. In addition, since sensors are components of a larger system design it is the combination of overall system acceptance with the germane sensor features that ultimately define the sensor's adoption rate. The challenge is to have a high enough performance sensor to engage a device manufacturer that does not already have a sensor supplier or to have an even higher performing sensor that will replace an incumbent sensor supplier.

Capture speed and accurate collection to support an acceptable consumer experience define the sensor performance. Capture speed with fingerprint sensors is pretty good today while capture for facial and iris has been slower due to the time taken to adjust the camera for optimal image capture. Innovations to improving the capture speed of these sensors continue to be developed and products including such improvement are starting to come to the market now.

Accuracy rates of biometrics are usually measured via False Match Rate (FMR) versus False Non-Match Rate (FNMR) and False Accept Rate (FAR) versus False Reject Rate (FRR). FMR and FNMR characterize the matching algorithm performance, and FAR and FRR characterize system performance. In words, FMR is the rate at which anybody else can be recognized as the intended person by the matching engine, FNMR is the rate at which the intended person is not recognized by the matching engine, FAR is the overall system rate related to FMR and FRR is the overall system rate related to FNMR. In the following equations, FTA is the Failure to Acquire rate, FTC is the Failure To Capture rate and FTX is the Failure To eXtract rate. These quantities are related to each other according to

$$FAR = FMR * (1 - FTA) \tag{3.1}$$

$$FRR = FTA + FNMR * (1 - FTA) \tag{3.2}$$

$$FTA = FTC + FTX * (1 - FTC) \tag{3.3}$$

where FTC and FTX are estimated as follows:

$$FTC = ((\text{terminated capture}) + (\text{not sufficient quality}))/(\text{capture attempts}) \tag{3.4}$$

$$FTX = (\text{feature extraction failure})/(\text{samples submitted to the feature extractor}) \tag{3.5}$$

The amounts in the right-hand side of (3.4) and (3.5) are the counts in a given test. Equation (3.3) relates to the sensor via FTC and to the environment's software via FTX.

FTX can also be affected by the quality of the biometric capture coming from the sensor. Hence, sensor output affects FAR and FRR in positive and negative ways.

Rather than discuss actual values for these performance metrics, the following challenges to optimize these values through sensor improvement are considered. While there are many important points to be made with respect to these performance metrics, the following points are particularly pertinent:

1. Using a common canonical language to compare one sensor or system to another is good IF the consumer cares about the relationship between usability and security. But companies don't usually publish this information as they embed all of the information into the user experience that is delivered by their product. As a result, if the user experience is good enough, then the product is acceptable.
2. As security requirements become more pronounced, these performance metrics will become more important because they statistically describe how often an unauthorized person can access your protected device and how often you, as the authorized user, will not be able to be authenticated on your device. From the industry's perspective, the liability of making a FAR error far outweighs a FRR error and a primary challenge here is to find a reasonable approach to managing this liability. After all, no biometric is 100 per cent accurate; a consumer's purchase of a device that uses biometrics must come with some acknowledgement that the consumer was made aware of potential risks.
3. When it comes to security, it is probably best to have an independent third party test that compares results between various vendors. Such tests remove any perceived bias that the company has in testing its own sensors when they deliver a performance statement to the market.
4. Since FTC, FTX, FMR, FNMR are not 0 today, there is room for improvement in every sensor but the law of diminishing returns applies here. What is the cost to benefit ratio as the sensors become better and better and even more importantly, are the current offerings acceptable to end users' needs?
5. Underneath the hood, these performance metrics are important to a company as they develop a sensor so they can objectively ascertain its performance and build out their product approach.
6. Improving the FAR and FRR of fingerprints means improving sensor size and resolution, improving FAR and FRR of iris matching means improving the image quality and improving the FAR and FRR of facial matching means improving image quality and the matching algorithms.

3.2.2.3   Integration

Improvements in integration between components in the smart phone have improved smart phones dramatically. Processor and co-processor support for capture and image processing, battery utilization, size and weight reduction, and user interfaces have improved dramatically. As newer technologies continue to be developed even deeper levels of integration challenge the sensor manufacturer to deliver more value to the mass market. Two examples of such integration come to mind: Synaptics' incorporation of a matching engine directly on the sensor [10] and capture of iris images on a smart phone with no more components required other than addition of an IR LED and a small mirror [11].

Innovative integration such as these two approaches could be the platform for the next step forward for sensors in mobile devices. Yet, researching and developing these enhancements to existing technology are not inexpensive and thus present a significant challenge.

3.2.2.4   Image quality

Discussion of image quality challenges begins with a common theme that has been mentioned several times already: mass-market biometric sensors are made to be as inexpensive as possible and as small as possible. There are, of course, premium versions of components built into some pricier devices, but these same two general principles apply within every smart phone and tablet sold. This comment is not made to imply that such cameras cannot capture good pictures. No, quite the contrary. A lot of resources have been placed into making imaging better over the years and very good pictures can result. The challenge is that the camera configurations used because of the price and size constraints work against optimal image quality.

Camera modules are made to be as simple as possible in these cases. As a result, lens configurations used typically have fewer optical elements, the apertures are fixed and the 35-mm focal length equivalent is a wide angle lens. Due to the small size that the camera must fit within, the image sensor is very close to the lens, and the image sensor therefore has smaller pixels which need more light in order to capture a good image. Thus, the aperture is typically on the range of f/2.0 to f/2.4. With this simple configuration, how these variables affect image quality are now investigated. Remember that in the discussion that follows, acceptable, not optimal, biometric samples can be captured on these platforms today and that the author is simply pointing out areas for sensor improvement that would improve the overall FAR and FRR values for the device.

Before launching into the discussion of image quality, an excellent read on the principles of geometric optics can be found in [15]. This book covers seemingly all of the technical details needed to understand and appreciate the optics involved in cameras.

One of the most common issues that arise is image distortion. You might have noticed when you take a straight-on picture of a scene with vertical or horizontal lines across the entire image that the lines in the image do not turn out parallel and, in fact, if you draw a straight edge across any single line, you would discover that the straight line in the scene is not straight in the image. Try this by capturing an image of the side of a clapboard house or of a building with windows that are perfectly parallel to one another. Make sure the picture is taken straight on so you can try to avoid perspective effects. These distortions arise because the lens is a wide angle lens with fewer elements, and these lenses are not typically what are known as flat field lenses – lenses that create high fidelity images wherein parallel lines in the scene become parallel lines in the image. These distortions introduce non-uniformities in biometric sample captures – a capture of the same face or iris can vary now from

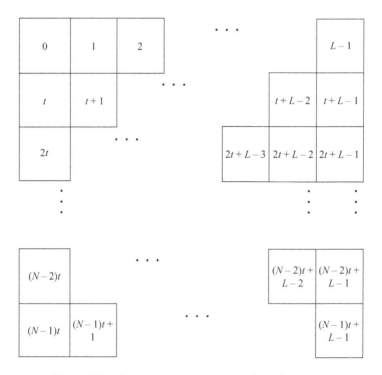

Figure 3.2   Progressive scanning pixel read-out timing

image to image because it is distorted in ways that depend on where the biometric sample is located in the image.

Another type of distortion the reader may be familiar with is motion blur. Motion blur is defocusing of areas of the image caused by the scene moving with respect to the camera while the camera is capturing the image. Motion blur renders it difficult to locate and recognize objects in a scene.

Let's dig deeper into what causes motion blur. The silicon chips typically used in these cameras are Complementary Metal-Oxide Semiconductor (CMOS) chips that use progressive scanning. The progressive scanning pixel read out approach is illustrated in Figure 3.2. Let's say that each pixel in the CMOS chip is exposed to light for a fixed amount of time, T. Progressive scanning begins pixel exposure at different times so as to allow each pixel to be read according to the clock sequence shown. The upper left pixel is read out at time 0, followed by the pixel to its right at time 1 and so on up to the end of the first pixel line. At the end of L clock cycles, the first line has been read out. After a total clock cycle delay of t cycles from the reading of the first pixel, reading the second line of pixels starts. This process repeats for all N lines in the sensor and after completion an image has been collected. In the time, it takes to scan an entire image from top to bottom, if the target scene and camera have a good probability of moving with respect to one another, blur can result because the light on different scan lines was collected at different times.

Charge Coupled Device (CCD) chips, on the other hand, typically use a global shutter to collect all the pixels in an image simultaneously. Given the same exposure time T as in the CMOS example above, the pixel exposures are synchronized to the same start and end time. While the pixels are not read out simultaneously, they did capture the light simultaneously and as such CCD chips lower the possibility of motion blur but the possibility still exists. CCD chips are not used in smart phones due to their increased cost.

Motion blur can be reduced by increasing the speed at which lines are scanned so as to give the object and sensor less time to move relative to one another. Limits exist on how fast scan lines can be captured due to the limited power budgets and sensor sizes. Importantly, the larger the sensor used, the more pixels that have to be cycled through and thus the lower the frame rate. Hence larger sensors are more susceptible to motion distortion.

Distortion and motion blur can be managed in several ways. This could be accomplished by fixing the relationship between the camera and the object using cameras that position the face (or iris) at the same position with respect to the camera, for example. In some applications, this is exactly what is done with a head piece that has the cameras mounted in it. These solutions fly in the face of what true mobility means – it would no longer be fun to capture a selfie! Algorithmic correct could be added to fix distortions but this must be done on each phone individually because each phone has its own unique imaging characteristics. Motion detection can be added to the system to only enable capture when minimal motion has been detected but, this may unacceptably delay capture. In fact, this is exactly is what is done on the author's smart phone and this leads to the inability to capture images when movement is in the scene – a bad result for capturing mobile biometrics on smart phones.

The image quality investigation continues by looking at image focus quality now. For the purposes of this section, it is assumed that the camera resolution is high enough to capture enough pixel data for biometric recognition. This is clearly the case given today's 10 megapixel and greater sensor sizes. That is a good start but what are the factors that affect focus quality? First and foremost, the quality of the lens determines, in large part, the best focus quality that can be achieved by the camera. Lower quality lenses are less expensive but more recent smart phones have put a lot of work to improve the lens quality – perhaps, you have observed that better pictures are now coming off of smart phones? The aperture and focal length also affect the focusing capability of the camera.

Since iris images and facial photos will likely be taken closer to one's face (as in capturing a selfie), this imaging situation places some burdens on the camera. These challenges relate to the concept known as depth of field – the range of distance from a camera that the image will be in focus. As a scene gets closer to a camera, the depth of field decreases for a given aperture and focal length. Consequently, if the distance from a person's tip of their nose to their ears is 15 cm there is a very good chance that the depth of field is smaller than 15 cm.[4] This means portions of the person's face

---

[4] Just think about macro-photography and how challenging it is to get close in scenes well focused. This is because the depth of field is small.

will not be in focus. The same idea applies to the iris of the eyes because ultimately the position of the lens is determined by the auto-focus engine the camera uses and the depth of field range quite possibly might not contain the eyes.

Another factor works against good image quality here as well. A smaller aperture (large f value) has a better depth of focus for the same lens but a smaller aperture also allows less light to pass to the sensor. This is one of the reasons that lenses in smart phones are so slow (low f value); there needs to be enough light to capture the scene properly. The lower the lens speed the worse the focus capability of the camera.

Taking a higher level view, today's smart phone cameras typically have auto-focus features. Have you ever noticed when you try to take a picture with your smart phone that there is a delay between when you click on capture and when the picture is actually captured? This time gap causes millions of missed Hallmark moments per day because you cannot hit the capture button 1–2 s before the Hallmark scene presented itself. The reason for this time gap is that the sensor is feeding images back to a processor, and the smart phone is calculating a degree of merit function for focusing. When this degree of merit function converges to the best value over a sequence of frames, the image is in focus and it is captured. During this time, the lens must be moved back or forth, a new image captured, and the figure of merit calculated and compared. This is the most common way to adjust focus on smart phones today because it is software centric and therefore less costly to implement over a large number of devices.

An alternative for focusing includes a radio frequency emitter to measure the distance to objects and this distance can be used to adjust the lens. Adjustment with this type of technology is much faster because it can be running in parallel with the streaming of the images from the sensor and the sensor basically always stays in focus. But since this approach is too costly for mass-market sensors, the discussion is relegated to the first method of focus.

Smart phone lenses are typically moved using a technology called Voice Coil Modulators (VCM). VCMs work by applying an electrical potential across a slide mechanism to move a lens in or out. Two of the problems with VCMs are the accuracy and speed in which they can move a lens. The lower the speed and accuracy, the slower and less accurate the focus response. In addition, VCMs tend to be power hungry and since the camera is the element of the smart that today takes the most power, reduction in the power budget would be a good feature to have in a new auto-focus engine.

Recent advances to these auto-focus issues are being sold today but they are not yet completely implemented in smart phones and tablets. Once such approach is a Micro-Electro-Mechanical (MEMS)-based auto-focus actuator that uses electrostatic forces to move the lens in smaller and more accurate increments to achieve the same effective change in focus. This auto-focus mechanism also significantly reduces power consumption from VCMs and can provide image stabilization. Since adjustment can be controlled to a finer level and the adjustments required are not so large, the effective time to focus an image is reduced by a factor of at least 2. Yet, even in this case, there may are delays caused by calculating the figure of merit. A summary overview of Ivensas' MEMS auto-focus lens holder can be found at [16]. In a nutshell, the auto-focus features on today's smart phones are getting better, but it is still an enemy of biometrics because of the time lag between clicking the capture button and the time the actual image is captured.

One issue commonly observed with smart phone cameras relates to the automatic nature of the image capture. When the auto-focus operates, what part of the scene is the camera focusing on? The focus engine likely calculates the figure of merit across the entire image and therefore if part of the capture scene is close in and part is far away, the auto-focus may actually end up not focusing on the near or the far scene but somewhere in between. Likewise, the automatic features of white balancing and contrast enhancement can also introduce non-uniformities in the image that make it harder to reliably match the biometric.

Some thoughts about utilizing a single camera for facial and iris capture in the same device are pertinent here. If the same camera is used to capture facial and iris images, the following points need to be addressed:

1. Zoom control or image resizing to ensure images meet the expected resolution.
2. If both iris images are collected careful attention must be paid to depth of field issues.
3. One's saccadic eye movements create a possibility for motion blur if the camera is too slow.
4. Iris imaging is done based on IR illumination because it makes iris images more uniform between people. Can the camera capture scenes with normal visible light and IR light? Does the camera have an IR source and a visible light source? Not unreasonably, typical smart phone cameras include an IR cut filter that excludes IR energy from the wavelengths that are used to create the image. How is the illumination filtered properly?

It could also be the case that a second camera can be used to capture iris images and this second camera can meet the use case for iris image much better that a universal camera. From a mass-market perspective, this will add cost to the platform.

Addressing all of these items simultaneously for multiple image capture scenarios is like standing on the edge of a sword with your bare feet and trying to balance. It is a gentle balance between all factors. So, indeed there are challenges to capturing the images with optimal image quality that could support the FAR and FRR rates that will need to be observed for more secure environments. Improvements clearly have been made in this area over time and the industry continues to make improvements.

The end of the technical challenge discussion has been reached. According to the NPD process, the output of the research and development phase is a product that is ready to be handed over to manufacturing. If the team has done a good job, the product has been extensively tested even though testing will continue. The team will have also met the modified use case specifications with the stakeholders' approvals. Presumably, the research and development team has set the product up for success because it can be manufactured at a quality level and cost point that enables the manufacturer to realize a successful product launch.

3.2.3 *Manufacturing*

Much work is yet to be done: the product is not ready to launch because a quality driven, cost-effective process for building the projected number of sensors must be engaged and qualified first. A contingency plan that guides the company for unforeseen demand fluctuation must also be in place prior to launch due to the high cost of raw materials.

The design team should have taken manufacturability requirements into consideration early during the design process. If the job was done properly, turning on manufacturing could be a simple matter of qualifying the process, machinery, contract manufacturer and other manufacturing aspects to ensure that high quality and consistent products result. For mass-market products, this is a huge challenge because every tenth of a cent saved makes a difference. Therefore, tweaks to the manufacturing system will be made as time and manufacturability goals permit before the manufacturing line is certified as production ready.

A company's profitability is reported on one of its financial statements named the Income Statement[5] which, at a very simple level, contains revenue and costs. The costs are broken down into the costs to manufacture the products the company sells and other non-manufacturing related expenses such as leases, taxes, payroll and so on. The net income, or profit, of a company is given by

$$\text{(Net Income (NI))} = \text{(Gross Margin (GM))} - \text{Expenses} \qquad (3.6)$$

where

$$\text{GM} = \text{Revenue} - \text{(Cost of Goods (CoGs))} \qquad (3.7)$$

Technically, the Gross Margin is a monetary amount but sometimes people refer to the Gross Margin as the rate defined by:

$$\text{(Gross Margin Rate (GMR))} = \text{(GM)/(Revenue)} \qquad (3.8)$$

For example, if $100,000 of revenue is realized with a CoGs of $45,000 then GM is $55,000 and the GMR is 55 per cent. When looking at Apple's and Alphabet's latest complete fiscal year GMRs, you find that Apple's is 40.1 per cent and Alphabet's is 61.1 per cent. The primary reason for the difference between these two GMRs is that Apple sells a lot of hardware whose CoGs must be subtracted from the Revenue, whereas Alphabet sells dramatically less hardware, i.e. the cost of manufacturing software is close to $0.

Depending on market competition, there may be leeway in the manufacturing cost of the sensor if the sensor has significant features that clearly make it preferable to the competition. The company may decide that it is more important to get the product to the market as quickly as possible rather than worry about the lowest cost. This is a decision made on the executive level after the executives are armed with market positioning and customer perception data. Once market launch occurs the on-going concern becomes improving market position and increasing GM by decreasing CoGs due to economies of scale and process improvement. Newer sensors

---

[5] Sometimes, the Income Statement is known as the Profit and Loss Statement.

sold into a market with existing competition will likely require lower gross margins unless the product is disruptive. Higher GMs fuel the new product development engine.

It is usually better to own at least the key enabling components of the sensor and their associated intellectual property. Ownership mitigates support and cost risks that a third party adds into their price. Purchasing from such a third party can reduce the GMR because the third party builds profit on their technology into their price. If the manufacturer owns the intellectual property, they have the flexibility to outsource the manufacture of the component if the business advantages outweigh the disadvantages. Business strategy determines the direction that will be taken. For example, a sensor manufacturer could decide that the risk of losing intellectual property is too high within the first manufacturing year and therefore decides to manufacture in-house, wherein a proprietary manufacturing process is refined. After the 1-year benchmark passes, then it might be time to move manufacturing outside to a lower cost producer because now the intellectual property threat is not as great. At the same time, a second component could start as an outsourced component, and after some time passes, the manufacturer may discover they can reduce their CoGs by manufacturing this component internally.

An important aspect of making the insource versus outsource decision is the capability of the supply chain: the sources and logistics of the raw materials needed to build the sensors. A company's ability to deliver volumes of sensors can only be as good as its supply chain. If an imaging chip source is exclusive and that source can only produce 50,000 units a month but the projected demand is 100,000 per month, then the sensor manufacturer has a problem. Product supply will not be able to keep up with the sensor demand. Consequently, when building a sensor, it is rather important to know the projected volume that will be needed and the time frames that they will be needed within. Because of the sensitive nature of the supply chains, supply contracts are often put into place that control the ability to increase and decrease component supplies so that both the manufacturer and the supplier are not severely damaged if product demand is much higher or lower than anticipated. To help mitigate supply risks, multiple suppliers of non-unique components can be used. This could be employed as a strategy to both keep costs down and quality up, if used properly.

It is impossible to overstate the effect that the supply chain can have on the manufacturer. If managed poorly, the supply chain can destroy a sensor's adoption, and in the worst case, it can put the company out of business. In the case where demand for the resulting sensor is too low, a poor supply agreement can lead to surplus inventory that costs the manufacturer all important cash. Maintaining a large inventory with slow turnover leads to poor usage of the company's resources, and it could end up bankrupting the company. On the other hand, if product is flying off the shelf and the suppliers can keep up with the component demands, the manufacturer must ensure they have enough financial backing to manage the ever increasing cash needs. If they cannot, the company can literally go bankrupt because they are too successful. Hence, careful planning and in-depth knowledge of the adoption rates are very important inputs to the plan for success.

Improving the GM revolves around streamlining processes that have acceptable quality outputs while minimizing the raw component costs. Improving the process often means configuring the manufacturing facility in an optimal fashion that supports a better workflow through the facility thereby minimizing the overall human time taken to build a product. Manufacturing constantly looks for opportunities to improve the quality and the GM because these affect the NI and the lowest cost producer is in a position to win the mass-market sensor game. The lowest cost producer has more latitude to adapt their approach and respond to competitive threats. Improving the yield rate is one significant way to improve the GM.

The yield rate is defined as the number of good units produced from a given quantity of raw materials divided by the number expected to have been produced from the same quantity of material. Manufacturers strive for 100 per cent yield because that means there is no manufacturing waste and this also lowers the average CoGs in (3.7). Yet, it is very difficult, if not impossible, to realize 100 per cent yield due to various factors including component defects, assembly errors, employee stealing (also known as shrinkage), logistics errors and other process errors. The quality control process provides feedback to the manufacturing process to maximize yield by identifying and correcting root causes of manufacturing defects. As a result, processes often incorporate automation that both improves manufacturing throughput and results in consistent quality output. Such quality and automation processes, while essential, can be quite costly to implement and they could be a significant decision factor in whether to insource or outsource.

Manufacturing plays a large-supporting role that is aligned with the remainder of the business strategy. If next-generation products and cost-reduction opportunities are not uncovered and investigated in a timely manner by the research-and-development parts of the NPD process, then continuity of any success the manufacturer realizes could indeed be momentary. In other words, the comprehensive strategic plan for the sensor needs to be communicated between the various stakeholders and effectively implemented through aligned commonly understood goals.

Once the challenges in manufacturing have been addressed, it is time to sell, integrate and support the sensor. Assuming the sales opportunities exist for the sensor, this sensor will have to be integrated and supported in the host environment.

3.2.4   Integration

The quality of the sensor's integration into a host system impacts both the cost and effectiveness of the end user experience. This was clearly shown in the case of the MEMS focus adjustment. Seamless integration occurs at the software and hardware levels. For example, today's microprocessor architectures allow high-speed data transfer from image sensors to memory where fast figure of merit calculations can be done for focus control. All of the parts, the hardware, firmware and processing software must work in synchrony to generate the best possible integration for the best user experience.

From an electronic perspective, resource requirements such as processing and battery power come into play. The smart phone revolution has helped propel biometrics into the mobile landscape in this respect because the need for fully integrated System on a Chip (SoC) architectures has advanced the underlying technology to the point where biometrics can address important use cases. While recent steps forward in SoC architectures lower power consumption, increasing computational demands need to be carefully balanced since battery power density improvements will likely take longer to evolve than the computational capabilities [17]. As a consequence, it is always good to think about minimizing the power budget while maximizing responsiveness through more sophisticated use of the microprocessor architecture by engaging with processing enhancements that may include items such as multi-threading or OpenGL.

An environment that supports robust testing of the integrated platform can be pivotal, and it must not be understated that the reliability of the final integration must be verified by extensive testing. The last thing the manufacturer or the consumer wants is for an overlooked problem to arise. Unfortunately, even extensive alpha and beta testing cannot ensure that all such problems are identified and solved. When such a problem does arise, the ability to respond to the issue can define the product and the brand name. For better or worse, the company's ability to respond to issues when they arise can be a defining moment.

3.2.5   Support

Companies that field mass-market biometric sensors face tremendous challenges when it comes to supporting their technologies. If even a small fraction of the units are not working because of the fingerprint sensor, the smart phone company is responsible to fix the manufacturer defect. The problem can lie with the smart-phone manufacturer or the sensor manufacturer or both. Fixing such problems, after the phones are distributed, can be very expensive if there is no software update that can be applied and, in this case, part of the cost of that liability may lie on the sensor manufacturer. Thus, the importance of pre-launch testing is critical to ensure that there are no hardware issues requiring physical intervention to repair the device. On the other hand, software updates are much easier and less expensive to deliver.

Even though the manufacturer has tested his sensors extensively, there is still the risk that the sensor has some fatal flaw. To address such cases, a contingency plan must exist in order to direct the response. An infrastructure system that supports this contingency properly helps the manufacturer pinpoint the root cause of the issue so it can be fixed. Historical manufacturing records may need to be evaluated during an engineer's root cause analysis in order to identify the batch and lot number of components or the processing step that caused the failure in the first place. In this way, a robust manufacturing system that collects and manages quality measurement and analysis supports the product. This is another link to the importance of good manufacturing processes. These manufacturing systems cost quite a bit of money and they require quite a bit of time to implement but the benefits of such a system can be tremendous when they help the company manage their way through a potentially catastrophic failure. This system (or set of systems) that supply this functionality should be in place and tested prior to the product launch precisely because one never knows when a problem will strike.

Here is a well-known secret about how larger companies tend to work when it comes to solving problems for mass-marketed products. Corporate counsel typically advises the company to adopt a policy to not admit they have a problem with their product until a fix has been identified and made available. This frustrates consumers because there is no admission that a failure exists, but from the company's perspective, the company is working to minimize financial risk. As long as they do not acknowledge there is a wide spread problem, they can focus on trying to solve the problem while postponing the liability of implementing a fix. Most likely, the problem will be solved sooner than later and the company can make the fix available as a product update. In the interim, the company avoided the extra burden of engaging the consumer when no solution has been identified yet. A caveat is that once the news of the problem becomes public, the company should take action to solve the problem as soon as possible.

A recent example of this hit home when one of the author's employees had a smart phone with two lost vertical pixel lines on the screen. Many conversations were made with the manufacturer about the resolution to this problem since the phone was only two months old and easily within warranty. After making little progress, an authorized reseller of the phone engaged with the manufacturer and it came out that this model of phone had an approximately 30 per cent failure rate due to this problem. The manufacturer agreed to replace the smart phone with a refurbished unit that had the same problem but had since been repaired. Imagine the financial and perception impact that this knowledge could have on a brand name smart phone if that failure had become widely published on the Internet and there had been no remedy.

3.2.6 Higher level considerations

Developing a sensor for a set of defined use cases can be very expensive due to the many process and technical pitfalls that have to be tackled. Taking a higher level view of these issues, from a business perspective manufacturers need to ask themselves questions like: if we are going to spend all of this capital developing a sensor, how can we leverage that investment into something more comprehensive and what does this mean that we need to accomplish technically? Strategic business and technology roadmaps are needed to envision additional returns on the company's investment. It could easily be the case that without such a vision, the investment required for the initial sensor development will not be approved by the executives. And, by the way, how much could the required investment be to bring the sensor to market? That question has not been answered yet.

3.2.6.1   Business roadmap: continuity strategy

Technology companies do not make large investments in building products without having envisioned a positive and substantial Return On Investment (ROI). That is, if the company invested $100,000 to develop a sensor, they expect to receive a large multiple of this in revenue over a period of time as their ROI. Extending this example, if it took 1 year to develop the sensor and the company's GMR on the product is 50 per cent, then it is likely that the company will expect at least $300,000 in sales the first year after launch. That amount recovers the investment, the CoGs and provides a ROI, hopefully. Remember that the $100,000 remaining needs to be applied, at least partially, to the corporate expenses according to (3.6). In practice, companies set prices based on a financial model that considers corporate strategy, projected GM, product pipeline, executive and shareholder expectations and expected ROI. Part of this model is amortization of the development cost over a period of years, not a single year as in the example above.

A larger ROI in currency amount is expected and justified when the original investment into product improvements or new products extends into a product family. Such a continuity strategy, or road map, can be critical. Components of this road map may include:

- Injecting feedback about existing products into the corporate NPD process will help define new directions or critical improvements. Often times, next-generation products are already in research and development and this feedback can be very useful in these efforts. Early adopters and focus groups provide useful commentary.
- As mentioned previously, if a sensor meets market expectations and it is the lowest cost to manufacture, then this product will have a much better chance at surviving any price war. Therefore, a premium is placed on cost minimization in part through product improvement. Considerations for raw materials, logistics, contract manufacturing, infrastructure capability, internal resources and capital resources must all be balanced.
- Consideration of the sensor lifetime so that continuity in product offerings is achieved. Would it not be deadly to have a product reach end of life with sales dropping to zero and the company does not have the next-generation product ready to go? Part of the art in this area is to be able to read the market two to three years ahead of time since often that is how long a next-generation product can take to bring to market.
- New use cases envisioned for this sensor that MAY include biometric applications. Could there be use cases for the technology in other markets that could benefit from the same base technology without the company losing focus or spending too much money to penetrate these new opportunities?

The company should consider the competitive market landscape and give objective internal critiques about where it desires to be in the market in given time frames. Strategically planning for short term, medium term and long term corporate growth sets the basis for success. As Alan Lakein said, '*Failing to plan is planning to fail*'.

3.2.6.2 Technology roadmap: next-generation products

Looking into the proverbial crystal ball, what other sorts of technology can apply to the search for improving or leap frogging the capabilities of mass-market biometric sensors? This information should be captured in a cohesive technology action plan that sets out the technological vision needed to execute the business plan for the next 5 years moving forward. This is the chance to take those white board concepts the engineers have been talking about and align them with the business roadmap. Rather than discussing how to do this, this section considers several technologies that offer insights into the future of mobile biometric sensors, maybe.

With plenoptic, or light field, imaging, a single-image sensor captures multiple functionalized views of a scene. With these differing views, images can be focused at different distances from the camera *after* the image is captured. There is even the ability to generate a three dimensional image that could perhaps be used for better facial and iris matching. Lytro [18] sells a commercial camera using this technology and according to Wikipedia [19] HTC introduced a smart phone in 2014 that mimics light field imaging. Several other techniques exist in commercial cameras that can take a sequence of images or multiple images at various focal distances and then use software to bring the desired image into focus. In each case when an in focus image is needed to do things like identify a person's face, the image can be manipulated using the multi-focus image set to get a properly focused image of that person. Likewise, using the same image set, a person in the background that may not be in focus will be able to be focused in a second generated image so they can be recognized as well.

Alternative methods can also be used to generate a three-dimensional image. For example, one can use spatial light modulation to illuminate a scene with light patterns. The pattern's shadows in the scene vary according to the surface profile and by understanding how the shadow varies, it is possible to reconstruct the scene in three dimensions with fairly good accuracy. Approaches using spatial light modulation have been applied to fingerprint and facial image capture and today a commercial facial capture system exists that uses this approach [20]. While not miniaturized to fit into a mobile biometric sensor, this technology, in conjunction with a robust processor and the appropriate software application, has some pretty interesting use cases.

Outside of the three-dimensional space we live in, sensors can also capture other 'dimensions' by incorporating scanning at different illumination wavelengths. The goal is to capture details that visible light systems cannot capture so that matching or processing improvements can be supported. An example of how this could apply to biometrics is already very well known: HID Global purchased Lumidigm in early 2014 because Lumidigm had multi-spectral fingerprint scanners that captured images traditional fingerprint capture scanners cannot. In [7], Zhang, Guo and Gong also describe iris and fingerprint modalities as well as others from a multi-spectral perspective.

The industry will undoubtedly have to turn to the science of the small simply because sensors need to be … smaller (and more functional). One such example described earlier was the MEMS auto-focus actuator that makes dramatic improvements on a miniaturized camera's focus capabilities. Other MEMS applications might directly impact biometric sensors in other ways. Another example of this impact is with thin film transistor (TFT)-based sensors. TFT applications are jumping from touchscreens and high definition television into biometrics with a very recent sensor announced [21].

Lest we forget, biometric dividends may be able to be achieved using nano-science – the study of items with size on the order of nano-meters. Can the photonic technology being discovered by Duke University [22] be used to remove ambient light effects within sensors, for example? While the world talks a lot about nanotechnology, once cost-effective, repeatable and manufactural technology has been developed perhaps the next generation sensors will use nanotechnology concepts extensively.

We would be remiss if we did not mention the role of software in the technology roadmap at this point. The reason is simple: sensors capture biometric signatures but these signatures must be processed in order to do something interesting with them. Mobile hardware architectures are very sophisticated today and it is fair to say that billions have been placed into maximizing the processing capability on these platforms. This processing capability is available to support advances in software that enable even more sophisticated biometric algorithms. For example, see the facial location algorithm example in [23] which dramatically improves the speed of unconstrained facial location on lower end processors. New and more sophisticated software will be a primary actor in overcoming challenges in these tightly integrated mobile environments.

3.2.6.3 Capital requirements

Plans incorporating the aforementioned insights must provide access to the capital needed in order to realize the mass-market potential. Exactly how much money could needed for mass-market success? It depends on the plan, how much of business process defined in Figure 3.1 the company wants to take on and the risk level the company is willing to assume. While no fixed guidelines exist, the following examples place parameters around what seem to be the most successful (or potentially successful) cases observed to date. As a cautionary note: be careful of the data found on the Internet as it is not always accurate. While all of the examples are of fingerprint sensors or facial recognition software (*not* a sensor) they serve as mile markers for the entry fee for success.

1. Apple paid $320 million for AuthenTec in 2008. A part of the technology purchased was the capacitance fingerprint sensors that Apple ultimately launched on their iPhones 5S in 2013. Prior to the launch, how much in additional investment did Apple make in order to achieve the level of success it has so far?
2. Qualcomm paid an undisclosed amount for Ultrascan in March 2013 but, to that point in time Ultrascan had advertised they had taken investments of about $30 million to build an ultrasound/capacitance based fingerprint sensor. Today, almost 3 years after the purchase of Ultrascan, the resulting product(s) have yet to completely launch but Qualcomm is attending industry shows and is clearly working hard to bring this product to the market.
3. Synaptics Inc. sells intuitive human interface solutions including fingerprint sensors that are built into smart phone products. While fingerprint sensors are a small portion of their product portfolio, Synaptics spent over $293 million on research and development in their latest complete fiscal year, 2015. This budget represents 17.2 per cent of its revenue and therefore, armed with the knowledge of the processes above, one conjectures that maintaining and extending the fingerprint sensor line costs quite a bit of money.
4. NEXT Biometrics, a Norwegian fingerprint sensor company, is a public entity that had retained earnings of about ($24) million in September 2015. This means

the cost to run the company, including development of their sensors, cost at least $24 million more than the revenue they have earned to date.

5. Facebook acquired a facial recognition software company face.com in June 2012 for a reported $55 to $60 million. face.com's expertise is in mobile facial recognition. Apple agreed to purchase facial analysis company Emotient in January 2016. This acquisition followed related Apple purchases of Perceptio, Faceshift and Polar Rose. Google acquired three facial recognition related companies: Viewdle, PittPatt and Neven Vision for undisclosed amounts. Evidently, facial analysis will become a large factor in mobile biometrics moving forward. The purchase prices for these last seven acquisitions have not been made public.

There are examples of companies that have not spent nearly as much capital as the above mentioned companies for small format biometric sensors. Some of these have been successful and some not but none of them seem to have made a large impact in the mobile mass-market biometric space yet. Success seems to play out when the combination of commitment to a strategic vision, access to capital, technology appropriateness and challenge abatement all synergistically combine. With these data points, hopefully the reader has a flavour for how much it costs to address the challenges to develop and deploy mobile mass-market biometric sensors.

3.3   Conclusion

It is expected that soon 50 per cent of all smart phones will have a fingerprint sensor on them by the end of 2019 [14] and many others will undoubtedly have facial and iris matching capabilities. The number of fingerprint sensors manufacturers that own most of the mobile sensor market right now is estimated at about 20 across the globe. Addition of iris and facial sensors into this mix adds even more mass-market biometric sensor providers. Lessons learned and current market conditions in smart phone experiences suggest that some of these suppliers will be able to navigate the challenges of bringing successful sensors to market and some will not. Eventually, it might be expected that only a handful of very successful sensor providers will be left. But, it certainly will be interesting watching and participating in the dramatic technological improvements that will be seen in this area in the upcoming years as the challenges are addressed and even better sensors come to market!

References

[1]   D. Day, "Biometric Applications, Overview," in *Encyclopedia of Biometrics*, S. Z. Li, Ed., New York, Springer Science+Business Media, 2009, pp. 76–80.

[2]   "IAFIS," Federal Bureau of Investigation, 7 April 2015. [Online]. Available: https://www.fbi.gov/about-us/cjis/fingerprints_biometrics/iafis. [Accessed January 2016].

[3]   D. Setlak, "Biometric Sample Acquisition," in *Encyclopedia of Biometrics*, S. Z. Li, Ed., New York, Springer Science +Business Media, 2009, pp. 96–100.

[4]   S. Z. Li, Ed., Encyclopedia of Biometrics, New York: Springer Science + Business Media, 2009.

[5]   "New Product Development," Wikipedia, 16 January 2016. [Online]. Available: https://en.wikipedia.org/wiki/New_product_development. [Accessed January 2016].

[6]   J.-F. Mainguet, "Biometrics," 11 December 2015. [Online]. Available: http://biometrics.mainguet.org/. [Accessed January 2016].

[7]   D. Zhang, Z. Guo and Y. Gong, Multispectral Biometrics: Systems and Applications, New York, New York: Springer, 2016.

[8]   "FIDO Alliance," 2016. [Online]. Available: https://fidoalliance.org/. [Accessed January 2016].

[9]   "GlobalPlatform," 2016. [Online]. Available: http://www.globalplatform. org/mediaguidetee.asp. [Accessed January 2016].

[10]   "Press Release Match-in-Sensor | Synaptics," Synaptics, 2016. [Online]. Available: http://www.synaptics.com/en/press-releases/match-in-sensor.php. [Accessed January 2016].

[11]   "Mobile Solution | Iris recognition Smartphone," IriTec, Inc., 2016. [Online]. Available: http://www.iritech.com/products/solutions/iris-mobile-solution. [Accessed January 2016].

[12]   "Key Security Factors," NEXT Biometrics, [Online]. Available: http:// nextbiometrics.com/security/key_security_factors/. [Accessed January 2016].

[13]   J. Steer, "The FireEye Mobile Threat Report," 27 February 2015. [Online]. Available: https://www.fireeye.com/blog/threat-research/2015/02/ the_fireeye_mobilet.html. [Accessed January 2016].

[14]   Y. Zhang, Z. Chen, H. Xue and T. Wei, "Fingerprints on Mobile Devices," August 2015. [Online]. Available: https://www.blackhat.com/docs/us-15/ materials/us-15-Zhang-Fingerprints-On-Mobile-Devices-Abusing-And-Leaking-wp.pdf. [Accessed January 2016].

[15]   E. Hecht, Optics, San Francisco: Addison Wesley, 2002.

[16]   V. Chandrasekharan and E. Tosaya, "MEMS Camera Auto Focus," Invensas, 21 May 2014. [Online]. Available: http://www.invensas.com/Company/ Documents/Invensas_MEPTEC2014_MEMSCameraAutofocus.pdf. [Accessed January 2016].

[17]   R. V. Noorden, "The rechargeable revolution: A better battery," *Nature,* vol. 507, no. 7490, pp. 26–28, 6 March 2014.

[18]   "Lytro," January 2016. [Online]. Available: https://www.lytro.com/. [Accessed January 2016].

[19]   "Lytro," Wikipedia, January 2016. [Online]. Available: https://en.wikipedia. org/wiki/Lytro. [Accessed January 2016].

[20]   "Morpho 3D Face Reader," Morpho, January 2016. [Online]. Available: http://www.morpho.com/en/biometric-terminals/access-control-terminals/ facial-terminals/morpho-3d-face-reader. [Accessed January 2016].

[21] "FlexEnable Sensor," FlexEnable, 19 January 2016. [Online]. Available: http://www.flexenable.com/Newsroom/flexenable-and-isorg-reveal-first-large-area-fingerprint-and-vein-sensor-on-plastic/. [Accessed January 2016].

[22] G. M. Akselrod, J. Huang, T. B. Hoang, *et al.*, "Large-Area Metasurface Perfect Absorbers from Visible to Near-Infrared," *Advanced Materials,* vol. 27, no. November (48), pp. 8028–8034, 2015.

[23] S. Liao, A. Jain and S. Li, "A Fast and Accurate Unconstrained Face Detector," *IEEE Transactions on Pattern Analysis and Machine Intelligence,* vol. 38, no. 2, pp. 211–223, 2016.

Chapter 4

Deep neural networks for mobile person recognition with audio-visual signals

M.R. Alam[1], M. Bennamoun[1], R. Togneri[2], and F. Sohel[3]

This chapter starts with a general and brief introduction of biometrics and audio-visual person recognition using mobile phone data. It begins with a discussion of what constitutes a biometric recognition system, and it then details the steps followed when audio-visual signals are used as inputs. This is followed by a review of the existing speaker and face recognition systems which have been evaluated on a mobile biometric database. We then discuss the key motivations of using deep neural network (DNN) for person recognition. We finally introduce a Deep Boltzmann Machine (DBM)–DNN, in short DBM–DNN, based framework for person recognition. An overview of the sections and sub-sections of this chapter is shown in Figure 4.1.

4.1 Biometric systems

4.1.1 What is biometrics?

The term 'biometrics' refers to the automatic recognition of persons using their behavioural and/or physiological data (e.g., fingerprint, retinal, facial image, speech or their combination). A biometric recognition system compares features extracted from input biometric samples (e.g., the face image) with pre-stored client model(s). This has been proven to be a more secure option than the traditional knowledge-based or token-based recognition approach: since knowledge or tokens, such as a personal identification number, password or an ID card, may be lost, forgotten, fabricated or stolen. A biometric system operates in one of the following two modes: *identification* or *verification*. A biometric 'verification' system may be deployed- in various real-life applications, e.g., account holder verification for automatic teller machine or making online payments, and customs and border protection. A biometric recognition system operates in the 'identification' mode, e.g., when an unidentified person is unlikely

[1] School of Computer Science and Software Engineering, The University of Western Australia, Australia
[2] School of Electrical, Electronics and Computer Engineering, The University of Western Australia, Australia
[3] School of Engineering and Information Technology, Murdoch University, Australia

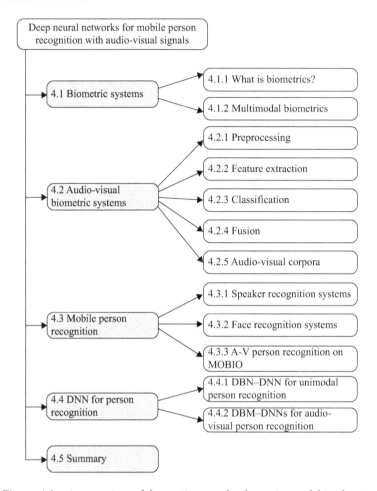

Figure 4.1   An overview of the sections and sub-sections of this chapter

to claim the identity of one of the registered clients. For example, crime scene and forensic investigations and access control.

Decision making in the identification mode is determined by two cases: a) *closed-set* and b) *open-set*. In closed-set identification, it is assumed that an unidentified person is already registered in the database and he/she is classified as one of the N registered persons. In open-set identification, however, the system tries to find whether the unknown person is registered in the database or not. Thus, the task is considered to be an $N + 1$ class identification problem including a reject class. In the verification mode (and in contrast to both forms of identification), the system tries to determine whether a person is who he/she is claiming to be. An unknown person's claim is accepted only if the matching score for the client model corresponding to the claimed identity is above a pre-determined threshold.

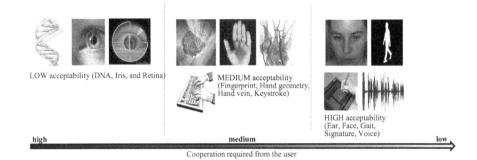

LOW acceptability (DNA, Iris, and Retina)

MEDIUM acceptability
(Fingerprint, Hand geometry,
Hand vein, Keystroke)

HIGH acceptability
(Ear, Face, Gait,
Signature, Voice)

high   medium   low

Cooperation required from the user

*Figure 4.2   Commonly used biometric traits and the level of cooperation required
from the end users during data acquisition*

4.1.2   Multimodal biometrics

A number of biometric traits are used for person recognition (see Figure 4.2). The
choice of a particular trait (or biometric) depends on a number of factors, including: the
uniqueness of the trait, the cost and size of the data acquisition sensor, the robustness
of the acquired data to noise and various other nuisances, and how easy it is for an
impostor to fake the biometric. A categorization of the commonly used biometric
traits based on the level of user cooperation (that is required from the user during data
acquisition) is shown in Figure 4.2.

   If a recognition system uses a single biometric trait then it is known as an unimodal system. The decision about the identity of a person is made using a single
classifier. However, the captured biometric sample may be of poor quality due
to (e.g., in the case of face recognition) variations in the pose, illumination and
background. Some biometric traits (e.g., voice and signature) are also vulnerable
to spoofing. Under these challenging conditions, a unimodal system may produce
unreliable recognition results, because it relies on a single input modality. A multimodal biometric system makes decisions by fusing information from multiple
modalities at different levels, e.g., fusion at the outcome level of each modality
(i.e., score-level fusion) or by using a single system with concatenated inputs (i.e.,
feature-level fusion). Such systems are considered more robust than their unimodal
counterparts.

   In [1], the following advantages of a multimodal biometric system were reported:
(a) it addresses the issue of non-universality, **(b)** it facilitates the indexing and filtering
process of large-scale biometric databases, **(c)** spoofing attack becomes increasingly
difficult and **(d)** it addresses the problem of noisy data by using a quality measure
of the sensed data during the fusion process. Furthermore, the level of difficulty
required to acquire some biometric traits (e.g., voice, facial and gait) may be less
than others (e.g., DNA and retinal image). Therefore, some multimodal biometric
systems (e.g., audio-visual) are more cost effective and easily deployable in a real-
life scenario than others (e.g., DNA + retinal image and fingerprint + palm print).

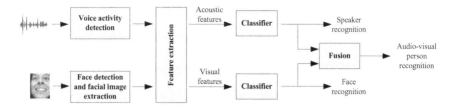

Figure 4.3   Block diagram of an audio-visual biometric recognition system

The following sections of this chapter will therefore focus on a standard audio-visual biometric recognition system (see Figure 4.3).

4.2   Audio-visual biometric systems

An audio-visual biometric system is designed to recognize unknown persons using their speech and facial images. Such a system (see Figure 4.3) is generally composed of the following parts: **(a)** a speaker sub-system and **(b)** a face sub-system. If sub-systems are fused at the score level, their matching scores are combined using a score fusion rule. If sub-systems are fused at the rank level, their ranked lists are combined using a rank fusion rule. Moreover, the modalities can also be fused at the feature level or data level. The functions of a standard audio-visual biometric recognition system are briefly described below.

4.2.1   *Preprocessing*

The captured audio and visual signals are commonly preprocessed before any meaningful features can be extracted. This stage enhances the quality of the captured data and it is critical in improving the overall recognition accuracy of the system.

- **Silence removal**: Although the type of required preprocessing is dependent on the intended features to be extracted (which are then used for classification), speech enhancement is used by most speaker recognition systems. It is usually performed with the help of a voice activity detection algorithm, e.g., [2], which removes any non-voice segments from the speech signals.
- **Face tracking**: Tracking by detection is performed on the video frames, based on the detection of the face from facial features, e.g., eyes and nose. The Viola–Jones algorithm [3] is the most commonly used algorithm for face detection due to its extremely fast feature computation, efficient feature selection, and scale and location invariant detection. Once a face part is detected, a region of interest (ROI) representing that face part is cropped. The ROI can represent various face parts, such as the eyes, nose or mouth (e.g., [4–6]) or the whole face (e.g., [7–11]). Then, this ROI is photometrically normalized (e.g., using the Tan–Triggs algorithm [12]) to minimize the effects of illumination and background variations.

4.2.2   Feature extraction

Features are extracted from the preprocessed audio and visual signals. The choice of a particular type of feature depends on factors such as the type of application, classifier, and the robustness to channel distortion and session variability. For example, for speaker recognition it is important to extract features that capture the speaker-specific characteristics.

4.2.2.1   Acoustic features

A speech utterance is first divided into overlapping fixed duration (e.g., ideally between 20 and 30 ms) segments, also known as frames. A frame is generated ideally after every 10 ms and multiplied by a window function to smooth the effect of using finite size segments. The following features are used by most speaker recognition systems:

- **Mel-frequency cepstral coefficients (MFCCs)**: A Fast Fourier Transform is applied on each frame to obtain a set of P complex spectral values. These coefficients are then converted to Q Mel-spaced filter bank values such that $Q \ll P$, using triangular filter banks spaced according to the Mel scale. Since human hearing exhibits logarithmic compression, the log of each filter bank output is taken. The final step is to convert these Q log filter bank values into R cepstral coefficients using the Discrete Cosine Transform (DCT). Ideally, $R = 13$ cepstral coefficients including c_0, which represents the average log-power, are exacted from each frame. The first- and second-order derivatives of these coefficients are augmented to form a 39-dimensional MFCC + delta + acceleration feature vector per frame of an utterance. MFCCs are ideally used in GMM-UBM-based systems where a client's Gaussian Mixture Model (GMM) [13], represented as λ_j, is built by adapting a global GMM model also known as the Universal Background Model (UBM). The feature vectors extracted from all the frames of the enrolment samples from a client j are used to adapt the UBM.
- **GMM super-vector (GSV)**: A GSV is defined as a large vector of length $N \times M$ formed by concatenating the N-dimensional means from each of the M mixtures of a GMM. If the UBM is adapted using the features (e.g., MFCCs) from all the frames of a single utterance, one can obtain a GMM for that utterance [14]. The means, μ_i, of the components of the *utterance-specific* GMM are obtained by adapting the means of the UBM. Therefore, a GSV representing an utterance, x, is formed by $\mu_x = [\mu_1'|\mu_2'|\dots|\mu_M']'$. They can be used to build linear models of the clients. In addition, the recent variability modelling techniques [15–17] are based on the GSV subspace.
- **i-vectors**: Inter-session variability (ISV) [15] and joint factor analysis (JFA) [16] are the session variability modelling techniques widely used in the literature. Both methods assume that the observations, $x_{i,j}$, from the ith sample of client j are drawn from the following distribution:

$$\mu_{i,j} = m + Uf_{i,j} + d_i, \tag{4.1}$$

where U is the low-dimensional session variability subspace trained with an expectation maximization (EM) algorithm and $f_{i,j}$ are latent factors, which are set with a standard prior [15]. ISV and JFA differ in their definition of the client-dependent offset. Given latent factors z_i and y_i, the offset is expressed as $d_i = Dz_i$ in ISV, where D is a function of the UBM covariance representing the between-class variability subspace. In JFA, the offset is expressed as $d_i = Vy_i + \hat{D}z_i$, where V is a low-dimensional within-class variability subspace. Both V and $\hat{D}$ are learnt using the EM algorithm. Due to the high dimensionality of GSV space, the JFA can fail to separate between-class and within-class variation in two different subspaces. A Total Variability Modelling (TVM) [17] method was proposed to overcome this issue by treating each sample in the training set as if it comes from a distinct client. The TVM training process assumes that the ith sample of client j can be represented by a GSV:

$$\mu_{i,j} = m + Tw_{i,j}, \tag{4.2}$$

where T is the low-dimensional total variability matrix and $w_{i,j}$ is the i-vector representation of the speech sample. A cosine similarity scoring [17] and the probabilistic linear discriminant analysis (PLDA) [18] are commonly used for evaluating the i-vector-based systems.

4.2.2.2　Visual features

Visual features are extracted from the normalized facial images. Unlike speech utterance, the dimensions of the facial images is fixed. Therefore, some recognition systems directly use the pixel intensity values as features. Other systems use more sophisticated hand-crafted features from the local segments of the image samples. The following feature types are commonly used by face recognition systems:

- **Appearance-based**: The transform vectors of the ROI pixel intensities are used as appearance-based features. The transform techniques used are principal component analysis (PCA), LDA, DCT, and Wavelet transforms. These are computationally inexpensive and can be extracted dynamically in real time. As pixel intensities are used for computation, the quality of the appearance-based features degrades under intense head-pose and lighting variations. If a grey-scale image of size $a \times b$ is represented by $x \in \mathbb{R}^{a \times b}$ then the image can be transformed to a feature vector such that $x \in \mathbb{R}^{a \times b} \to \hat{x} \in \mathbb{R}^{q \times 1}$, where $q = a \times b$. This type of feature is suitable for classifiers which create dictionaries or templates of the clients and use the Euclidean distance between a test feature vector and the dictionaries/templates for scoring.
- **Parts-based**: In this approach, each face image is divided into regions and then features are extracted from each region. Examples of features include the local binary patterns (LBP) presented in [19]. In its simplest form, $LBP_{8,1}$ (with a radius of 1 pixel and 8 sampling points), patterns are extracted from each pixel of an image by thresholding the values of the 3×3 neighbourhood of the pixel against the central pixel. The extracted patterns are then interpreted as binary numbers. An LBP operator with a radius of 2 pixels, 8 sampling points and uniform patterns (i.e., if it contains at most two 0-1 or 1-0 transitions when viewed as a circular bit

string) is represented as $LBP_{8,2}^{u2}$ and is commonly used in face recognition research. Typically, a histogram of LBP is computed in each region after applying the $LBP_{8,2}^{u2}$ operator to the face image. Since only 58 of the 256 possible 8-bit patterns are uniform, the $LBP_{8,2}^{u2}$ operator provides a compact representation while building the LBP histograms. The histograms obtained from all the regions of a face image are concatenated to form a feature vector.

- **i-vectors**: Recently, the TVM method has been used for extracting i-vectors representations for the visual modality inputs [20]. First, feature vectors are extracted from the overlapping blocks in the normalized facial images. Feature extraction is carried by extracting say $c \times c$ blocks of pixels using an exhaustive overlap. If an image is of size $a \times b$ then a total of $k = (a - c - 1) \times (b - c - 1)$ blocks are extracted from that image. In [21], different values of c were evaluated and it was shown that a lower error rate can be achieved when c is set to 12. The pixel values of each block are normalized to zero mean and unit variance. Then, the $l + 1$ lowest frequency 2D-DCT [22] coefficients are extracted from each block removing the zero frequency coefficient. The value of l is commonly set equal to 44. The resultant 2D-DCT coefficients are also normalized to zero mean and unit variance in each dimension with respect to the other feature vectors obtained from the same image. The feature vectors obtained from all the training images can be used to build the UBM. Then, the TVM method (Equation (4.2)) is applied to extract an i-vector given the observations of an image.

4.2.3 Classification

The role of a classifier is to build the client models and score them in the test phase. In this section, a brief description of two of the most widely used classifiers in recent years is presented.

4.2.3.1 Support vector machine (SVM)

The SVM is a two-class classifier which fits a separating hyperplane between the two classes. During the training phase, an optimal hyperplane is chosen such that it maximizes the Euclidean distance to the nearest data points on each side of the plane. The nearest data points on either side of the hyperplane are known as support vectors. An SVM classifier is constructed from the sums of kernel functions [23]:

$$S(x) = \sum_{i=1}^{L} \alpha_i c_i K(x, x_i) + d, \qquad (4.3)$$

where α_i is the Lagrange multiplier associated with the ith support vector, c_i is the corresponding classification label (i.e., $c_i = +1$ if x belongs to the user and $c_i = -1$ if x belongs to the impostor), d is a learned constant, L is the number of support vectors, $\sum_{i=1}^{L} \alpha_i c_i = 0$, $\alpha_i > 0$ and x_i are the support vectors obtained from the training set. A kernel that is constrained to satisfy the Mercer condition is expressed by:

$$K(x, y) = b(x)^t b(y), \qquad (4.4)$$

where $b(x)$ is the mapping from the input space to a higher dimensional separating space. Finally, a verification decision is made based on whether the value of $S(x)$ is above or below a threshold.

Although the SVM classifier is suitable for verification, multiclass identification can be performed by designing a *one-against-all* (OAA) SVM for each of the N registered clients. For example, the SVM for an unknown person claiming to be a registered client j is a two-class system, where class 0 represents the training data from client j and class 1 represents the training data from all the other $(N - 1)$ clients. Identification is thus performed by carrying out a total of N two-class classifications and selecting the SVM with the maximum decision function value. The functional operation of the OAA SVM classifier for a client j can be expressed by:

$$S_j(x) = \sum_{i=1}^{L_j} \alpha_i^j c_i^j K(x, x_i^j) + d^j, \tag{4.5}$$

where $(\{\alpha_i^j, x_i^j\}_{i=1}^N, d^j)$ are obtained from the SVM optimization algorithm. A rank-1 identification is performed by determining the SVM classifier with the maximum decision value:

$$\hat{j} = \underset{1 \leq j \leq N}{\max} \, S_j(x). \tag{4.6}$$

4.2.3.2   Linear regression-based classifier (LRC)

Suppose there are N classes with L_j number of training images from the jth class. A class specific model X_j is developed by stacking the q-dimensional feature vectors:

$$X_j = [x_j^{(1)} x_j^{(2)} \cdots x_j^{(L_j)}] \in \mathbb{R}^{q \times t_j}, \, j = 1, 2, \ldots, N. \tag{4.7}$$

During the training phase, each class j is represented by a subspace X_j. When a test feature vector $y \in \mathbb{R}^{q \times 1}$ is presented, y should be represented as a linear combination of the training samples of the class it belongs to [24]. This is expressed by: $y = X_j \beta_j$, where $\beta_j \in \mathbb{R}^{L_j \times 1}$ is the vector of parameters. The vector β_j can be estimated using least-squares estimation:

$$\hat{\beta}_j = (X_j^T X_j)^{-1} X_j^T y. \tag{4.8}$$

The estimated vector of parameters, β_j, along with the predictors X_j are used to predict the response vector for each class j by using $\hat{y}_j = X_j \hat{\beta}_j$, where $\hat{y}_j$ is the predicted response vector in the jth subspace. The distance measure between the predicted response vector $\hat{y}_j$ and the original response vector y is then calculated:

$$d_j(y) = \|y - \hat{y}_j\|_2 \tag{4.9}$$

and the rank-1 identification decision is given in favour of the class with minimum distance:

$$\hat{j} = \underset{1 \leq j \leq N}{\min} \, d_j(y). \tag{4.10}$$

4.2.4 Fusion

The Fusion of multiple modalities may be carried out at an early stage (i.e., feature fusion) or a late stage (i.e., score- or rank-fusion). Since feature-level fusion may result in large feature vectors which slow down the recognition process, fusion at the score level is commonly adopted in most biometric systems. If the matching scores obtained from multiple sub-systems are heterogeneous (e.g., posterior probability and Euclidean distance), many fusion rules first transform these into a common domain before fusion can take place. This process of transforming the scores is known as score normalization [25]. Once multiple score sets are normalized they can be combined into one score using the following fusion approaches.

4.2.4.1 Non-adaptive fusion

If the weights of the sub-systems are not adjusted based on some specific criteria then it is referred to as **non-adaptive** fusion. This approach may be used if it assumed that the data from both modalities are clean. Fusion is performed by combining the matching scores obtained from the classifiers of the sub-systems using an appropriate rule (e.g., product, sum, min or max rule). A theoretical framework of the rule-based fusion methods is presented in [26]. If an input vector x_m is presented to the mth classifier then the matching score for the jth class is $P(\lambda_j|x_m)$, which is the probability of x_m belonging to class j. Let $c \in \{1, 2, \ldots, N\}$ be the class to which the input is finally assigned. The identity c can be determined using the following rules:

- **Product rule**:

$$c = \operatorname*{argmax}_{j} \prod_{m=1}^{M} P(\lambda_j|x_m) \tag{4.11}$$

- **Sum rule**:

$$c = \operatorname*{argmax}_{j} \sum_{m=1}^{M} P(\lambda_j|x_m) \tag{4.12}$$

- **Max rule**:

$$c = \operatorname*{argmax}_{j} \max_{m} \sum_{m=1}^{M} P(\lambda_j|x_m) \tag{4.13}$$

- **Min rule**:

$$c = \operatorname*{argmax}_{j} \min_{m} \sum_{m=1}^{M} P(\lambda_j|x_m) \tag{4.14}$$

4.2.4.2 Adaptive fusion

In a real-life scenario, data from a modality may contain noise due to variations in the input data (e.g., illumination, background and pose variations on the face images; additive noise and channel distortion, and session variability on the speech samples). Therefore, an **adaptive** fusion approach would be essential where the sub-systems are weighted based on a criteria such as the quality of the sensed data or some statistics of the matching score distribution. The following adaptive rules are used to determine c:

- **Adaptive product rule**:

$$c = \operatorname*{argmax}_j \prod_{m=1}^{M} P(\lambda_j|x_m)^{\alpha_m} \qquad (4.15)$$

- **Adaptive product rule**:

$$c = \operatorname*{argmax}_j \sum_{m=1}^{M} \alpha_m P(\lambda_j|x_m) \qquad (4.16)$$

where α_m is the weight assigned to the classifier of the mth modality and $\alpha_1 + \alpha_2 + \cdots + \alpha_M = 1$. The statistics of the matching score distributions are commonly used in the literature to calculate the modality weights. For example, the dispersion of the log-likelihoods [27], the entropy of the posteriors [28] and the C-ratio [29,30].

4.2.4.3 Linear logistic regression (logReg) fusion

The **logReg** fusion approach combines a set of M classifiers using the sum rule. Let $f_m(x_m, \lambda_j)$ represent the score assigned by classifier $m \in M$ between its input vector x_m and the jth client model, λ_j. Therefore, the fused score for the jth client model is obtained using a linear combination:

$$f_\beta(x_m, \lambda_j) = \beta_0 + \sum_{m=1}^{M} \beta_m f_m(x_m, \lambda_j), \qquad (4.17)$$

where $\beta = [\beta_0, \beta_1, \ldots, \beta_M]$ are the fusion weights also known as the regression coefficients. These coefficients are calculated by using the maximum likelihood estimation of the logistic regression model on the scores of a development set [31]. Let $\mathcal{Y}_{cli}$ be the set of pairs $y = \{x_m, \lambda_j\}$, where the identity of the test sample and that of the client is the same. Similarly, let $\mathcal{Y}_{imp}$ be the sets of pairs $z = \{x_m, \lambda_j\}$, where the identity of the test sample is different from the identity of the client. Then, the objective function to maximize is:

$$L(\beta) = -\sum_{y \in \mathcal{Y}_{imp}} \log\left(1 + \exp\left(f_\beta(y, \beta)\right)\right) - \sum_{y \in \mathcal{Y}_{cli}} \log\left(1 + \exp\left(-f_\beta(y, \beta)\right)\right).$$

$$(4.18)$$

The maximum likelihood procedure converges to a global maximum. The optimization is carried out using the conjugate-gradient algorithm [32]. This fusion approach has been used to combine heterogeneous speaker [33,34], face [35] and bimodal recognition systems [20,36].

4.2.5 Audio-visual corporation

There exists only a few audio-visual databases that are freely available for biometrics research. This is because the field is just fledgling and the data collection process poses some challenges such as the synchronization between the audio and video data, storage and the privacy of the subjects. Existing audio-visual databases vary in size, in the types of recorded utterances and in the recording environment. In this section, a brief review of existing audio-visual databases is presented.

4.2.5.1 BANCA

The BANCA database [37] was captured in four languages (e.g., English, French, Italian and Spanish) using high- and low-quality microphones and cameras. The subjects were recorded over a period of three months and in three different scenarios such as controlled, degraded and adverse. There are 208 recorded subjects with an equal number of males and females. Each subject recorded 12 sessions with 2 recordings (1 true *client access* and 11 informed *impostor attack*) per session. The subjects were asked to say a random 12 digit number, name, address and date of birth. A cheap analogue web cam and a high-quality camera were used to record the video data. Similarly, both cheap and high-quality microphones were used to capture the audio data. Thus, the BANCA database contains realistic and challenging conditions and allows for robustness comparison of different systems.

4.2.5.2 M2VTS and XM2VTS

The M2VTS [38] contains audio-visual data of 37 subjects, uttering digits 0 to 9. The data were captured in five sessions, with a gap of at least one week between the sessions. An extended M2VTS (XM2VTS) database was created with the audio-visual data of 295 subjects [39]. In XMT2VTS, data collection was carried out in four sessions uniformly distributed over a period of five months, in order to capture the variabilities that are due to the changes in appearance, mood and physical conditions. The database was acquired using a Sony VX1000E digital camcorder and DHR1000UX digital VCR. The subjects were asked to read twice through three sentences written on a board positioned just below the camera. The subjects were also asked to rotate their head while they were being recorded (refer to Figure 4.4 for the sequence head movements).

4.2.5.3 VidTIMIT

The VidTIMIT [40] database consists of videos (with speech) of 43 subjects (19 females and 24 males) reciting short sentences and rotating their heads under facial expressions. The sentences were selected from the test portion of the TIMIT corpus. This database is useful for lip reading and multi-view face and speech/speaker recognition research. A broadcast quality digital video camera was used to record the subjects in the database. There are 10 videos per subject collected in 3 sessions. The first six videos were captured in Session 1, the next two in Session 2 and the other two in Session 3. There was an average delay of 7 days between Sessions 1 and 2 and 6 days between Sessions 2 and 3. The first two sentences for all subjects were the same, while the remaining eight were different for each subject. Each subject performed a sequence of head rotations as shown in Figure 4.4.

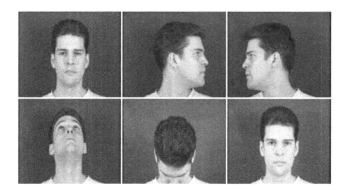

*Figure 4.4   A sequence of rotating head movements included in the M2VTS,
            XM2VTS and VidTIMIT databases (examples are taken from
            XM2VTS)*

Table 4.1   AusTalk subjects represent variations in geography, dialect and emotion

State	Capitals	Subjects	Regional	Subject	Other	Subjects	Total
NSW	UNSW	48	–	–	Emotion	36	84
	USYD	48	–	–	Disordered	16	64
	–	–	Armidale (UNE)	48	–	–	48
	–	–	Bathurst (CSU)	48	–	–	48
QLD	UQ	120	Townsville (UQ)	48		–	168
VIC	UMELB	120	Castlemaine (UMELB)	48	–	–	168
SA	Flinders	96	–	–	–	–	96
NT	–	–	Darwin (CDU)	24	Aboriginal English (CDU)	48	96
			Alice Spring (CDU)	24			
WA	UWA	96	–	–	–	–	96
TAS	UTAS	48	–	–	–	–	48
ACT	UC	36	–	–	–	–	36
	ANU	48	–	–	–	–	48
Total	–	660	–	240	–	100	1,000

4.2.5.4   AusTalk

The AusTalk [41–43] dataset consists of a large collection of audio-visual data
acquired from 15 different locations in all states and territories of Australia (see
Table 4.1). The project started in 2011 and was funded by an Australian Research
Council grant. Twelve identical stand-alone recording set-ups (see Figure 4.5) were
built and shipped to 17 different collection points at 15 locations. Each subject
in the database was recorded in three sessions at intervals of at least one week.

Figure 4.5   AusTalk data capturing environment

Table 4.2   AusTalk data collection protocol (time in minutes)

Session 1		Session 2		Session 3	
Task	**Time**	**Task**	**Time**	**Task**	**Time**
Opening Yes/No	3	Opening Yes/No	2	Opening Yes/No	2
Words	10	**Words**	10	**Words**	10
Read narrative	5	*Interview*	15	*Map task* (First run)	20
Re-told narrative	10	–	–	–	–
Read digits	5	**Read digits**	5	*Map task* (Second run)	20
–	–	**Read sentences**	8	*Conversation*	5
Closing Yes/No	2	Closing Yes/No	2	Closing Yes/No	2
–	35	–	42	–	59

The database contains spoken words, digits, sentences and paragraphs (see Table 4.2) from 1,000 subjects representing regional and social diversity and linguistic variation of Australian English.

The participants' speech was recorded using five microphones and two stereo cameras. Each of the three sessions includes different subsets of the read and spontaneous speech tasks. The participants were prompted to read aloud a list of words, digits and sentences (bold faced in Table 4.2) presented on a computer screen in front of them. These recordings are ideally suited for speaker recognition (e.g., [30,44,45]). The 'Read narrative' and 'Re-told narrative' tasks (underlined in Table 4.2) in Session 1 provide materials for the study of differences between spontaneous language. The 'Interview,' 'Map task' and 'Conversation' tasks (italicized in Table 4.2) provide materials for speech analysis. Finally, a set of 'Yes/No' questions recorded at the beginning and end of each session provides a range of positive and negative answers.

Table 4.3   A summary of the MOBIO database

Site	Phase I			Phase II		
	# subjects (female/male)	# sessions	# shots	# subjects (female/male)	# sessions	# shots
BUT	33 (15/18)	6	21	32 (15/17)	6	11
IDIAP	28 (7/21)	6	21	26 (5/21)	6	11
LIA	27 (9/18)	6	21	26 (8/18)	6	11
UMAN	27 (12/15)	6	21	25 (11/14)	6	11
UNIS	26 (6/20)	6	21	24 (5/19)	6	11
UOULU	20 (8/12)	6	21	17 (7/10)	6	11

4.2.5.5   MOBIO

MOBIO is a challenging bimodal database containing 61 h of audio-visual data recorded in 12 sessions. It includes 192 unique audio-visual signals from each of the 150 participants, acquired at 6 different sites over a period of one and a half years. It is a challenging database because data were captured with real noise. For example, the video frames contain uncontrolled illumination, expression, near-frontal pose and occlusion, while speech signals are relatively short. Hence, the MOBIO database is suitable for evaluating systems that would operate in an uncontrolled environment.

The MOBIO data collection was carried out using two mobile devices: a Nokia N93i mobile phone and a 2008 MacBook laptop. Only the first session data was captured using the laptop, while in the other sessions the mobile phones were used. Data acquisition was carried out in two phases: Phase I and Phase II. A Dialogue Manager (DM) was installed on the mobile phones for each site and for each user. During data recording the DM prompted the participants with pre-defined and random short response questions, free speech questions and to read a pre-defined text. For the short response questions, the DM also supplied pre-defined and fictitious answers to the participants. In addition, the responses to the free speech questions were fictitious and hence did not necessarily relate to the question. The following pre-defined short response questions were considered: (a) What is your name? (b) What is your address? (c) What is your birth date? (d) What is your licence number? and (e) What is your credit card number? A summary of the MOBIO database is shown in Table 4.3.

4.3   Mobile person recognition

The recent popularity of smart phones, tablet and laptop computers has laid the platform for various mobile-friendly applications. Many of these applications are required to handle the personal information of their users. Hence, it is important that access is only given to the registered users. Recently, person recognition using mobile phone data has been extensively studied for developing more robust biometric systems. Two evaluation competitions were also held on face and speaker recognition using MOBIO

Table 4.4   The unbiased MOBIO evaluation protocol used at ICB 2013

Set	Phase-I		Phase-II	Samples/ subject	Subjects		
	Session 1	Sessions 2–6	Sessions 7–12		TR	DEV	EVAL
Training	5*p*, 10*f*, 5*r*, 1*l*	5*p*, 10*f*, 5*r*, 1*l*	5*p*, 5*f*, 1*l*	192	50	–	–
Development	5*p*	–	–	5	–	42	58
Evaluation	–	10*f*, 5*r*	5*f*	105	–	42	58

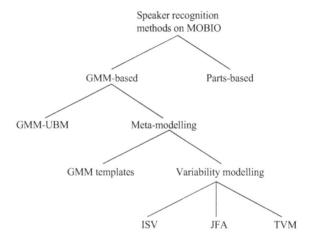

Figure 4.6   A taxonomy of existing speaker recognition methods using MOBIO

at ICPR 2010 [46] and at ICB 2013 [47,48]. At ICB 2013, an unbiased protocol (see Table 4.4) was used to assure a fair evaluation and comparable results. The protocol divides the MOBIO database into three disjoint sets: training, development and evaluation. The training set is used for the learning of the background modelling and/or for score normalization, while the development set can be used for tuning the meta-parameters (e.g., number of Gaussians). Client models are built and biometric systems are evaluated using the evaluation set. The following performance measures are used: equal error rate (EER), half-total-error rate (HTER) and detection error trade-off. These measures rely on the false acceptance rate (FAR) and the false rejection rate (FRR), which are calculated for the development and evaluation sets independently.

4.3.1 Speaker recognition systems

A number of speaker recognition systems using different modelling approaches have been evaluated on MOBIO (see Figure 4.6). Based on client modelling the speaker recognition systems using the MOBIO corpus can be categorized as follows.

Table 4.5   Speaker recognition systems evaluated using MOBIO

Reference		Modelling	Scoring	Female		Male	
				EER	**HTER**	**EER**	**HTER**
Khoury	Alpineon	TVM	PLDA	7.98	10.67	5.04	7.07
et al. [48]	ATVS	TVM	PLDA	16.83	17.85	14.88	15.42
	CDTA	TVM	cosine	19.47	22.64	12.73	19.40
	CPqD (sub-I)	GMM-UBM	LLR				
	CPqD (sub-II)	TVM	cosine				
	CPqD (Fusion)	Fusion	logReg	14.34	15.98	11.82	10.21
	EHU	TVM	PLDA	17.93	19.51	11.31	10.05
	GIAPSI	GMM-UBM	LLR	11.59	12.81	9.68	8.86
	IDIAP	ISV	linear [49]	12.01	14.26	9.96	10.03
	L2F (sub-G)	GMM-UBM	LLR				
	L2F (sub-S)	GSV	SVM				
	L2F (sub-I)	TVM	cosine				
	L2F (Fusion)	Fusion	logReg	13.48	22.14	10.59	11.12
	L2F-EHU		logReg	11.00	17.26	7.88	8.19
	Mines-Telecom	GMM-UBM	LLR	11.42	11.63	10.19	9.10
	Phonexia	TVM	PLDA	8.36	14.18	9.60	10.77
	RUN	TVM	PLDA	25.40	23.11	24.64	22.52
Khemiri		GMM-UBM	3N	–	7.01	–	4.87
et al. [50]							
[1]Khoury		GMM-UBM	LLR	17.94	17.68	13.41	12.12
et al. [51]							
		ISV	linear	12.22	16.23	10.40	10.36
		TVM	linear	12.59	17.36	11.31	11.11
		Fusion	logReg	9.21	14.65	7.31	7.89
Boulkenafet		TVM (LDA)	cosine	14.40	28.61	12.33	20.98
et al. [52]							
		TVM (CEA [53])	cosine	12.47	25.29	10.99	20.65
[2]Roy *et al.* [54]		Slice feature	BSC	–	15.50	–	18.90

[1]Using the **mobile-0** protocol [20].
[2]Using the **MOBIO Phase I** database [46].

4.3.1.1   GMM-based methods

GMM-based modelling is the most commonly used method for client modelling in speaker recognition research. Recent variability modelling methods (e.g., ISV, JFA and i-vector) are also built upon the GMM-based modelling to compensate for the within-class and between-class variations.

- **GMM-UBM:** A client model, λ_i, is built by adapting the UBM using all the utterances from client i and the maximum *a posteriori* adaptation method [13]. Scoring is carried out by estimating the log-likelihood ratio (LLR) of a test sample with regard to the client models. In Table 4.5, a list of speaker recognition systems evaluated on the MOBIO database is presented. In [48], the GMM-UBM was used by the systems identified as: **CPqD (sub-I)**, **GIAPSI**, **L2F (sub-G)** and **Mines-Telecom**. These systems built gender-dependent UBMs with 512 or 1,024 Gaussian components. The evaluation results showed that the GIAPSI system achieved the best HTER of 12.81% for the male clients among the systems which used a single decision maker. While none of the CPqD (sub-I), GIAPSI, L2F (sub-G) and Mines-Telecom performed score normalization, a Nearest Neighbour Normalization (3N) technique was recently proposed [50], which improved the HTERs of the GMM-UBM-based systems (see Table 4.5). In this approach, the test utterance is compared with the claimed identity model as well as the other models stored in the database. Then the difference between the LLR of the claimed identity and the maximal LLR of the other models is calculated. The claim is accepted only if the difference is above a predefined threshold.

- **Meta-modelling**: In this approach, the GMM-UBM method is utilised for two purposes: a) to generate a utterance-specific GMM (GSV) and b) for variability modelling.
  - **GMM templates**: A GSV, $\mu_{i,j}$, corresponding to an utterance is formed by concatenating the means of a GMM $\lambda_{i,j}$, representing only that utterance. Such super-vectors are used to build client templates with an SVM, referred to as a GMM-SVM in [55]. A sub-system of the **L2F** system, identified as L2F (sub-S) in [48], submitted to the 2013 speaker recognition evaluation used a GMM-SVM.
  - **Variability modelling**: These methods aim to estimate and eliminate the effects of within-class and between-class variations. For example, the ISV [15] and JFA [16] methods attempt to eliminate the within-class variabilities commonly caused by the sensor (i.e., microphone) and the environment (i.e., background) noises. It is assumed that session variability results in an additive offset to the GSV. The JFA method can fail to separate within-class and between-class variations in two different subspaces, potentially due to the high dimensionality of the GSVs. The TVM method learns a lower dimensional total variability subspace (T) from which the i-vectors are extracted. In [48], the systems identified as **Alpenion**, **ATVS**, **CDTA**, **CPqD (sub-II)**, **EHU**, **L2F (sub-I)**, **Phonexia** and **RUN** used i-vectors. Although the Alpenion system achieved the best HTER among all the submitted systems, it used a combination of 9 different TVM-based sub-systems. Among the single systems, the EHU achieved the lowest HTER for both male and female clients. However, an ISV-based system (IDIAP) performed better than EHU. Similar results were reported in [51], where a comparison between the baseline GMM-UBM, ISV and the i-vector-based methods was presented (see Table 4.5). Moreover, a Conformal Embedding Analysis (CEA) [52] was presented as an alternative to the LDA subspace learning used in the TVM, but the HTERs achieved were not better than the TVM-based systems presented in [48].

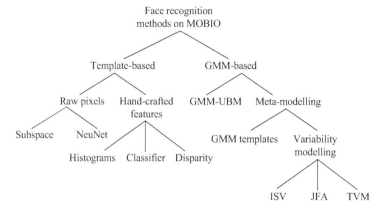

Figure 4.7   A taxonomy of the existing face recognition methods using MOBIO

4.3.1.2   Parts-based method

A fast parts-based method was presented in [54] using a Boosted Slice Classifier (BSC) and a novel set of features, called 'slice features', extracted from the speech spectra. The proposed method was inspired from the following object detection algorithms in the computer vision domain: **(a)** rapid object detection using a boosted cascade of Haar features [3,56], **(b)** fast key-point recognition using random Fern features [57] and **(c)** face detection and verification using LBP [56]. The 1-D spectral vectors derived from speech are considered equivalent to 2D images. The classifier measures the spectral magnitudes at pairs of frequency points. The most discriminative classifiers is selected using the Discrete Adaboost algorithm. The BSC system was compared with the 17 systems presented by five research groups at the ICPR 2010 [46] for the first speaker recognition evaluation on the MOBIO Phase I database. Although the BSC system did not achieve the best HTERs for the male and female clients (18.9% and 15.5%, respectively), it was computationally less complex than the GMM-UBM methods (see Table 4.5).

4.3.2   Face recognition systems

Different hand-crafted features such as the Patterns of Oriented Edge Magnitude (POEM), Gabor features, LBP, Local Phase Quantization (LPQ) and texture information have been utilised by the face recognition systems using MOBIO. Moreover, the session variability modelling techniques have been recently used in face recognition with some degree of success. In Figure 4.7, a taxonomy of existing face recognition methods on MOBIO is presented.

4.3.2.1   Template-based

In this approach, raw or hand-crafted features are extracted from the enrolment images. Then, client models are built using one of the following: **(a)** an average LBP histogram,

Table 4.6   Template-based methods of face recognition on MOBIO

Participant	Feature	Image	Block	Method	HTER Female	Male
Baseline	Raw	64 × 80	–	Subspace (PCA+LDA)	20.94%	17.11%
UC-HU	Raw	200 × 200	–	NeuNet + LDA Subspace	10.83%	6.21%
CDTA	LBP	64 × 80	8 × 8	Histogram	28.48%	11.92%
TUT	LBPHS	140 × 154	49 × 40	Classifier (PLS)	13.91%	11.54%
Idiap	Gabor	120 × 150	20 × 25	Disparity	12.50%	10.29%
UTS	Gabor wavelets	unknown	8 × 8	Histogram		
	LPQ	unknown	7 × 7		13.56%	11.95%
GRADIANT	POEM	125 × 140	10 × 10	Disparity		
	Gabor	85 × 100	10 × 10		12.27%	9.52%
CPqD	LBP	108 × 108	8 × 8	Classifier (SVM)		
	dLBP	68 × 68	16 blocks			
	MSLBP-68	68 × 68	16 blocks		11.20%	7.66%
	MSLBP-108	108 × 108	16 blocks	Histogram		
UNIJ-ALP	3 cropping × 3 features	–	–	Subspace (PCA)	10.45%	7.45%

(b) a Partial Least Square (PLS) classifier, **(c)** an SVM classifier or **(d)** a subspace learning (e.g., PCA and/or LDA analysis).

- **Raw pixel values**: In [47], the baseline system was developed by finding a PCA + LDA [58] projection matrix on the raw pixel values taken from histogram equalized images of order 64 × 80 pixels (see Table 4.6). The dimensionality of the PCA and LDA subspaces was limited to 200 and 199, respectively. The cosine similarity between the projected features of a model and probe image was used as the score. The **baseline** system achieved HTERs of 20.94% and 17.11% for the female and male clients, respectively. In addition, the system identified as **UC-HU** in [47] learned features from grey scale images of order 200 × 200 pixels using a Convolutional Neural Network similar to the one in [59]. Then, the Fisher LDA approach was used in order to adapt the learned features to the discriminant face aspects of the individuals in the training set. Person-specific linear models were learnt by taking into consideration the samples of the person being enrolled as the positive class and all the other samples in the training set for the negative class. The dot product between the model and the probe samples was used as a score. The UC-HU system achieved the lowest HTER, 10.83% and 6.21% for the female and male clients, respectively, among all the simple systems which participated in the face recognition evaluation at ICB 2013.

- **Hand-crafted features**: Table 4.6 shows that most of the systems which participated in the 2013 MOBIO face recognition evaluation [47] used one or more of
  the following hand-crafted features: **(a)** LBP, **(b)** Gabor wavelet responses (i.e.,
  Gabor Phases), **(c)** POEM and **(d)** colour information. Among all the template-
  based systems, the best performance was achieved by the **UNIJ-ALP** system
  which combined nine sub-systems by extracting three different features (intensity, Gabor and LBP) from each of the three styles of cropped faces (i.e., tight,
  normal and broadly cropped). This was also referred to as the *representation
  plurality* method in [60]. However, a much simpler (single system) Gabor feature-
  based approach was used by the **Idiap** system and achieved HTERs of 12.50%
  and 10.29% for the female and male clients, respectively. It is also observed in
  Table 4.6 that a lower HETR can be achieved using larger images and blocks.
  Moreover, the fusion of multiple SVM classifiers (e.g., **CPqD**) proved to be a
  better approach for the LBP-based features.

4.3.2.2 GMM based

The success of the GMM-based method for speaker recognition inspired a number
of face recognition systems to apply it to the MOBIO dataset. These systems can be
divided into two groups: **(a) GMM-UBM** modelling and **(b) Meta-modelling**. The
meta-modelling methods can be further divided into two groups: **(i) GMM template**
modelling and **(ii) Variability modelling** methods. In [61], a comparison between the
GMM-UBM and GMM template-based methods was performed using the MOBIO
face verification evaluation protocol [46]. It was shown that the GMM template-
based methods performed significantly better than the GMM-UBM approach for both
female and male clients. On the other hand, different **variability modelling**-based
methods have also been applied to face recognition. For example, the ISV and JFA
methods were compared for face verification in [21]. It was reported that the HTERs
achieved with the ISV modelling were 11.4% and 8.3% for female and male clients,
while the corresponding HTERs of the JFA were 13.0% and 7.3%, respectively. On
an average, the ISV approach performed better than the JFA, which is consistent with
the speaker recognition evaluation. Recently, a score calibration technique [62] and
a local ISV [63] method were proposed in an attempt to improve the performance of
ISV. In addition, an i-vector-based face recognition approach using the TVM was presented in [64]. A comparison of several session compensation and scoring techniques
was evaluated. It was reported that a combination of the Within-Class Covariance
Normalization (WCCN) along with the Cosine Kernel Normalization (C-norm) performed better (HTERs of 15.2% and 8.7% for female and male clients, respectively)
than any other session compensation methods considered using the MOBIO database.

4.3.3 Audio-visual person recognition on MOBIO

Besides the unimodal speaker/face recognition systems, several audio-visual systems
have been proposed. A brief overview of existing audio-visual systems is presented
in Table 4.7. One common aspect between all these systems is their use of the logReg
fusion. In [65], the GMM-UBM and LBP histogram methods were, respectively, used

Table 4.7   Audio-visual person recognition systems on MOBIO

Reference	Methods A: Audio modality V: Audio modality	HTER					
		Audio		Visual		Fusion (logReg)	
		Female	Male	Female	Male	Female	Male
[3]Shen *et al.* [65]	**A**: GMM-UBM **V**: LBP histograms	33.1%	33.7%	26.7%	27.9%	19.3%	22.7%
McCool *et al.* [66]	**A**: TVM + PLDA **V**: LBP histograms	17.7%	18.2%	28.2%	24.1%	13.3%	11.9%
Motlicek *et al.* [36]	**A**: ISV **V**: ISV	15.3%	8.9%	12.2%	7.5%	9.7%	2.6%
Khoury *et al.* [20]	**A**: S-TVM + PLDA **V**: F-TVM + cosine	17.36%	11.11%	16.16%	8.9%	9.93%	3.77%
	A: S-All **V**: F-All	14.64%	7.89%	11.62%	**6.06%**	6.30%	1.89%
Khoury *et al.* [31]	**A**: S-1 + S-2 + $\cdots$ + S-11 **V**: F-1 + F-2 + $\cdots$ + F-8	6.87%	4.63%	**8.47%**	6.27%	3.80%	**1.78%**
[4]DBM–DNN	**A**: DBM–DNN$_{S\text{-}GSV}$ **V**: DBM–DNN$_{F\text{-}LBP}$	11.54%	9.68%	11.52%	9.75%	5.08%	3.55%
	A: DBM–DNN$_{S\text{-}TVM}$ **V**: DBM–DNN$_{F\text{-}TVM}$	15.61%	13.77%	14.01%	14.37%	8.69%	6.67%
	A: DBM–DNN$_{S\text{-}All}$ **V**: DBM–DNN$_{F\text{-}All}$	**5.08%**	**3.55%**	8.69%	6.76%	**3.38%**	2.29%

[3]Protocol used in the ICPR 2010 face and speech competition [46].
[4]A DBM–DNN is a special kind of **Deep Neural Network** (DNN) which is initialized as a generative **Deep Boltzmann Machine** (DBM) and trained using the standard back-propagation method (see Section 4.4) for details.

for the audio and visual modalities. Their system used the evaluation protocol from ICPR 2010 [67], while all the other systems which are listed in Table 4.7 were based on the protocol of ICB 2013 [47,48]. A combination of i-vector and LBP histogram-based recognition systems was studied in [66]. However, much improved results were reported in [36] based on the use of ISV for both modalities. Furthermore, a comparison study (in that same paper) confirmed that ISV modelling achieved lower HTERs than the JFA and the traditional GMM-based approaches. Moreover, the TVM method was used for both modalities in [20]. Session compensation, modelling and scoring were carried out using the PLDA [18]. In addition, the use of the cosine similarity measure [17] between the enrolment and test i-vectors was also used in [20]. In the same paper, three different sub-systems corresponding to the face and speaker were combined using the logReg fusion to obtain F-All (= F-GMM+F-ISV+ F-TVM) and S-All (= S-GMM+S-ISV+S-TVM), respectively. Finally, these combined unimodal systems were fused to obtain HTERs of 6.30% and 1.89%, respectively, for the female and male subjects of B-All (= S-All+F-All). The best visual modality HTER of 6.06% for the male clients was obtained by the F-All in [20].

The best HTER for the female clients (submitted to the ICB 2013 evaluation) was obtained by fusing eight speaker recognition systems [31]. However, the best audio modality HTERs 3.55% and 5.08%, respectively, for the male and female clients were achieved with the unimodal fusion of the DBM–DNN-based sub-systems of that modality (i.e., $DBM–DNN_{S-All} = DBM–DNN_{S-GSV} + DBM–DNN_{S-TVM}$ and $DBM–DNN_{F-All} = DBM–DNN_{F-LBP} + DBM–DNN_{F-TVM}$). Moreover, the bimodal fusion of all the DBM–DNN-based sub-systems (i.e., $DBM–DNN_{S-All} + DBM–DNN_{F-All}$) resulted in the best HTER of 3.38% for the female clients, while the best fused HTER of 1.78% for the male clients was achieved by fusing the 19 sub-systems in [20]. In summary, the results in Table 4.7 provide evidence that the DBM–DNN-based systems are able to play an important role in audio-visual person recognition using mobile phone data. A detailed description of DBM–DNN is presented in the next section.

4.4   Deep neural networks for person recognition

A DNN is a feed-forward Artificial Neural Network (ANN) with multiple layers of hidden units between its input and output layers. Such networks can be discriminatively trained by back-propagating the derivative of the mismatch between the target outputs and the actual outputs [68]. In the training phase, the initial weights of a DNN can be set to small random values. However, a better way of initialization is to generatively pre-train the DNN as a DBN or as a Deep Boltzmann Machine (DBM) and then fine-tune using the enrolment samples [69].

4.4.1   A DBN–DNN for unimodal person recognition

A DNN which is pre-trained generatively as a DBN is referred to as a DBN–DNN [68]. In a DBN, the top two layers are undirected but the lower layers have top-down directed connections (see Figure 4.8). Undirected models such as restricted Boltzmann machines (RBMs) are ideal for layer-wise pre-training [68]. An RBM

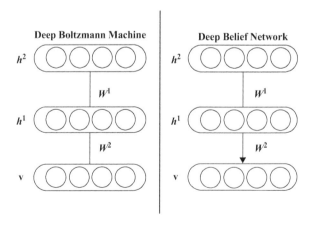

*Figure 4.8   **left***: *DBM;* ***right***: *DBN*

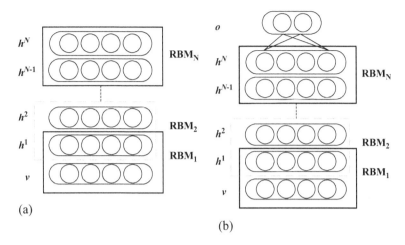

Figure 4.9   (a) Generative DBN and (b) discriminative DBN–DNN

is a type of Markov random field with a bipartite connectivity graph, no sharing of weights between different units and a subset of unobserved variables. Multiple RBMs can be stacked to form a DBN.

Recently, a DBN–DNN approach for speaker recognition was presented in [70]. First, a global DBN model (Figure 4.9(a)) is built by using unlabelled samples from the training set. Then, an impostor selection algorithm and a clustering method are used for each client to achieve the balance between the genuine and impostor samples. When the number of positive and negative samples is balanced the DBN parameters are adapted to each client. The adaptation process starts by pre-training each network initialized by the DBN parameters and using few numbers of iterations to avoid over-fitting. Once the adaptation process is completed, a network is fine-tuned by adding a label layer (with two units) on top and then using the stochastic back-propagation (Figure 4.9(b)). The connection weights between the top label layer and the adjacent layer below are initialized randomly and then pre-trained by back-propagating the error one layer for few iterations. Finally, full back-propagation is carried on the whole network. If the genuine and impostor feature vectors are represented as $(l_1 = 1, l_2 = 0)$ and $(l_1 = 0, l_2 = 1)$ during the training process, the final output score in the testing phase in the LLR format can be given as follows:

$$LLR = \log(o_1) - \log(o_2) \qquad (4.19)$$

where o_1 and o_2 represent the outputs of the top layer units. Although the DBN model was designed for a verification scenario, it can be modified and used for identification. For example, during the fine-tuning process, N units (corresponding to all the targets) instead of two may be added on top of a network initialized by the DBN parameters. Then, back-propagation can be carried out on the whole network by using only the genuine samples of all the target persons. The unit with the maximum value is declared the winner. The DBN–DNN approach presented in [70] was evaluated for speaker recognition on the NIST SRE 2006 corpora. A similar approach can also be adapted and used for audio-visual person recognition.

4.4.2 A DBM–DNN for person recognition

A DBM is a variant of the Boltzmann machine which not only retains the multi-layer architecture but also incorporates the top-down feedback (see Figure 4.8). Hence, a DBM has the potential of learning complex internal representations and dealing more robustly with ambiguous inputs (e.g., image or speech) [71].

4.4.2.1 DBM training

Consider a two-layer DBM with no within-layer connections, Gaussian visible units $\mathbf{v}_i \in \mathbb{R}^D$ and binary hidden units $\mathbf{h}_j^1, \mathbf{h}_k^2 \in \{0, 1\}^P$. A state of the DBM can be represented by a vector $x = \{v, h^1, h^2\}$, where $v = [\mathbf{v}_i]_{i=1,...,U}$ represent the units in the visible layer, and $h^1 = [\mathbf{h}_j^1]_{j=1,...,P_1}$ and $h^2 = [\mathbf{h}_k^2]_{k=1,...,P_2}$, respectively, represent the units in the first and the second hidden layer. Then, the energy of the state of the DBM is given by:

$$E(x|\theta) = \sum_{i=1}^{D} \frac{(v_i - b_i)^2}{2\sigma_i^2} - \sum_{i=1}^{D}\sum_{j=1}^{P_1} \frac{v_i}{\sigma_i^2}\mathbf{h}_j^1 \mathbf{w}_{ij}^1 - \sum_{l=1}^{2}\sum_{j=1}^{P_l} \mathbf{c}_j^l \mathbf{h}_j^n - \sum_{j=1}^{P_1}\sum_{k=1}^{P_2} \mathbf{h}_j^1 \mathbf{h}_k^2 \mathbf{w}_{jk}^2,$$

(4.20)

where the terms σ_i represent the standard deviation of the units in the visible layer, whereas $\mathbf{b}_i$ and $\mathbf{c}_j^l$, respectively, represent the biases of the units in the visible and the lth hidden layer. In addition, the symmetric interaction terms between the visible-to-hidden and hidden-to-hidden units are contained in $W^1 = \{\mathbf{w}_{ij}^1\}$ and $W^2 = \{\mathbf{w}_{jk}^2\}$, respectively.

DBMs can be trained with the stochastic maximization of the log-likelihood function. The partial-derivative of the log-likelihood function is:

$$\frac{\partial \mathscr{L}(\theta|v)}{\partial \theta} = \left\langle \frac{\partial E(v^{(t)}, h|\theta)}{\partial \theta} \right\rangle_{data} - \left\langle \frac{\partial E(v, h|\theta)}{\partial \theta} \right\rangle_{model},$$

(4.21)

where $h = [h^1, h^2]$, $\langle . \rangle_{data}$ and $\langle . \rangle_{model}$ denote the expectation over the data distribution $P(h|\{v^{(t)}\}, \theta)$ and the model distribution $P(v, h|\theta)$, respectively. Here, $\{v^{(t)}\}$ is the set containing all the training samples. Although the update rules are well defined, it is intractable to exactly compute them.

The variational approximation is commonly used to compute the expectation over the data distribution (the first term of (4.21)). The variational parameters for the lth hidden layer, μ_j^l, are estimated by:

$$\mu_j^l \leftarrow f\left(\sum_{i=1}^{P_{l-1}} \mu_i^{l-1} + \sum_{k=1}^{P_{l+1}} \mu_k^{l+1}\mathbf{w}_{kj}^l + \mathbf{c}_j^l \right)$$

(4.22)

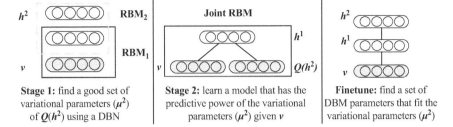

Stage 1: find a good set of variational parameters (μ^2) of $Q(h^2)$ using a DBN

Stage 2: learn a model that has the predictive power of the variational parameters (μ^2) given v

Finetune: find a set of DBM parameters that fit the variational parameters (μ^2)

Figure 4.10   Two-stage pre-training of a DBM with two hidden layers. Shaded nodes indicate clamped variables, while white nodes indicate free variables

where $f(.)$ is a sigmoid function, $\mu_i^0 = \mathbf{v}_i$ and $P_{l+1} = 0$. The variational approximation method provides the values of the variational parameters which maximize the following lower-bound with respect to the current parameters:

$$p(v|\theta) \geq \mathbb{E}_{Q(h)}[-E(v,h)] + \mathscr{H}(Q) - \log Z(\theta), \tag{4.23}$$

where

$$\mathscr{H}(Q) = -\sum_{l=1}^{2}\sum_{j=1}^{P_l}\left(\mu_j^l \log \mu_j^l + (1 - \mu_j^l)\log(1 - \mu_j^l)\right). \tag{4.24}$$

is an entropy functional. Hence, the gradient update step increases the variational lower-bound of the log-likelihood.

Subsequently, different persistent sampling methods [71,72] can be used to compute the expectation over the model distribution (the first term of (4.21)). The simplest approach is the Gibbs sampling which closely resembles the variational expectation-maximization (EM) algorithm [73]. Learning is carried out by alternating between: **(a)** finding variational parameters and **(b)** updating the DBM parameters using the stochastic gradient method. The objective of updating the DBM parameters is to maximize the variational lower-bound.

4.4.2.2   Person recognition using DBM–DNN

Similar to the DBN–DNN, a DBM can be converted into a discriminative network, which is referred to as a DBM–DNN [74]. At this stage, two DBMs (one for each modality) are trained. For example, the DBM$_{F-LBP}$ and DBM$_{F-TVM}$ for the face modality, and DBM$_{S-GSV}$ and DBM$_{S-TVM}$ for the speech modality. The steps involved in DBM–DNN-based person recognition are detailed below.

- **Generative pretraining of DBM**: It is not trivial to start the training from randomly initialized parameters [75]. Hence, the DBMs are pre-trained using the two-stage algorithm shown in Figure 4.10. In this approach, posterior distributions over the hidden units and DBM parameters are obtained separately. The two stages are detailed below:

- **Stage 1**: At this stage the objective is to find a good set of variational parameters regardless of the parameter values of the DBM. This is performed by taking the posteriors of the hidden units from another model such as a DBN or a Deep Auto Encoder. A DBN can be trained efficiently [76] to find a good approximate distribution over units in the even-numbered hidden layer. Hence, a set of initial variational parameters (μ^2) for the second hidden layer is found from a DBN (left panel of Figure 4.10).
- **Stage 2**: A joint distribution over the visible vector and variational parameters is learned using another RBM (central panel of Figure 4.10). The visible layer of the joint RBM corresponds to the visible layer and the even-numbered hidden layer of the DBM that is being pretrained. The connections between the layers of the joint RBM are bidirectional like those of the DBM. Finally, when the joint RBM is trained, the learned parameters are used as initializations for the training of the DBM (right panel of Figure 4.10) which corresponds to freeing h^2 from its variational posterior distribution $Q(h^2)$ obtained in *Stage 1*.

- **Discriminative fine-tuning of DBM–DNN**: Once the parameters of a DBM are learnt (see the right panel of Figure 4.10), they can be used to initialize the hidden layers of a corresponding feed-forward DNN. For a bimodal application (e.g., audio-visual person recognition), two such DBMs are generatively trained. These DBMs are then converted into discriminative DBM–DNNs such as the DBM–DNN$_{F-LBP}$ and DBM–DNN$_{S-GSV}$ [77], or the DBM–DNN$_{F-TVM}$ and DBM–DNN$_{S-TVM}$ [78]. This is done by first adding a softmax layer on top of each DBM and then fine-tuning them with the enrolment data by using the standard back-propagation algorithm (see Figure 4.11). When a set of probes is presented to the system, they are clamped to the visible layer of the corresponding DBM–DNN and the resultant softmax layer is used to generate the scores.

- **Decision making**: The outputs of the DBM–DNNs are combined using the sum rule of fusion (see Figure 4.11), which for a claimed identity j is given by the following equation:

$$f_j(v_F, v_S) = \sum_m (o_{m,j}),$$ (4.25)

where $m = \{audio, visual\}$ and $o_{m,j}$ represents the probability of the inputs v_F and v_S belonging to person j. An unknown person's claim j is accepted if $f_j(v_F, v_S)$ is above a predetermined threshold.

- **Evaluation**: The EER and the HTER are used for the evaluation of the DBM–DNNs in a held-out dataset scenario (e.g., MOBIO evaluation):

$$EER = \frac{FAR_{dev}(\theta_{dev}) + FRR_{dev}(\theta_{dev})}{2}$$ (4.26)

$$HTER = \frac{FAR_{eval}(\theta_{dev}) + FRR_{eval}(\theta_{dev})}{2}$$ (4.27)

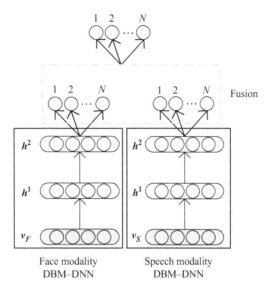

Figure 4.11   DBM–DNNs are initialized with DBM weights and discriminatively
fine-tuned. The output scores are fused before reaching a final
decision

These measures rely on the FAR and the FRR, which are calculated for the development and evaluation sets independently:

$$FAR(\theta) = \frac{|\{s_{imp}|s_{imp} \geq \theta\}|}{|\{s_{imp}\}|} \qquad (4.28)$$

$$FRR(\theta) = \frac{|\{s_{cli}|s_{cli} \geq \theta\}|}{|\{s_{cli}\}|} \qquad (4.29)$$

where s_{cli} are the client scores and s_{imp} are the impostor scores. A score threshold based on the EER of the development set is calculated as:

$$\theta_{dev} = \arg\min_{\theta} |FAR_{dev}(\theta) - FRR_{dev}(\theta)| \qquad (4.30)$$

and the HTER is calculated using this threshold.

4.5   Summary

This chapter has first presented an overview of the existing audio-visual person recognition systems, especially those evaluated in mobile environments. The speaker and face recognition systems using MOBIO are categorized based on their approach for client modelling. A number of other systems focusing on the bimodal fusion of the speaker and face sub-systems have been compared in the context of an unbiased protocol. Finally, the application of DNN-based frameworks such as the DBN–DNN and DBM–DNN is discussed. Competitive results in terms of HTER for both female and male clients have been reported using the DBM–DNNs. However, the DBM–DNN discussed in this chapter is only applicable in the context of a held out database. New clients can be enrolled and scoring can be performed by respectively following the adaptation and LLR scoring strategies used by the DBN–DNN. Despite the promising results reported here using DNNs, it should be emphasized that there are still scopes of further investigations, e.g., to improve the performance of the learning algorithms and the design of novel architectures using DNNs that are tailored to multimodal person recognition.

References

[1]   Ross A, Nandakumar K, and Jain AK. Introduction to multibiometrics. In: Handbook of Biometrics. Springer; 2008. p. 271–292.

[2]   Brookes M. Voicebox: Speech processing toolbox for MATLAB®. Software, available [Mar 2011] from www.ee.ic.ac.uk/hp/staff/dmb/voicebox/voicebox.html 1997.

[3]   Viola P, and Jones M. Rapid object detection using a boosted cascade of simple features. In: Computer Vision and Pattern Recognition, 2001. Proceedings of the 2001 IEEE Computer Society Conference. vol. 1. IEEE; 2001. p. I–511.

[4]   Fox NA, Gross R, Cohn JF, and Reilly RB. Robust biometric person identification using automatic classifier fusion of speech, mouth, and face experts. IEEE Transactions on Multimedia. 2007;9(4):701–714.

[5]   Brunelli R, and Falavigna D. Person identification using multiple cues. IEEE Transactions on Pattern Analysis and Machine Intelligence. 1995;17(10):955–966.

[6]   Dieckmann U, Plankensteiner P, and Wagner T. Sesam. A biometric person identification system using sensor fusion. Pattern Recognition Letters. 1997;18(9):827–833.

[7]   Chaudhari UV, Ramaswamy GN, Potamianos G, and Neti C. Audio-visual speaker recognition using time-varying stream reliability prediction. In: IEEE International Conference on Acoustics, Speech, and Signal Processing. vol. 5. IEEE; 2003. p. V–712.

[8]   Ben-Yacoub S, Luttin J, Jonsson K, Matas J, and Kittler J. Audio-visual person verification. In: IEEE Computer Society Conference on Computer Vision and Pattern Recognition. vol. 1. IEEE; 1999.

[9]   Sanderson C, and Paliwal KK. Identity verification using speech and face information. Digital Signal Processing. 2004;14(5):449–480.

[10]   Erzin E, Yemez Y, and Tekalp AM. Multimodal speaker identification using an adaptive classifier cascade based on modality reliability. IEEE Transactions on Multimedia. 2005;7(5):840–852.

[11] Sahoo SK, and Prasanna SRM. Bimodal biometric person authentication using speech and face under degraded condition. In: 2011 National Conference on Communications (NCC). IEEE; 2011. p. 1–5.

[12] Tan X, and Triggs B. Enhanced local texture feature sets for face recognition under difficult lighting conditions. IEEE Transactions on Image Processing. 2010;19(6):1635–1650.

[13] Reynolds DA, Quatieri TF, and Dunn RB. Speaker verification using adapted Gaussian mixture models. Digital Signal Processing. 2000;10(1):19–41.

[14] Campbell WM, Sturim DE, and Reynolds DA. Support vector machines using GMM supervectors for speaker verification. Signal Processing Letters, IEEE. 2006;13(5):308–311.

[15] Vogt R, and Sridharan S. Explicit modelling of session variability for speaker verification. Computer Speech and Language. 2008;22(1):17–38.

[16] Kenny P, Boulianne G, Ouellet P, and Dumouchel P. Joint factor analysis versus eigenchannels in speaker recognition. IEEE Transactions on Audio, Speech, and Language Processing. 2007;15(4):1435–1447.

[17] Dehak N, Kenny P, Dehak R, Dumouchel P, and Ouellet P. Front-end factor analysis for speaker verification. IEEE Transactions on Audio, Speech, and Language Processing. 2011;19(4):788–798.

[18] Prince SJ, and Elder JH. Probabilistic linear discriminant analysis for inferences about identity. In: Computer Vision, 2007. IEEE 11th International Conference on ICCV. IEEE; 2007. p. 1–8.

[19] Ahonen T, Hadid A, and Pietikäinen M. Face recognition with local binary patterns. In: Computer vision-eccv 2004. Springer; 2004. p. 469–481.

[20] Khoury E, El Shafey L, McCool C, Günther M, and Marcel S. Bi-modal biometric authentication on mobile phones in challenging conditions. Image and Vision Computing. 2014;32(12):1147–1160.

[21] Wallace R, McLaren M, McCool C, and Marcel S. Inter-session variability modelling and joint factor analysis for face authentication. In: 2011 International Joint Conference on Biometrics (IJCB). IEEE; 2011. p. 1–8.

[22] Sanderson C, and Paliwal KK. Fast features for face authentication under illumination direction changes. Pattern Recognition Letters. 2003;24(14): 2409–2419.

[23] Hearst MA, Dumais ST, Osman E, Platt J, and Scholkopf B. Support vector machines. Intelligent Systems and their Applications, IEEE. 1998;13(4): 18–28.

[24] Naseem I, Togneri R, and Bennamoun M. Linear regression for face recognition. IEEE Transactions on Pattern Analysis and Machine Intelligence. 2010;32(11):2106–2112.

[25] Jain A, Nandakumar K, and Ross A. Score normalization in multimodal biometric systems. Pattern Recognition. 2005;38(12):2270–2285.

[26] Kittler J, Hatef M, Duin RP, and Matas J. On combining classifiers. IEEE Transactions on Pattern Analysis and Machine Intelligence. 1998;20(3): 226–239.

[27]   Adjoudani A, and Benoit C. On the integration of auditory and visual param-
       eters in an HMM-based ASR. In: Speechreading by Humans and Machines.
       Springer; 1996. p. 461–471.
[28]   Heckmann M, Berthommier F, and Kroschel K. Noise adaptive stream weight-
       ing in audio-visual speech recognition. EURASIP Journal on Applied Signal
       Processing. 2002;2002(1):1260–1273.
[29]   Alam MR, Bennamoun M, Togneri R, and Sohel F. Confidence-based rank-
       level fusion for audio-visual person identification system. In: 3rd International
       Conference on Pattern Recognition Applications and Methods (ICPRAM
       2014). 2014.
[30]   Alam MR, Bennamoun M, Togneri R, and Sohel F. A confidence-based
       late fusion framework for audio-visual biometric identification. Pattern
       Recognition Letters. 2014;52:65–71.
[31]   Khoury E, Günther M, El Shafey L, and Marcel S. On the improvements of
       uni-modal and bi-modal fusions of speaker and face recognition for mobile
       biometrics. Idiap; 2013.
[32]   Minka TP. Algorithms for maximum-likelihood logistic regression. Technical
       Report at Research Showcase, Carnegie Mellon University, 2003.
[33]   Pigeon S, Druyts P, and Verlinde P. Applying logistic regression to the
       fusion of the NIST'99 1-speaker submissions. In: Digital Signal Processing.
       2000;10(1):237–248.
[34]   Brummer N, Burget L, Cernocky JH, *et al.* Fusion of heterogeneous speaker
       recognition systems in the STBU submission for the NIST speaker recognition
       evaluation 2006. In: IEEE Transactions on Audio, Speech, and Language
       Processing. 2007;15(7):2072–2084.
[35]   McCool C, and Marcel S. Parts-based face verification using local frequency
       bands. In: Advances in Biometrics. Springer; 2009. p. 259–268.
[36]   Motlicek P, Shafey L, Wallace R, McCool C, and Marcel S. Bi-modal authen-
       tication in mobile environments using session variability modelling. In: 2012
       21st International Conference on Pattern Recognition (ICPR). IEEE; 2012.
       p. 1100–1103.
[37]   Bailly-Bailliére E, Bengio S, Bimbot F, *et al.* The BANCA database and eval-
       uation protocol. In: Audio-and Video-Based Biometric Person Authentication.
       Springer; 2003. p. 625–638.
[38]   Pigeon S, and Vandendorpe L. The M2VTS multimodal face database (release
       1.00). In: Audio-and Video-Based Biometric Person Authentication. Springer;
       1997. p. 403–409.
[39]   Messer K, Matas J, Kittler J, Luettin J, and Maitre G. XM2VTSDB: The
       extended M2VTS database. In: Second international conference on audio
       and video-based biometric person authentication. vol. 964. Citeseer; 1999.
       p. 965–966.
[40]   Sanderson C, and Lovell BC. Multi-region probabilistic histograms for robust
       and scalable identity inference. In: Advances in Biometrics. Springer; 2009.
       p. 199–208.

[41] Burnham D, Estival D, Fazio S, *et al.* Building an Audio-Visual Corpus of Australian English: Large Corpus Collection with an Economical Portable and Replicable Black Box. In: INTERSPEECH; 2011. p. 841–844.

[42] Burnham D, Ambikairajah E, Arciuli J, *et al.* A blueprint for a comprehensive Australian English auditory-visual speech corpus. In: Selected Proceedings of the 2008 HCSNet Workshop on Designing the Australian National Corpus, ed. Michael Haugh *et al.*; 2009. p. 96–107.

[43] Wagner M, Tran D, Togneri R, *et al.* The big australian speech corpus (the big asc). In: SST 2010, Thirteenth Australasian International Conference on Speech Science and Technology. ASSTA; 2011. p. 166–170.

[44] Alam MR, Togneri R, Sohel F, Bennamoun M, and Naseem I. Linear Regression-based Classifier for audio visual person identification. In: 2013 1st International Conference on Communications, Signal Processing, and their Applications (ICCSPA). IEEE; 2013. p. 1–5.

[45] Alam MR, Bennamoun M, Togneri R, and Sohel F. An efficient reliability estimation technique for audio-visual person identification. In: 2013 8th IEEE Conference on Industrial Electronics and Applications (ICIEA). IEEE; 2013. p. 1631–1635.

[46] Marcel S, McCool C, Matějka P, *et al.* On the results of the first mobile biometry (MOBIO) face and speaker verification evaluation. In: Recognizing Patterns in Signals, Speech, Images and Videos. Springer; 2010. p. 210–225.

[47] Gunther M, Costa-Pazo A, Ding C, *et al.* The 2013 face recognition evaluation in mobile environment. In: 2013 International Conference on Biometrics (ICB). IEEE; 2013. p. 1–7.

[48] Khoury E, Vesnicer B, Franco-Pedroso J, *et al.* The 2013 speaker recognition evaluation in mobile environment. In: 2013 International Conference on Biometrics (ICB). IEEE; 2013. p. 1–8.

[49] Glembek O, Burget L, Dehak N, Brümmer N, and Kenny P. Comparison of scoring methods used in speaker recognition with joint factor analysis. In: IEEE International Conference on Acoustics, Speech and Signal Processing (ICASSP). IEEE; 2009. p. 4057–4060.

[50] Khemiri H, Usoltsev A, Legout MC, Petrovska-Delacrétaz D, and Chollet G. Automatic speaker verification using nearest neighbour normalization (3N) on an iPad tablet. In: 2014 International Conference of the Biometrics Special Interest Group (BIOSIG). IEEE; 2014. p. 1–8.

[51] Khoury E, El Shafey L, and Marcel S. Spear: An open source toolbox for speaker recognition based on Bob. In: 2014 IEEE International Conference on Acoustics, Speech and Signal Processing (ICASSP). IEEE; 2014. p. 1655–1659.

[52] Boulkenafet Z, Bengherabi M, Nouali O, and Cheriet M. Using the conformal embedding analysis to compensate the channel effect in the I-vector based speaker verification system. In: 2013 International Conference of the Biometrics Special Interest Group (BIOSIG). IEEE; 2013. p. 1–8.

[53] Fu Y, Liu M, and Huang TS. Conformal embedding analysis with local graph modeling on the unit hypersphere. In: CVPR'07. 2007 IEEE Conference on Computer Vision and Pattern Recognition. IEEE; 2007. p. 1–6.

[54] Roy A, Doss MM, and Marcel S. A fast parts-based approach to speaker verification using boosted slice classifiers. IEEE Transactions on Information Forensics and Security. 2012;7(1):241–254.

[55] Togneri R, and Pullella D. An overview of speaker identification: Accuracy and robustness issues. Circuits and Systems Magazine, IEEE. 2011;11(2): 23–61.

[56] Rodriguez Y. Face detection and verification using local binary patterns. Ecole Polytechnique Fédérale de Lausanne; 2006.

[57] Özuysal M, Calonder M, Lepetit V, and Fua P. Fast keypoint recognition using random ferns. IEEE Transactions on Pattern Analysis and Machine Intelligence. 2010;32(3):448–461.

[58] Zhao W, Krishnaswamy A, Chellappa R, Swets DL, and Weng J. Discriminant analysis of principal components for face recognition. In: Face Recognition. Springer; 1998. p. 73–85.

[59] Cox D, and Pinto N. Beyond simple features: A large-scale feature search approach to unconstrained face recognition. In: 2011 IEEE International Conference on Automatic Face & Gesture Recognition and Workshops (FG). IEEE; 2011. p. 8–15.

[60] Štruc V, Gros JZ, Dobrišek S, and Pavešic N. Exploiting representation plurality for robust and efficient face recognition. In: Proceedings of the 22nd International Electrotechnical and Computer Science Conference (ERK13); 2013. p. 121–124.

[61] Pereira TF, Angeloni MA, Simões FO, and Silva JEC. Video-Based face verification with local binary patterns and SVM using GMM supervectors. In: Computational Science and Its Applications–ICCSA 2012. Springer; 2012. p. 240–252.

[62] Mandasari MI, Günther M, Wallace R, Saeidi R, Marcel S, and van Leeuwen DA. Score calibration in face recognition. IET Biometrics. 2014;3(4):246–256.

[63] Anantharajah K, Ge Z, McCool C, *et al.* Local inter-session variability modelling for object classification. In: 2014 IEEE Winter Conference on Applications of Computer Vision (WACV). IEEE; 2014. p. 309–316.

[64] Wallace R, and McLaren M. Total variability modelling for face verification. Biometrics, IET. 2012;1(4):188–199.

[65] Shen L, Zheng S, Zheng S, and Li W. Secure mobile services by face and speech based personal authentication. In: 2010 IEEE International Conference on Intelligent Computing and Intelligent Systems (ICIS). vol. 3. IEEE; 2010. p. 97–100.

[66] McCool C, Marcel S, Hadid A, *et al.* Bi-modal person recognition on a mobile phone: using mobile phone data. In: 2012 IEEE International Conference on Multimedia and Expo Workshops (ICMEW). IEEE; 2012. p. 635–640.

[67] McCool C, and Marcel S. Mobio database for the ICPR 2010 face and speech competition. Idiap; 2009.

[68] Hinton G, Deng L, Yu D, *et al.* Deep neural networks for acoustic model-ing in speech recognition: The shared views of four research groups. Signal Processing Magazine, IEEE. 2012;29(6):82–97.

[69] Hinton GE, and Salakhutdinov RR. Reducing the dimensionality of data with neural networks. Science. 2006;313(5786):504–507.

[70] Ghahabi O, and Hernando J. Deep belief networks for i-vector based speaker recognition. In: 2014 IEEE International Conference on Acoustics, Speech and Signal Processing (ICASSP). IEEE; 2014. p. 1700–1704.

[71] Salakhutdinov R, and Hinton GE. Deep Boltzmann machines. In: International Conference on Artificial Intelligence and Statistics; 2009. p. 448–455.

[72] Desjardins G, Courville AC, Bengio Y, Vincent P, and Delalleau O. Tempered Markov chain Monte Carlo for training of restricted Boltzmann machines. In: International Conference on Artificial Intelligence and Statistics; 2010. p. 145–152.

[73] Bishop CM. Pattern recognition and machine learning. Springer; 2006.

[74] You Z, Wang X, and Xu B. Investigation of deep Boltzmann machines for phone recognition. In: 2013 IEEE International Conference on Acoustics, Speech and Signal Processing (ICASSP). IEEE; 2013. p. 7600–7603.

[75] Cho K, Raiko T, Ilin A, and Karhunen J. A two-stage pretraining algorithm for deep boltzmann machines. In: Artificial Neural Networks and Machine Learning–ICANN 2013. Springer; 2013. p. 106–113.

[76] Bengio Y, Courville A, and Vincent P. Representation learning: A review and new perspectives. IEEE Transactions on Pattern Analysis and Machine Intelligence. 2013;35(8):1798–1828.

[77] Alam MR, Bennamoun M, Togneri R, and Sohel F. A Deep Neural Network for Audio-Visual Person Recognition. In: 7th IEEE International Conference on Biometrics: Theory, Applications and Systems (BTAS 2015). IEEE; 2015.

[78] Alam MR, Bennamoun M, Togneri R, and Sohel F. Deep Boltzmann machines for i-vector based audio-visual person identification. In: 7th Pacific Rim Symposium on Image and Video Technology (PSIVT); 2015.

Chapter 5

Active authentication using facial attributes

Pouya Samangouei[1], Emily Hand[1], Vishal M. Patel[2], and Rama Chellappa[1]

5.1 Introduction

Passwords are no longer a reliable form of authentication. They have been the default authentication method for multiuser computer systems since the 1960s, when MIT's Compatible Time Sharing System [1] was born. They are easy to crack if they are simple and short, and otherwise, they are difficult for a user to remember. Once a malicious user somehow acquires a password, he/she has full control over the system. In order to avoid a password leak, one needs to update their password regularly, which adds to the frustration of maintaining the password. These are just a few drawbacks of using password-based authentication methods.

Studies have shown that one out of five mobile device users choose a simple, easily guessed password like "12345," "abc1234" or even "password" to protect their data [1]. As a result, hackers could easily break into many accounts just by trying the most commonly used passwords. In the case when secret patterns are used for gaining initial access on the mobile devices, users tend to use the same pattern over and over again. As a result, they leave oily residues or smudges on the screen of the mobile device. It has been shown that with special lighting and high-resolution photo, one can easily deduce the secret pattern [2].

To overcome the shortcomings of a password-authenticated system, researchers have focused on an alternative authentication approach, namely "Biometric identification" [3]. This approach includes the use of fingerprints [4], retinal scans [5], voice [6], facial image [7], etc. to validate users. Biometrics is more secure as they are unique for each user and very difficult to replicate. However, there are many challenges in determining robust biometrics, recognizing them accurately, and also efficiently implementing the identification algorithms on mobile devices with limited computational resources.

Mobile devices like smartphones, tablets, and wearable technologies are still mostly protected by password-based authentication methods. The devices usually

[1]University of Maryland Institute for Advanced Computer Studies, USA
[2]Electrical and Computer Engineering, Rutgers University, USA

contain vital information such as emails, bank account information, and private content. With such valuable data, they have increasingly been under different kinds of attacks [8]. Typically, mobile devices incorporate no mechanisms to verify continuously that the user in control is the same original user of the device. Thus, if the initial authentication is bypassed, the unauthorized individual gets access to the personal information of the original user of the device. As opposed to the password-based methods, biometric identification algorithms facilitates continuous authentication on these devices [9,10]. However, it still remains a problem to find an algorithm for continuous biometric authentication that is both accurate and computationally efficient.

The focus of this chapter is on face-based biometric authentication methods. These methods first use the camera sensor images to detect the face of the users. Next, they extract low-level features from the face images, and then apply their algorithm to the extracted features to authenticate the user. These algorithms have access to some model of the enrolled user for comparison. In [11], Hadid *et al.* use Haar cascade and Adaboost of [12] to detect face components and then use Local Binary Pattern (LBP) histograms [13] and nearest-neighbor thresholding for authentication. In [7], Fathy *et al.* extract two types of intensity features from the full face and facial parts and compare four still image-based verification algorithms with four image set-based methods. The common pitfall of most of these algorithms is that they are very sensitive to changes in the low-level feature domain. They are sensitive in the sense that if two face images are under the same pose and lighting condition, they can perform well, but in unconstrained settings they become very inaccurate.

In this chapter, it will be shown how algorithms based on a large number of facial attributes—human describable features—can address both efficiency and accuracy in the task of continuous authentication on mobile devices. Human attributes are also referred to as "soft biometrics" in literature [14] and have been shown to increase the performance of identity recognition algorithms. These features include gender, ethnicity, age, and other attributes that can be determined from the facial appearance of the user. The most important quality of facial attributes is their robustness. They are different from one person to another, and they are constant for the same person under different conditions. Attributes can also be fused with low-level features to boost performance. In [14], Jain *et al.* combined height, race, and gender information with fingerprint features to improve the recognition accuracy on an in-house dataset.

In addition to their robustness, attributes are also technically advantageous. First, the training of the attribute classifiers can be done offline. This is important since training accurate models on a mobile device is computationally infeasible. As the training is done offline, one can use large datasets containing images with various face poses and environmental conditions. Second, it will be seen in this chapter that as attributes are easier targets to learn, simple models can be trained for most of them with a good accuracy. Consequently, these models can be run efficiently and continuously on an average mobile device.

These attribute features have benefits regarding authentication and user experience. First, the attribute features are compact representations, meaning that if we have n binary attributes, the probability that two people have the same attributes is

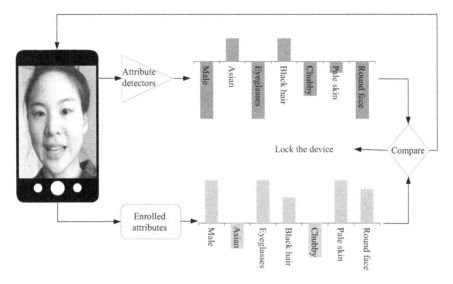

Figure 5.1 *The overview of active authentication based on facial attributes on mobile devices*

$1/2^n$. Therefore ideally by comparing the predicted attributes with the enrolled ones, authentication can be done accurately. Also, the enrollment of attributes can be done by asking them directly from the user, which decouples the enrollment from being dependent on the initial enrollment lighting and pose conditions. The overview of the approach taken in [15–17] is shown in Figure 5.1.

In this chapter, we first discuss the training of the attribute classifiers in Section 5.2 where a large-scale dataset and two learning approaches are employed: a linear model for each attribute in Section 5.2.1, and a deep convolutional neural network (CNN) for each attribute in Section 5.2.2. The former being the simplest model, and the latter being one of the most complex models one can learn for detecting the presence of each attribute. Then the performance of these two extremes is compared in Section 5.3 in different authentication scenarios. Finally, in Section 5.4, platform evaluations are conducted to show the feasibility of the different approaches for learning the attribute models. Finally, the chapter is concluded in Section 5.5.

5.2 Facial attribute classifiers

In computer vision, almost in all problems, the very first step is to extract features from a given visual signal. The initial feature extraction algorithms had been performed in the low-level pixel domain, until attributes as higher order features were introduced in Content-Based Image Retrieval where they are presented as a solution to decrease the semantic gap [18–20]. Attribute features fall under the category of "intermediate features," which first appeared in [20] as low-level semantic features but high-level image features.

Later applications of the intermediate features were in the task of object recognition and human identification domain. In [21], Ferrari and Zisserman learn visual attributes for objects such as "dotted" or "striped." In [22], Farhadi *et al.* use L1-regularized logistic regression on SVM outputs of visual attributes to learn object attributes such as "has wheels" or "metallic" from the images of PASCAL VOC 2008 [23]. Then they use these attribute features to describe the objects in a given image. Lampert *et al.* [24] learn the object attributes via kernel Support Vector Machines (SVMs) [25] to perform object recognition. They show good results on their Animal and Attributes dataset. There are many other areas that attribute features have been shown to be useful including zero-shot learning [26], scene classification [27], and action recognition [28].

Human attributes like gender, ethnicity, hair color, etc. have also been exploited in different recognition and verification tasks. In [29], Kumar *et al.* train 73 attribute SVMs by incrementally concatenating features from different face components. They test their attribute classifiers on the PubFig [25] dataset. Bourdev *et al.* [30] decompose the pedestrian image into a set of parts containing salient patterns, they extract Histogram of Oriented Gradients (HOG) [31] that feature from these parts and train linear SVMs per part per attribute to create feature vector from SVMs' responses. Then they train attribute classifiers on these feature vectors. Zhang *et al.* [32] train a CNN on full body images to detect attributes. They show good results on the Berkeley Attributes of People dataset and the Attributes 25k [30] dataset. In [33], Berg *et al.* use labeled images with part locations as their training set. They learn one SVM per class pairs and part pairs to take into account the class relationship and part relationship, then they create a feature vector out of these classifiers. The feature vector is used to train the final attribute classifiers. In [34], Kumar *et al.* trained local SVMs and used Adaboost to optimize for the best SVMs for the ten attributes of their FaceTracer dataset. In [35], Liu *et al.* propose a deep learning framework for attribute prediction where different layers have different pretraining initialization and they test their attribute classifiers on their CelebA dataset.

In this chapter, to capture the spectrum of attribute models, two types of attribute classifiers are trained. One is a simple linear model using regularized logistic regression per attribute in Section 5.2.1, the other is a deep CNN per attribute in Section 5.2.2 on the CelebA [35] dataset to compare these models for the task of authentication in Section 5.3.

CelebA dataset. CelebA dataset is a large-scale face attribute dataset. It contains 200k celebrity photos from 10,177 different identities. It has labels for 40 attributes per image along with 5 landmark location on faces. Recall that one of the advantages of using facial attributes is that the models can be trained with large datasets.

5.2.1 *Linear attribute classifiers*

The simplest model that one can learn for detecting binary attributes is a linear model. The L2-Regularized Logistic Regression method is chosen as the first approach to train the linear classifiers, as the training data set is large and in this approach the linear model can be updated using one batch of the data at a time. Let $P(y = 1|x,w) = \sigma(w^T\hat{x}) = 1/1 + e^{-w^T\hat{x}}$, where $\hat{x} = [x, \ 1]^T$ is the probability that the attribute $y \in \{0, 1\}$ exists in $x \in R^n$, which is the feature vector extracted from the face image, given the weights $w \in R^{n+1}$. The last weight w_{n+1} is the linear model offset. Then, the likelihood function can be written as follows:

$$p(y|x,w) = p(y = 1|x,w)^y (1 - p(y = 1|x,w))^{1-y}.$$

Consequently, the negative log likelihood cost will be

$$L(w) = E_x[-y \log p(y = 1|x,w) - (1 - y)\log(1 - p(y = 1|x,w))].$$

To prevent overfitting to the training data, the regularization term is added to the cost function, so the final objective becomes:

$$\underset{w}{\operatorname{argmin}} J(w) = L(w) + \frac{1}{2}\lambda\|w\|^2.$$

To minimize this objective function, batch gradient descent which is a variant of stochastic gradient descent (SGD) [36] is used. It both helps us in dealing with large data and avoiding local minimas. The batch gradient descent algorithm updates the weight vector with

$$w^n = w^{n-1} - \alpha\nabla_w J(w^{n-1}),$$

where α is found using backtracking for batch n. The batch size is set to 50. The hyper-parameter λ is selected using the provided development set of the CelebA dataset. The identities in the development set do not overlap with the training set.

To extract features from the images, first the face images of CelebA dataset are aligned to a canonical coordinate and size of 78×78 pixels. Then, extract uniform LBPs histograms are extracted with a cell size of 6 on the intensity channel as well as RGB color channels and concatenate them together to create a feature vector $\in R^{25056}$.

5.2.2 Convolutional neural network attribute model

With the increase in popularity of CNNs for feature extraction and classification in recent years, there have been many software packages introduced for implementing CNNs. Theano [37], Torch [38], Tensorflow [39], Caffe [40], and DeepLearnToolbox [41] are some of the more popular packages for CNNs.

In order to train a CNN, one must specify a network architecture. This is no easy task, and there has been much debate over the years about what constitutes the "best" architecture for a given problem. There is still no principled way of determining the optimal architecture for a problem, and so, they are generally chosen through experimentation. The network architecture is described below. For more details about CNNs, see [42].

The network weights are learned for each attribute using Caffe. The architecture [43] contains three convolutional layers and three fully connected layers. The architecture can be seen in Figure 5.2. The input to the network is 256×256 mean-subtracted color images, containing full faces. The network takes a random crop of 227×227

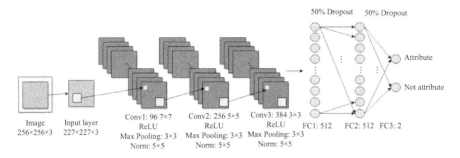

Figure 5.2   CNN architecture for attribute feature extraction and classification

from the color images during training. This is done for two reasons: to avoid over-fitting and to manufacture additional training data. Overfitting is described in more detail in a later section. During testing, the crop is taken from the center of the image. The first convolution layer (Conv1) contains 96 7×7 filters, that is 96 separate convolution masks of size 7×7, which are applied to the cropped input image. This results in 96 feature maps, which are the input to the layers that follow. Conv1 is followed by a Rectified Linear Unit (ReLU), which simply passes the nonnegative values through to the next layer, making the negative values 0. Following this is a max pooling layer, which takes the maximum value from each 3×3 window in the 96 feature maps. The Max Pooling layer is followed by a normalization layer, which takes the norm of each 5×5 window in each of the 96 maps. The second convolution layer (Conv2) follows a similar pattern with 256 5×5 filters being applied to the 96 maps from the previous layer. Conv2 is followed by a ReLU layer, a max pooling layer with a 3×3 filter and a normalization layer with a 5×5 filter. Conv3 has 384 5×5 convolution filters again followed by ReLU, max pooling with a 3×3 filter, and normalization with a 5×5 filter. The first fully connected layer (FC1) follows Conv3. FC1 contains 512 units, and it is fully connected to Conv3. FC1 is fully connected to FC2 – the second fully connected layer. There is a 50% dropout between the two layers [44]. Dropout is used to prevent overfitting. It randomly sets 50% of the weights between the two layers to 0 during training. FC2 is fully connected to FC3 with a 50% dropout. FC3 has two units – one for the presence of an attribute and one for the absence of the attribute.

An interesting property of CNNs is that the lower layers (the ones closest to the input layer) learn low-level features, where the layers closer to the output learn high-level features. One can think of the lower layers as learning things like edges and the higher layers as learning more complex shapes. This can give some direction to the programmer as to the number of layers to include in their network. One important consequence of this is that face alignment is not necessarily a required preprocessing step when using CNNs. The transformation required to align the faces can be learned in the first layers of the network, and thus is not required as a preprocessing step.

In order to train the CNN, the weights must be learned for a given problem. In the case of attribute classification, the problem is determining the presence or absence of an attribute in a face image. Given the input and its corresponding label, it learns the weights that produce the target output. If a mistake is made, the error is back propagated through the network and the weights are changed using stochastic gradient descent. Thus, the CNN is trained through the method of backpropagation.

5.2.2.1   Overfitting

Deep networks have so many parameters that it is easy for them to learn the training data so well that they do not generalize to outside data – also known as overfitting. There are many different techniques to avoid this. One way to prevent overfitting is to create a very diverse training set. The more the training set represents the real world, the better the network will perform on outside data. Data augmentation is typical with CNNs, and the two common ways of doing this are mirroring the input and taking random crops of the input. Mirroring the input doubles the size of the training set and allows the network to be more robust to this type of image flipping. Random cropping of the input image during training allows for more data augmentation, and for different information to be highlighted with each crop. A random crop is taken from each image in each batch during training, so a given image will be presented to the network in many different ways. This is a type of data augmentation. It also keeps the network from learning features that are specific to the center of an input image. Both mirroring and random cropping are used for data augmentation when training the deep CNN.

5.2.2.2   Hyperparameters

A big part of training a CNN is determining the hyperparameters for the network. The hyperparameters determine whether or not training will converge and how long it will take to do so. Some hyperparameters, which the programmer must specify in Caffe, include the learning rate, the learning rate policy, Γ, momentum, step size, and weight decay. The learning rate is the step size for the gradient descent procedure. As training goes on, the learning rate is lowered until convergence. The learning rate policy specifies how the learning rate is decreased during training. There are several options for this, including exponential decay and a step function. Exponential decay reduces the learning rate according to an exponential function, and the step function reduces the learning rate by the value specified in Γ every step size iterations. In order for the CNN to continue learning in the face of plateaus and local minima in the weight space, the current weight update contains a fraction of the previous weight update. This fraction is specified by the momentum parameter. In training, we want the weights to exponentially decay to zero, and the rate at which they decay is specified by the weight decay parameter. All of these parameters are problem specific, much like the CNN architecture itself. There is no principled way of specifying these parameters, so again they are determined through experimentation.

The following hyperparameter values were used in the experiments (Table 5.1):

The CNN works as a dimensionality reduction tool, or a feature extractor. It takes the input – a $256 \times 256 \times 3$ image – and transforms it into a 512 dimensional feature vector (FC2). FC3 is used for classification. A softmax operation is applied to the units in FC3, resulting in a probability distribution over the classes for a given input.

Table 5.1   Hyperparameters of the attribute CNNs

Learning rate	0.001
Learning rate policy	Step
Step size	10,000
Momentum	0.9
Weight decay	0.0005

Training is performed in batches of size 50. This means that 50 images are presented to the network before weights are adjusted according to the errors produced. Each iteration consists of presenting 50 images to the network and adjusting the weights. The network is trained for 50,000 iterations.

Rarely are CNNs trained until convergence. Instead, a validation set – a set of data separate from the training and testing set – is used to determine when training should stop. For each attribute, a separate CNN is trained, using the architecture described above. After every 1,000 iterations, the model is tested on a validation set, and that, accuracy is used to determine which model to use for each attribute.

For classification, an image is passed to the network, subtracting the mean from the training data, and using the weights computed during training, the network gives a probability distribution over the classes as a result. The class with the highest probability is taken as the classification result for a given input.

CNNs have gained popularity very quickly in recent years due to the availability of GPUs and CNN software packages, which allow for quick and easy architecture specification and model training. As CNNs perform feature extraction, there is no need for hand-derived features specific to a given problem. Because of this, the same CNN architecture can be used for each attribute, and different features are learned depending on the problem. This is one of the many reasons that CNNs are being used for many of the big problems in computer vision today.

5.2.3   Performance of the attribute classifiers

From Table 5.2, it can be seen that overall, the CNN attribute classifiers do very well, with an average of over 90%. Linear attribute classifiers are moderately good with an average of 80%.

The big lips, oval face, and pointy nose CNN models were unable to break 80% testing accuracy. This is likely due to the subjective nature of attributes. Some attributes, like wearing hat, are very obvious, and a "yes" or "no" answer can be given easily, but some attributes are much more difficult to determine. Whether or not, someone has Big Lips is very subjective. This is true of oval face, and pointy nose as well. Also, lower accuracies in some of the other more subjective attributes can be seen. Arched eyebrows, attractive, and bags under eyes have lower accuracies that are much more subjective than the attributes with very high accuracies, like wearing eyeglasses, bald, and male.

Table 5.2   Accuracies of CNN and Logistic Regression on the test portion of CelebA dataset

Attribute	CNN	LogisticReg	Attribute	CNN	LogisticReg
5 o'clock shadow	0.9433	0.8251	Male	0.9826	0.9568
Arched eyebrows	0.8353	0.739	Mouth slightly open	0.9387	0.8267
Attractive	0.823	0.7561	Mustache	0.9666	0.8141
Bags under eyes	0.8507	0.6996	Narrow eyes	0.8704	0.7323
Bald	0.9882	0.9212	No beard	0.9607	0.8755
Bangs	0.9599	0.93	Oval face	0.7474	0.6013
Big lips	0.706	0.5902	Pale skin	0.8972	0.7388
Big nose	0.8388	0.6801	Pointy nose	0.7727	0.631
Black hair	0.897	0.7521	Receding hairline	0.9343	0.7636
Blond hair	0.9605	0.8734	Rosy cheeks	0.9502	0.8289
Blurry	0.9616	0.7855	Sideburns	0.9782	0.9039
Brown hair	0.8899	0.6284	Smiling	0.9262	0.8977
Bushy eyebrows	0.9258	0.7736	Straight hair	0.8262	0.6401
Chubby	0.957	0.8343	Wavy hair	0.8261	0.7461
Double chin	0.9638	0.8591	Wearing earrings	0.9052	0.7004
Eyeglasses	0.9966	0.9612	Wearing hat	0.9898	0.9354
Goatee	0.9714	0.8587	Wearing lipstick	0.938	0.9058
Gray hair	0.9816	0.8872	Wearing necklace	0.8645	0.5995
Heavy makeup	0.9112	0.8614	Wearing necktie	0.9666	0.779
High cheekbones	0.8731	0.8303	Young	0.8794	0.7898

Both models performed roughly equally on the development set as they did on the testing data. This means that the training, validation, and test data all come from the same, or a very similar distribution.

5.3 Authentication

In this section, the performance of the trained attribute models in the task of authentication is explored using two publicly available datasets, active authentication (AA) [7] and MOBIO [6]. These datasets consist of videos of the users in different sessions. The training data per person for the authentication task is limited to the enrollment videos of that person. Two challenges arise as a result. One is that there are no negative training samples since at first, each device sees its original user, the other one is that the enrollment video contains little number of frames which leads to overfitting in most of the models.

To perform authentication, the faces from all the frames are cropped and aligned via [45] by Asthana *et al.* to the same canonical coordinate used for training the linear classifiers. To extract the attribute features, face images are fed to both the CNN and the Linear attribute models that are trained in the previous section, concatenating their responses to form the feature vector. As a baseline, the LBP histogram features are extracted from each face image, which have proved to be useful for the face recognition task in [46]. To extract one single feature vector per video, LBP histogram and attribute features are calculated per frame and then averaged out.

After extracting features from videos, the similarity score of two given videos are simply calculated as the inverse of the Euclidean distance of their representations. Using these scores, the similarity matrix is formed between the enrollment and the test videos. Next, to evaluate the authentication performance of each model and also the baseline low-level features, the well-known Receiver Operating Characteristic (ROC) curves are plotted for each experiment. The ROC curve shows the True Acceptance Rate (TAR) on the y axis versus False Acceptance Rate (FAR) on the x axis which can be derived from the similarity matrix between the gallery and the probe. Also, the Equal Error Rate (EER), which is the rate where FAR equals False Rejection Rate (FRR), of all the approaches are compared together. This value is the point where ROC curve hits the diagonal line passing through $(1, 0)$ and $(0, 1)$ on the axes of the curve's coordinate system.

As a proof-of-concept method, to show that the fusion of low-level and attribute feature boosts the authentication performance, mean-variance normalized similarity scores of each method are summed and authentication is performed using these fused scores.

The experiments are divided into two categories:

1. Short-term authentication: The appearances of the user, like hair model and make up, are the same in the enrollment and testing videos.
2. Long-term authentication: The appearance of the user may be different from the enrollment video.

5.3.1   *Short-term authentication*

There are two types of short-term authentication scenarios. In the first one, the user is enrolled at the login time, then continuously authenticated after that. This is important as someone may take on the device after the user's initial login. Thus, this type has two features:

1. The appearance of the user does not change in the enrollment and testing video.
2. Environmental condition like illumination is same in enrollment and testing video.

The second type of the short-term scenario happens when the appearance of the user is still the same, but the environment changes.

The performance of facial attributes can be evaluated in these two settings of the short-term authentication using the AA [7] dataset. This dataset consists of videos of 50 subjects who participated in 3 sessions of experiments. All the sessions are taken in the same day for each individual and are different from one another in their illumination. The first session's videos are taken in an environment with office light, the second session is in a low-light environment, and the third session's videos have natural lighting. The videos are all captured by the front camera of an iPhone 5. Within each session, there are 5 videos. One corresponding to the enrollment stage, and the other four to testing. In each of the testing videos, the user is performing

*Figure 5.3   Sample frames of the AA [7] dataset. The first row shows the frames
           from the office light session. The second row shows the low light
           session and the third row corresponds to the natural light session*

some predefined tasks, while the touch data alongside the front camera videos are
recorded. The touch data is not used in the experiments. Sample video frames from
this dataset are shown in Figure 5.3.

For the first scenario, the enrollment and testing videos are taken from the same
session. This guarantees that both the appearance and the environment conditions are
the same. The ROC curves depicting the first short-term scenario on the AA dataset
are shown in the left column of Figure 5.4. It can be seen that the attribute features
perform better than the low-level LBP features.

To evaluate the effectiveness of the facial attributes in the second scenario, where
environment changes, the enrollment video of a user is taken from one session and
the testing videos from the other two sessions. The ROC curves showing the performance of the facial attributes are plotted in the second column of Figure 5.4. Again,
the attribute features perform better than the low-level features.

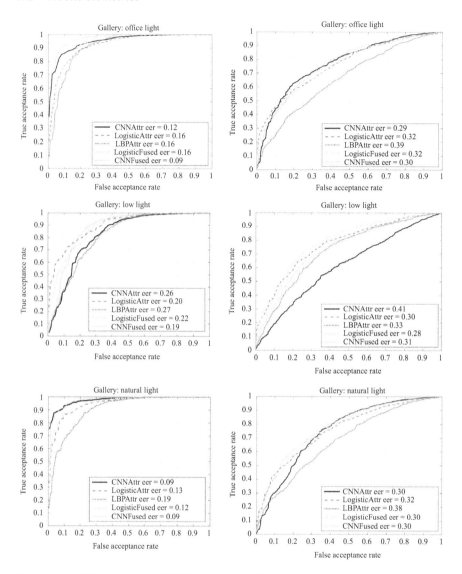

Figure 5.4   ROC curves and EER values of different scenarios of AA dataset. The left column shows the experiments where testing and enrollment environment are the same. The right column contains ROCs of enrollment in one environment and testing in the remaining two

To compare these two scenarios, the performance of all the algorithms drop when the enrollment video is from a different session; however, the attribute features are more robust than LBP features and still perform better. Furthermore, it is seen that the CNN attribute model performs very poorly when the enrollment videos are from

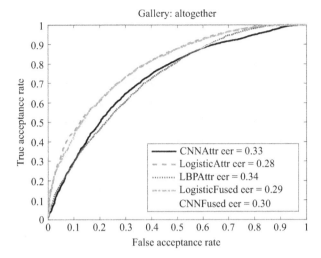

Figure 5.5   The ROC curves and EER values of the authentication methods on all sessions of AA dataset together. The gallery consists of three enrollment videos per subject

the low light environment. This is due to overfitting of the CNN attribute model to CelebA dataset distribution. The celebrity images are mostly in nicely illuminated environments. Also, this comparison suggests that even attribute features from a linear model can produce better results than low-level features, and not much worse than attribute features from the complex CNN models.

Figure 5.5 shows the ROC curve on the entire dataset. In this setting, 3 enrollment videos per person are in the gallery and the rest of 12 testing videos of that person are in the probe. Again, the attribute classifiers performed better than LBP features. As the enrollment videos of the low light session are present in the gallery videos, the performance of CNN features is not much better than the LBP features. This reminds us of the effect of the large number of parameters of the deep models in the training time and the required diversity of the data.

5.3.2   Long-term authentication

In the long-term authentication scenario, the time interval between enrollment and testing are long. It can be due to a one-time enrollment requirement of the system. This may lead to significant changes in the appearance of the person from the enrollment model. For example, the user may get a different haircut, or wear sunglasses. Also, the enrollment may be done with a different camera sensor rather than the device that the user is going to use.

The dataset suitable for evaluating the attribute features in the long-term setting is the MOBIO dataset. It contains 152 subjects with 13 sessions for each person.

Figure 5.6   Sample images of the MOBIO dataset. The first row is different people in different environments, the second row shows image pairs of the same identity and their appearance changes across the sessions

Videos of one of the sessions are taken by a MacBook 2008 laptop, and the rest are taken by a Nokia N93i. The videos are taken in six different sites and on different days, resulting in appearance changes. Also, the illumination of the sessions is not consistent and may vary from one session to another due to the location of the camera sensor at the time of the video capturing. However, the environment changes are less within the same site. Sample images from this dataset are shown in Figure 5.6.

The 12th mobile video session for each subject is chosen as the enrollment video as this session is the mostly available session across the dataset. The rest of the mobile videos are taken as the test videos. No laptop videos are taken for enrollment or testing videos. Also, all the videos within one session are assumed to form a single video to make sure that the long-term protocol is not violated. In [6], they take a single video of one session as enrollment which leads to having videos from the same session in the probe which combines the short-term authentication and long-term authentication scenarios.

In Figure 5.7, the performance of all the algorithms on the whole mobile videos of the MOBIO dataset are shown. Also, in Table 5.3, the EER values for different authentication methods for each site are shown. As it can be seen, the EER values are notably better for attribute features. This is due to lower quality of videos in this dataset compared to AA dataset as the videos are captured with an older device. The CNN model and Logistic model perform closely well; however, in most of the times, the best result comes from the fusion of the CNN attribute features with the low-level LBP features.

Next, the performance of the attributes in the second category of the long-term authentication setting is evaluated, which is enrolling the user with a different camera sensor from the testing time. To do so, all the laptop videos of the MOBIO subjects are taken as the enrollment videos and the rest of the mobile videos are considered as the test videos. The ROC curve of this experiment is plotted in Figure 5.8. As it can

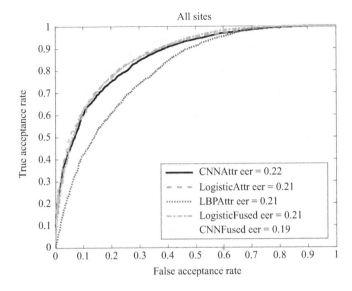

Figure 5.7 The ROC curve of the MOBIO [6] dataset for different authentication methods

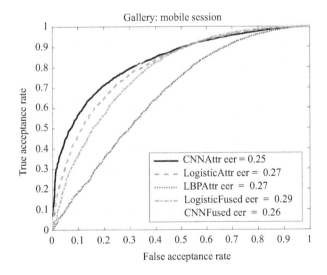

Figure 5.8 The ROC curve and EER values for the long-term authentication scenario under varying camera sensors on MOBIO [6] dataset

Table 5.3   EER values of all the experiments on AA and MOBIO dataset

Session	M1	M2	M3	M4	M5	M6	MA	ML	A1	A1C	A2	A2C	A3	A3C	AA
LBP	0.36	0.35	0.26	0.33	0.27	0.23	0.28	0.38	0.16	0.39	0.27	0.33	0.19	0.38	0.34
LogisticAttrs	0.26	**0.22**	<u>0.24</u>	**0.23**	0.20	0.24	<u>0.21</u>	0.27	0.16	0.32	<u>0.20</u>	<u>0.30</u>	0.13	0.32	**0.28**
CNNAttrs	<u>0.25</u>	<u>0.28</u>	0.26	0.24	<u>0.11</u>	<u>0.18</u>	0.22	**0.25**	<u>0.12</u>	**0.29**	0.26	0.41	<u>0.09</u>	<u>0.30</u>	0.33
LogisticFused	0.27	0.30	0.26	0.24	0.19	0.22	<u>0.21</u>	0.29	0.16	0.32	0.22	**0.28**	0.12	<u>0.30</u>	<u>0.29</u>
CNNFused	**0.24**	<u>0.28</u>	**0.22**	<u>0.24</u>	**0.09**	**0.16**	**0.19**	<u>0.26</u>	**0.09**	<u>0.30</u>	**0.19**	0.31	<u>0.09</u>	<u>0.30</u>	0.30

The bold numbers are the best performing method in each column. The second best is underscored. Each column is for a setting as follows. M1: MOBIO BUT, M2: MOBIO IDIAP, M3: MOBIO LIA, M4: MOBIO UMAN, M5: MOBIO UNIS, M6: MOBIO OUOLU, MA: All of the mobile sessions of all sites, ML: MOBIO laptop camera sensor videos as enrollment, the rest of mobile videos as test, A1: AA office light A1C: AA office light cross session, A2: AA low light, A2C: AA low light cross session, A3: AA natural light, A3C: natural light cross session, AA: Enrollment videos of all sessions as Gallery, the rest as probe.

be seen, the attribute features perform significantly better than the low-level features because low-level features are sensitive to the sensor change.

5.3.3   Discussion

The EER values of all the experiments can be found in Table 5.3. The performance of the attribute classifiers is better than the low-level LBP features. Comparing the CNN and the linear attribute models, the CNN models perform better as expected in most of the cases; however, the linear attribute features also show promising results, for example, in the overall MOBIO (MM column of Table 5.3) and AA (AA column of Table 5.3) experiments, they have better EER values than the CNN attribute features.

As it is seen, the fusion of CNN attributes and LBP features produces the best or the second best result. It means that the CNN attributes scores and LBP feature scores complement each other. This was also the case in AA dataset as seen in Figure 5.5; however, with the AA dataset, the performance of both together was worse than the Linear model alone.

The linear model is a first-order function of the LBP features; however, CNN attributes are higher order functions of the low-level features, and hence, they capture higher order dependencies. Thus, it is more probable that the lower level information gets lost, and a simple similarity score from the low-level features can help recover a part of that information lost in the network.

5.4   Platform implementation feasibility

In the continuous biometric authentication field, a set of challenges are related to the computational complexity and the memory usage of algorithms. The limited resources of a mobile phone are shared among the many processes that run at the same time on the device. The authentication process must have the highest priority among the running programs and all of them must wait until the authentication is done. So if the algorithm's computational complexity is high, it can stall the device which results in frustration of the user. In addition, as the authentication program runs continuously on the device, it can drain the battery quickly.

To find out which attribute feature detection model can work efficiently on a mobile device, a linear and a kernel SVM are trained for each attribute on a Google Nexus 5 phone with the Android operating system. The phone has 2GB of RAM and a quad core CPU clocked at 2.2 GHz. OpenCV [47] library is used for most of the steps of the algorithms including feature extraction, training the classifiers, and testing on the device.

The CNN attribute models are not suitable to be implemented on the phone. Although on workstations, they give the outputs for a test image in a couple of milliseconds while running on GPUs, they take over 100 ms on a 3-GHz CPU. On mobile devices where GPUs are not always available and CPUs are generally less powerful, they will take more time to produce results. Keeping in mind that it is need to extract all of the 40 attributes at once, it can take more than four seconds per frame to extract the attribute feature vector with the current implementations of CNNs on mobile devices. However, if a feasible CNN implementation is at hand, they are the most accurate models to be used on the device.

To test other models a kernel SVM model is trained per attribute. This model is similar to those which save the training data itself or a function of the data as a part of themselves, for example, Nearest Neighbor classification or dictionary-based methods like approaches using Principle Component Analysis for dimensionality reduction or Sparse Representation-based models. The resource usage of the kernel SVM model is a lower bound for these types of algorithms as the amount of saved data is comparably low to the training data, while the margin between the two classes are maximized. Linear SVM attribute models represent the linear models such as the ones learned with Logistic Regression in Section 5.3.1.

As the SVM classifier training is not scalable, they are trained on LFW [48] which is a smaller dataset than CelebA. A linear and a kernel SVM for each attribute are trained. The scale 1 image size is 128×168 pixels which leads to a feature vector of size 76,800.

At testing time on the phone, the Haar cascade face detector [12] is used which is available in OpenCV to detect the face of the user from the stream of frames. To align the face image with the canonical coordinate the facial landmarks are extracted via [49] by Kazemi *et al.* in DLib [50]. The same LBP features are extracted as in Section 5.3.1., but with OpenCV's implementation, which is a different implementation than VLFeat [51] on the intensity and RGB channels and concatenate them.

5.4.1 *Memory*

The memory capacity of the mobile devices are also very limited. If the model or the algorithm occupies a lot of memory, the operating system is forced to place the running programs that do not fit in RAM into the swap storage. The swap storage is both slower and less power efficient as it is a secondary storage. Thus, this can lead to less responsiveness and shorter battery life of the mobile device. Therefore, the less memory the authentication program consumes the better.

To calculate the memory usage, the RAM usage is measured before and after loading the classifiers into the memory. Then, the average of memory usage for each model over all the attributes is taken. For kernel SVM classifiers, the average memory usage is 52 MB per classifier. This number makes sense, because each classifier stores around 300 to 400 support vectors of dimension 76,000 for each attribute class. This is not a high memory usage if there is just have one SVM classifier, but there are one classifier for each attribute, summing up to 40 classifiers. Therefore, it will take about 2 GB of memory if all of them are loaded at once. One solution is to load them one by one into the memory and compute the features; however, this approach will result in a significant slowdown as there will be many secondary storage transactions.

The linear SVM model on the other hand takes less than 1 MB of memory in total as it does not have any support vectors stored in it. With the linear SVM model, it is feasible to keep all 40 classifiers in memory.

5.4.2 Computation efficiency and power consumption

So far, the only attribute method that is still feasible to implement on the device is the linear SVM model. To perform the computation efficiency test, two modes are considered. The first one is when face detection and alignment need to be performed, and the other is simply extracting the attributes from the aligned and cropped face image. A sample face image is loaded on the phone and it runs through a loop of 5,000 attribute feature extraction. In each iteration, all the 40 attributes are calculated. The device's clock is used to measure the frame per second speed of the models. In addition to frame per second speed, the energy efficiency of the attribute feature classifiers is measured with the power profiler application PowerTutor from [52]. To also see the effect of image size on the performance of the algorithm, the SVM classifiers are learned on four scales 0.3, 0.5, 0.7, 1.

To estimate the power usage, the PowerTutor provides the separate power profiles for each running application. For each application, PowerTutor shows how much energy it has consumed from the time it started running with a margin of 0.1 J. To make the measurement more precise, immediately after the 5,000th frame's attribute features are extracted the program gets terminated.

To calculate the endurance of the algorithm, the amount of energy consumption per frame is estimated. Then, assuming it is needed to do authentication every second, the endurance of the authentication algorithm is calculated with respect to the battery capacity and power profile of the device without running the authentication.

On the Nexus 5 device, the battery capacity is 2,300 mAh and the average working voltage is 4.3 V which can be verified by the power profiler. This means that in total it has about 35,604 J. To get the average power usage of the phone when the authentication algorithm is not running, the power profiler is run for 5 min when the phone is in idle mode. It shows the average power usage of the phone as 520 mW.

Table 5.4   *The speed and power consumption of linear attribute models for different image sizes. The W/O column is for the experiment that had no alignment and detection*

Scale	0.3		0.5		0.7		1	
Size/dim	32 × 48/3,840		64 × 80/16,128		88 × 112/33,280		128 × 168/76,800	
D/A	W/O	W/	W/O	W/	W/O	W/	W/O	W/
FPS	114	20	31	16	13	8	5	4
Energy (J)	26.8	128.9	93.5	201.2	207	269.1	524.9	603
Energy/frame (mJ)	5.4	25.8	18.7	40.2	41.4	73.8	105	120.6
Endurance(h)	18.8	18.1	18.3	17.6	17.6	16.6	15.8	15.4

Assuming that authentication rate is 1 per second, and having the energy consumed per frame for attribute feature extraction, the endurance of the linear model facial attribute classifiers are calculated in Table 5.4.

From Table 5.4, it is seen that the speed is decreased as the image size increases; however, still at scale 1, authentication can be done with a speed of more than one frame per second. The linear models that are learned in Section 5.3.1 had 17,600 dimensions, so the models can run on a Google Nexus 5 for about 17.5 h at a rate of 1 authentication per second.

5.5   Summary and discussion

In this chapter, facial attribute features are examined as a means of authentication on mobile devices. A linear model is trained via Logistic regression as the simplest attribute model one can learn. Also, a complex deep CNN model is trained per attribute which produces very good attribute recognition results. The performance of the other models on the authentication task will be roughly between these two methodologies. Both of the models are tested on two challenging authentication datasets and compared the results of the low-dimensional attribute features only using low-level features. It is seen that in both short- and long-term authentication scenarios, both models are more robust than the low-level features. It is also seen that a simple-score-level fusion with the low-level features can boost the authentication performance. At the end, the feasibility of different models is discussed for implementation on a mobile platform. The most feasible model with respect to memory usage, computational complexity, and power consumption is the linear model which also has produced reasonable authentication results.

Finally, there are also a set of challenges related to the authentication solely based on the facial attributes. First of all, training robust attribute classifiers is not a trivial task and itself is an open problem. Second, even if the attribute classifiers are ideal, the person attributes may change over time. This can be problematic if we are in the long-term AA setting, i.e. the user is enrolled once and is authenticated afterwards with the same model.

Acknowledgments

The work of Pouya Samangouei and Vishal Patel is supported by the DARPA Active Authentication Project under cooperative agreement FA8750-13-2-0279. Sections on deep features for attribute classifiers contributed by Emily Hand and Rama Chellappa are based upon work supported by the Office of the Director of National Intelligence (ODNI), Intelligence Advanced Research Projects Activity (IARPA), via IARPA R&D Contract No. 2014-14071600012. The views and conclusions contained herein are those of the authors and should not be interpreted as necessarily representing the official policies or endorsements, either expressed or implied, of the ODNI, IARPA, or the U.S. Government. The U.S. Government is authorized to reproduce and distribute reprints for Governmental purposes notwithstanding any copyright annotation thereon.

References

[1]   A. Vance, "If your password is 123456, just make it hackme," *New York Times*, vol. 20, 2010.

[2]   A. J. Aviv, K. Gibson, E. Mossop, M. Blaze, and J. M. Smith, "Smudge Attacks on Smartphone Touch Screens.," *WOOT*, vol. 10, pp. 1–7, 2010.

[3]   A. Jain, L. Hong, and S. Pankanti, "Biometric identification," *Commun. ACM*, vol. 43, no. 2, pp. 90–98, 2000.

[4]   P. Gupta, S. Ravi, A. Raghunathan, and N. Jha, "Efficient fingerprint-based user authentication for embedded systems," in *Proceedings. 42nd Design Automation Conference, 2005.*, 2005, pp. 244–247.

[5]   R. P. Wildes, "Iris recognition: an emerging biometric technology," *Proc. IEEE*, vol. 85, no. 9, pp. 1348–1363, 1997.

[6]   C. McCool, S. Marcel, A. Hadid, *et al.*, "Bi-Modal Person Recognition on a Mobile Phone: using mobile phone data," in *IEEE ICME Workshop on Hot Topics in Mobile Multimedia*, 2012.

[7]   M. E. Fathy, V. M. Patel, and R. Chellappa, "Face-based active authentication on mobile devices," in *IEEE International Conference on Acoustics, Speech and Signal Processing*, 2015.

[8]   CNBC, "Smartphones: Hackers' target for 2015." 2015.

[9]   T. Sim, S. Zhang, R. Janakiraman, and S. Kumar, "Continuous verification using multimodal biometrics," *Pattern Anal. Mach. Intell. IEEE Trans.*, vol. 29, no. 4, pp. 687–700, 2007.

[10]   M. Frank, R. Biedert, E. Ma, I. Martinovic, and D. Song, "Touchalytics: on the applicability of touchscreen input as a behavioral biometric for continuous

authentication," *IEEE Trans. Inf. Forensics Secur.*, vol. 8, no. 1, pp. 136–148, 2013.

[11] A. Hadid, J. Y. Heikkila, O. Silven, and M. Pietikainen, "Face and eye detection for person authentication in mobile phones," in *ACM/IEEE International Conference on Distributed Smart Cameras*, 2007, pp. 101–108.

[12] P. Viola, M. J. Jones, and D. Snow, "Detecting pedestrians using patterns of motion and appearance," *Int. J. Comput. Vis.*, vol. 63, no. 2, pp. 153–161, 2005.

[13] T. Ojala, M. Pietikäinen, and T. Mäenpää, "Multiresolution Gray Scale and Rotation Invariant Texture Classification with Local Binary Patterns," pp. 1–35.

[14] A. K. Jain, S. C. Dass, and K. Nandakumar, "Soft biometric traits for personal recognition systems," in *Biometric Authentication*, Springer, 2004, pp. 731–738.

[15] P. Samangouei, V. M. Patel, and R. Chellappa, "Attribute-based continuous user authentication on mobile devices," in *Biometrics Theory, Applications and Systems (BTAS), 2015 IEEE 7th International Conference on*, 2015, pp. 1–8.

[16] Samangouei, Pouya, and Rama Chellappa. "Convolutional neural networks for attribute-based active authentication on mobile devices." Biometrics Theory, Applications and Systems (BTAS), 2016 IEEE 8th International Conference on. IEEE, 2016.

[17] Samangouei, Pouya, Vishal M. Patel, and Rama Chellappa. "Facial Attributes for Active Authentication on Mobile Devices." Image and Vision Computing 58 (2017): 181–192.

[18] Y. Liu, D. Zhang, G. Lu, and W.-Y. Ma, "A survey of content-based image retrieval with high-level semantics," *Pattern Recognit.*, vol. 40, no. 1, pp. 262–282, 2007.

[19] R. Datta, J. Li, and J. Z. Wang, "Content-based image retrieval: approaches and trends of the new age," in *Proceedings of the 7th ACM SIGMM international workshop on Multimedia information retrieval*, 2005, pp. 253–262.

[20] M. Obeid, B. Jedynak, and M. Daoudi, "Image indexing & retrieval using intermediate features," in *Proceedings of the ninth ACM international conference on Multimedia*, 2001, pp. 531–533.

[21] V. Ferrari and A. Zisserman, "Learning visual attributes," in *Advances in Neural Information Processing Systems*, 2007, pp. 433–440.

[22] A. Farhadi, I. Endres, D. Hoiem, and D. Forsyth, "Describing objects by their attributes," in *Computer Vision and Pattern Recognition, 2009. CVPR 2009. IEEE Conference on*, 2009, pp. 1778–1785.

[23] M. Everingham, L. Van Gool, C. Williams, J. Winn, and A. Zisserman, "The PASCAL visual object classes challenge 2008," 2008.

[24] C. H. Lampert, H. Nickisch, and S. Harmeling, "Attribute-based classification for zero-shot visual object categorization," *Pattern Anal. Mach. Intell. IEEE Trans.*, vol. 36, no. 3, pp. 453–465, 2014.

[25]   C. Cortes and V. Vapnik, "Support-vector networks," *Mach. Learn.*, vol. 20, no. 3, pp. 273–297, 1995.

[26]   M. Liu, D. Zhang, and S. Chen, "Attribute relation learning for zero-shot classification," *Neurocomputing*, vol. 139, pp. 34–46, 2014.

[27]   G. Patterson and J. Hays, "Sun attribute database: Discovering, annotating, and recognizing scene attributes," in *Computer Vision and Pattern Recognition (CVPR), 2012 IEEE Conference on*, 2012, pp. 2751–2758.

[28]   J. Liu, B. Kuipers, and S. Savarese, "Recognizing human actions by attributes," in *Computer Vision and Pattern Recognition (CVPR), 2011 IEEE Conference on*, 2011, pp. 3337–3344.

[29]   N. Kumar, A. C. Berg, P. N. Belhumeur, and S. K. Nayar, "{A}ttribute and {S}imile {C}lassifiers for {F}ace {V}erification," in *IEEE International Conference on Computer Vision (ICCV)*, 2009.

[30]   L. Bourdev, S. Maji, and J. Malik, "Describing people: A poselet-based approach to attribute classification," in *Computer Vision (ICCV), 2011 IEEE International Conference on*, 2011, pp. 1543–1550.

[31]   N. Dalal and B. Triggs, "Histograms of oriented gradients for human detection," in *Computer Vision and Pattern Recognition, 2005. CVPR 2005. IEEE Computer Society Conference on*, 2005, vol. 1, pp. 886–893.

[32]   N. Zhang, M. Paluri, M. Ranzato, T. Darrell, and L. Bourdev, "Panda: Pose aligned networks for deep attribute modeling," in *Computer Vision and Pattern Recognition (CVPR), 2014 IEEE Conference on*, 2014, pp. 1637–1644.

[33]   T. Berg and P. N. Belhumeur, "POOF: Part-based one-vs.-one features for fine-grained categorization, face verification, and attribute estimation," in *Computer Vision and Pattern Recognition (CVPR), 2013 IEEE Conference on*, 2013, pp. 955–962.

[34]   N. Kumar, P. Belhumeur, and S. Nayar, "FaceTracer: A Search Engine for Large Collections of Images with Faces," pp. 1–14.

[35]   Z. Liu, P. Luo, X. Wang, and X. Tang, "Deep Learning Face Attributes in the Wild," vol. 1, Nov. 2014.

[36]   L. Bottou, "Large-Scale Machine Learning with Stochastic Gradient Descent," *Proc. COMPSTAT'2010*, pp. 177–186, 2010.

[37]   J. Bergstra, O. Breuleux, F. Bastien *et al.*, "Theano: a CPU and GPU math compiler in Python," in *9th Python in Science Conference*, 2010, no. Scipy, pp. 1–7.

[38]   R. Collobert, "Torch7: A Matlab-like environment for machine learning," *BigLearn, NIPS Work.*, pp. 1–6, 2011.

[39]   M. Abadi, A. Agarwal, P. Barham *et al.*, "TensorFlow: Large-scale machine learning on heterogeneous systems, 2015," *Softw. available from tensorflow.org*.

[40]   Y. Jia, E. Shelhamer, J. Donahue *et al.*, "Caffe," in *Proceedings of the ACM International Conference on Multimedia—MM'14*, 2014, pp. 675–678.

[41]   R. B. Palm, "Deeplearntoolbox, a matlab toolbox for deep learning," *Online. Dispon{í}vel em https//github.com/rasmusbergpalm/DeepLearnToolbox*.

[42]   Y. LeCun, L. Bottou, Y. Bengio, and P. Haffner, "Gradient-based learning applied to document recognition," *Proc. IEEE*, vol. 86, no. 11, pp. 2278–2323, 1998.

[43]   G. Levi and T. Hassner, "Age and Gender Classification Using Convolutional Neural Networks," in *IEEE Conf. on Computer Vision and Pattern Recognition (CVPR) workshops*, 2015.

[44]   G. Hinton, "Dropout: a simple way to prevent neural networks from overfitting," *J. Mach. Learn. Res.*, vol. 15, pp. 1929–1958, 2014.

[45]   A. Asthana, S. Zafeiriou, S. Cheng, and M. Pantic, "Robust discriminative response map fitting with constrained local models," in *Computer Vision and Pattern Recognition (CVPR), 2013 IEEE Conference on*, 2013, pp. 3444–3451.

[46]   T. Ahonen, A. Hadid, and M. Pietikainen, "Face description with local binary patterns: application to face recognition," *Pattern Anal. Mach. Intell. IEEE Trans.*, vol. 28, no. 12, pp. 2037–2041, 2006.

[47]   G. Bradski, "No Title," *Dr. Dobb's J. Softw. Tools*, 2000.

[48]   G. B. Huang, M. Ramesh, T. Berg, and E. Learned-Miller, "Labeled Faces in the Wild: A Database for Studying Face Recognition in Unconstrained Environments," Oct. 2007.

[49]   V. Kazemi and J. Sullivan, "One millisecond face alignment with an ensemble of regression trees," in *Computer Vision and Pattern Recognition (CVPR), 2014 IEEE Conference on*, 2014, pp. 1867–1874.

[50]   D. E. King, "Dlib-ml: a machine Learning Toolkit," *J. Mach. Learn. Res.*, vol. 10, pp. 1755–1758, 2009.

[51]   A. Vedaldi and B. Fulkerson, "{VLFeat}: An Open and Portable Library of Computer Vision Algorithms." 2008.

[52]   L. Zhang, B. Tiwana, Z. Qian *et al.*, "Accurate online power estimation and automatic battery behavior based power model generation for smartphones," in *Proceedings of the eighth IEEE/ACM/IFIP international conference on Hardware/software codesign and system synthesis*, 2010, pp. 105–114.

Chapter 6

Fusion of shape and texture features for lip biometry in mobile devices

*Rahul Raman[1], Pankaj K. Sa[2], Banshidhar Majhi[3],
and Sambit Bakshi[2]*

6.1 Introduction

Each individual in this planet is claimed to be inimitably identified; as they carry certain features in their body that uniquely ascertain them. Study of such properties is called biometrics [1] and the features that can exclusively identify an individual are called biometric traits. Fingerprint, palmprint, hand geometry, iris patterns, retinal patterns, face, facial and body thermogram, heartbeat, ear, finger and hand vein pattern, voice, gait pattern, keystroke dynamics, online and offline signature, odour, dental patterns, and lips are some of the established biometric traits.

The different biometric traits are categorised into two classes, behavioural biometric and physiological biometric. Behavioural biometrics is the field of study related to the measure of distinctively identifying and determinate patterns in human activities such as gait patterns, signature, and keystroke dynamics. The term is in contrasts with physical or physiological biometrics. Physiological biometrics is the one which involves innate characteristics of an individual encoded in the body such as fingerprint, palmprint, iris, and lips. Like fingerprints and iris, unique terrains are present over lips as well, which makes lip as a potential biometric trait with high uniqueness.

Although the study of Cheiloscopy has gained much importance in recent times, the concept was pioneered and proposed in 1968 by Tsuchihasi [2] and Suzuki and Tsuchihasi [3]. They studied the lip prints of people of all ages and concluded that lip characteristics are unique and stable for a human being (further worked upon by [4]). Their work also established that lips are unique and carry stable features. Lip prints have been used for gender classification [5]. Lip prints characteristics have been mostly used in forensics and criminal police practice (cheiloscopy) [6]. Such approach to identity an individual is used by the police and forensic experts and is included as a sub-discipline of dactyloscopy.

[1]School of Computer Science of Engineering, VIT University, India
[2]Department of Computer Science and Engineering National Institute of Technology Rourkela, India
[3]Indian Institute of Information Technology Design and Manufacturing, India

6.1.1   Evolution of lip as biometric trait

Lip is yet not so established and used widely as a biometric trait due to certain reasons like difficulty in image acquisition, proper segmentation and preprocessing to achieve noiseless features. Not sufficient database is available dedicated for lip biometric as well. Accurate segmentation of region of interest is a major challenge in lip biometric system, which limits the performance of subsequent feature extraction and matching phases. Lievin and Luthon [7] have devised a novel approach for lip segmentation by exploiting the colour information from video of a speaker. First a logarithmic colour transform is performed from RGB (Red, Green, and Blue) to HSI (Hue, Saturation, and Intensity) colour space. Subsequently a statistical approach to Markov random field modelling is deployed to determine the red hue prevailing region and motion in spatiotemporal neighbourhood. Leung *et al.* [8] have proposed spatial fuzzy clustering algorithm for lip segmentation. The distribution of data in feature space and the spatial interactions between neighbouring pixels during clustering are taken into consideration for this implementation. A Particle Swarm Optimization–Support Vector Machine (PSO–SVM) lips recognition method based on active basis model has been presented in [9].

Different researchers have used lip as a biometric trait. In [10], lip has been identified as physiological as well as behavioural biometric. Movement of lip while speaking has got unique pattern and the study towards lip reading-based recognition describes it as a behavioural biometric. Lip has also got strong geometric features as its shape and texture and they form it as a physiological biometric. This proposes the immense potential of lip as a biometric trait both behavioural and physiological. In [11], Bakshi *et al.* have extracted local features from greyscale images and achieved high accuracy ($>90\%$), using scale invariant feature transform [12] and speeded up robust features [13] algorithms for local feature matching. Choras has established lip as a biometric over and again by his different researches [14,15] and proved that lip can be used as a strong biometric trait. In [16], groove patterns over human lip are studied for successful identification purpose. With the aim of applying lip print recognition for criminal identification, Smacki and Wrobel [17] have considered mean difference similarity measure to discriminate the feature obtained from Hough transformed binary images of lip prints. Kim *et al.* [18] describe a lip print recognition for security systems by multi-resolution architecture. In another work [19], Wrobel and Doroz have used generalised Hough transform over fragments of lip images. Authors have claimed correct recognition rate of 80% over a population of 30 lip print images acquired from 10 individuals. In the subsequent work [20], Wrobel *et al.* have applied the same methodology over a larger database and presented an elaborated result with error rate ranging between 27% and 17% as the length of lip-sections are varied from $40px$ to $10px$.

Lip as a physiological biometric trait is similar to popularly used biometric traits such as fingerprint, palmprint, and iris patterns. Lip also carries many strong geometrical features such as lines, bifurcations, bridges, pentagons, dots, lakes, crossings, triangles, etc. This anthropometric property has fuelled a lot of motivation to the researchers for investigating the potential of lip as a dependable biometric trait; and lip has emerged as a member of multimodal biometrics as reported by numerous current researches. There is ample research being conducted, where lip is used as a part of multimodal biometrics. Wang and Liew [10] have explored the potential of lip as behavioural as well as physiological biometrics and are combined with different traits to make a robust biometric trait. Jain and Hong [21] have combined fingerprint, face and speech together for authentication exploiting the behavioural aspect of lip. Ichino *et al.* [22] have also worked over multimodal biometrics. In their research, they have modelled lip movement along with voice using kernel Fisher discriminant analysis. Scientists have also worked for lip tracking for the purpose of speaker authentication [23,24].

6.1.2   Why lip among other biometric traits?

A typical biometric trait is expected to satisfy certain *procedural criteria* like universality, uniqueness, permanence, measurability, performance, and acceptability. Lip is among those biometric traits, which have high universality, uniqueness, permanence, and measurability, and its performance and acceptability as a biometric trait will increase as more and more research is exercised in the domain. Further, a biometric trait can practically be used if it satisfies certain *social criteria* like ease of use, error incidences, accuracy, and user acceptance. Fingerprint and palmprint have ease of use that made them prevalent, although factors like dryness, dirt and age can increase error incidences. They also have high accuracy and user acceptance. Other biometric traits with high-performance measures like retina, hand geometry, and face also suffers with issues of social criteria, for example, hand geometry has high error incidences due to hand injury and age; retina has error incidences due to acquisition with spectacles; and face has error incidences with illumination, age, spectacles, and hair. On the other hand, lip has less social issues. It too has got error incidences in the form of chapping of lips due to seasonal and health related-issues and hindrance of acquisition of image due to the presence of moustaches and lip paints. Yet, these issues are temporal over short span of time and hence can be acceptable and dealt with. Acquisition of lip, its enrolment and authentication are not preferable with fixed acquisition devices, that is unlike fingerprints and palmprints, the acquisition position cannot be fixed for lip print acquisition. This limits its use in public authentication domain. However, lip acquisition can be very effortlessly done with handheld devices. This fact advocates lip print as a preferable biometric for authentication in handheld devices over other biometric traits. Lip, as a biometric trait, can be considered descent on the measures of procedural criteria. Moreover, the ease of acquisition of lip through handheld device, without much need of conscious cooperation of the user, encourages users to prefer lip as a biometric for handheld device.

6.1.3   Biometric authentication for handheld devices

Handheld devices are major targets for theft. These devices are providing features useful in daily life, and users are getting dependent on it for different purposes. Internet access, personal digital assistance, banking, e-commerce, data storage, remote work and monitoring, and entertainment are to name a few. As the devices are getting more dependable, more and more confidential data are getting stored on these devices. However, the question that remains an issue is that, what is protecting such confidential data stored on these handheld devices? For ages, the passwords have been a secured means of authentication. As the power of our computers increased, so did the power of the computers used by the database hackers and code crackers. Today basic password takes no time to break through. A string of letters and numbers known by authenticated user does not seem enough to keep an account secure. In the future how are we supposed to keep all of our data safe and secure? Studies have shown that more than 50% of handheld device users are not willing to use PIN security feature though they are aware of it. The reason roots to the lack of confidence in using the PIN system or the lack of convenience. A large majority of those users feel the need of an alternative approach of authentication [25]. As these handheld devices are giving impact both in our personal and professional life, there is always a constant battle between security and convenience. A 32 digit alphanumeric password with symbols and characters in both the cases and a unique sequence of them would be a highly secure password but it is highly inconvenient for a handheld device considering its frequent usage. On the other hand, a password with only four digits can be cracked in seconds, although it is relatively convenient. The role of authentication medium is very crucial at this point as to let these handheld devices remain easy to operate yet embedding a high level of security in it.

Handheld devices are frequently used devices and hence require a convenient way for authentication. However, smart handheld devices carry important personal as well as official data and hence the authentication mechanism needs to be robust as well. Biometrics has come as an urgent solution for authentication of handheld devices with very high standards of security as well as convenience of use. Advantages of biometrics over other authentication paradigms have made it omnipresent, be it for authentication in networked devices for large population, or for handheld and mobile devices for limited number of users. Some of the biometric traits like fingerprint, pulse, and iris are already studied as biometric traits for mobile devices.

Saevanee and Bhatarakosol [26] have exploited the behavioural aspect of biometric like finger pressure and keystroke dynamics for authentication over mobile devices. Palmprint [27] and knuckle biometrics [28] are also used for authentication. Shabeer and Suganthi [29] have used voice and fingerprints for authentication, whereas in [30] voice is combined with face biometric for authentication over mobile devices.

6.1.4   *Suitability of lip biometric for handheld devices*

Advantages and disadvantages of lip over other biometric traits have already been discussed. Referring those points, the advantages and disadvantages of lip over other biometric traits for handheld devices can be understood. The handheld devices have front camera that suits the acquisition distance and angle needed for capturing lip prints. Every time a handheld device is needed to be accessed, it has to be brought in front of the face. This, naturally places lip in the field of view of camera of the handheld device, making this preferable over many other biometric traits. Moreover, other suitably placed biometric traits like retina and iris requires specific camera setup to acquire iris and retina features, and suffer from the hindrance in acquisition due to the presence of spectacles. Although, seasonal chapping and lip paints may drop the recognition rate of lip prints, yet this biometric trait has sufficient motivating reasons to be used for handheld devices.

6.2 Motivation

Potential of lip as a strong biometric trait and its suitability to be used for handheld devices are the key motivation behind the research. With the presence of strong features analogous to fingerprint, palmprint, and iris, it qualifies as a strong candidate as a biometric trait. It also does not require high precision camera for acquisition of its feature. Moreover, during the usage of handheld devices, particularly at the time of authentication, lip would be naturally positioned exactly before front camera of the handheld device. This provides another level of convenience to use as no extra step towards authentication like entering PIN or providing fingerprint is needed. With high security as well as convenience, we are encouraged to carry forward our research in the proposed direction.

6.3 Anatomy of lip biometric system

Lip as a biometric trait has diverse features. The movement of lip while speaking called lip motion feature is a behavioural biometric feature. Intensity variations inside the outer lip contour, referred as texture feature of lip and shape- and contour-related features called lip shape feature, collectively form its physiological feature. Many existing article have worked over either of the physiological features. Articles [31–33] have exploited shape as feature for recognition, whereas, few other articles [34–37] have used its texture and colour information as feature. Authors in [38] have attempted to fuse both shape and colour features of lip images by means of fuzzy clustering, though the combined feature is applied to segment lip portions from given image. With the goal of recognition and with the motive of robust authentication, this chapter describes a work over combination of both the physiological features of lip: shape and texture.

The authentication of lip needs a certain steps of preprocessing that starts from image acquisition followed by the proper segmentation of lip from the acquired image. A number of research have been carried out towards achieving proper segmentation of lip as already presented in the earlier section. In the present work, we have used two separate versions of NITRLip databases [39,40], which have segmented versions of lip available. The segmented lip-map is subjected to distance transform to find the centre of the lip. The contour of the lip is traversed 2π angle. For every angle θ (where $0 < \theta < 2\pi$), the pixels falling over the straight line connecting the centre pixel and the corresponding contour pixel are noted. Hence, a 2D feature vector having size $w_{\text{lip}}/2 \times 360$ would be formed, where $w_{\text{lip}}/2$ is the maximum distance between centre pixel and contour among all possible 360 values. This feature vector is scaled down into 1D feature by retaining the Coefficient of Variation of values for every

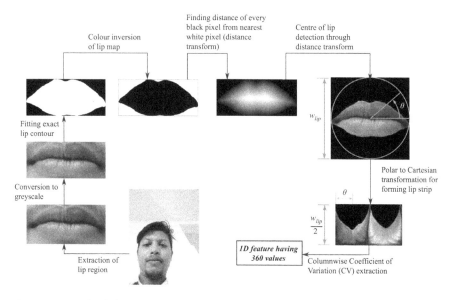

Figure 6.1   Block diagram of lip acquisition, segmentation, and feature extraction

360 columns. Finally, a 1×360 feature vector is formulated. The statistical measure *coefficient of variation* ($= standard\_deviation/mean$) makes the feature normalised and hence robust against illumination variance. Figure 6.1 illustrates this process in a block diagram. Recognition is performed on this feature template. The aggregated feature is proposed to contain the information of both shape and texture of lip, which is invariant to rotation and scale and also to the illumination variation. These features make the proposed model robust to lighting conditions; distance and angle of the handheld device for the user, which is a desirable feature for an authentication system of handheld device.

The resulting feature for each segmented image is in the form of a 1D array with 360 elements, where each element is related to its immediate neighbouring element forming a Markovian chain, and thus, Hidden Markov Model (HMM) is the best suitable for modelling such relations. As the shape and texture features of each of the entries are related to its proximity, hence the aggregated features which are entries in the array are apposite to be used for HMM-based modelling. The features of a lip, thus found must be able to differentiate each subject from another. The machine-learning method learns the feature patterns for limited number of users to classify them among the number of classes equal to the number of users of that handheld device. After a sufficient number of learning data sample from each of the classes are provided to the HMM training module, a consistency is established by the algorithm representing successful training of the samples. After successful training of samples, testing needs to be performed. When a live sample is captured from the front camera of the handheld device for authentication, it is classified among the existing classes that represents authenticated users of the device or a class for all other non-authenticated users. Based on this classification of the users, they can be authenticated or can be granted different levels of access. Following section describes the machine-learning method used here, that is, about the HMM in detail.

6.3.1 HMM-based modelling

To model this problem to HMM, their observation state, hidden state, transition, and emission probabilities are required to be defined. Following are the definitions with respect to the proposed model. An observed state or visible state denoted by $v(\theta)$ is a visible features, that is, accessible from an event sequence to be modelled. A sequence of observed state is denoted as:

$$V^{\Theta} = \{v(1), v(2), \ldots, v(\Theta)\}$$

In the proposed model, 360 lines originating from the centre of lip segment image and ending at the end of the lip contour are taken. Each line is angularly 1° apart from the previous line, thus making a complete rotation with 360 lines for 360°. For each of the line, gradient values representing the texture feature are averaged with the length of the line contributing to the shape feature. Thus, the aggregated feature has contribution of both shape and texture of lip. Each value in an array is the visible feature and related only to its neighbour, and forms a set of observed state sequence for HMM-based modelling.

Hidden state denoted by $\omega(\theta)$ are the states not observed directly rather needs interpretation from observed sequence. The perceiver does not have access to the hidden state, instead the algorithm can measure and exploit some properties of the observed state to infer hidden state. In the proposed model, four hidden states are taken considering the horizontal and vertical symmetry of the lip contour from the centre. The HMM-based modelling for this problem has also been carried out with 3 and 5 hidden states, when the training lasted for more than 100 iterations as compared to 50–60 iterations in case of 4 hidden states. This empirically justifies the selection of number of hidden states.

As depicted in Figure 6.2, the transition and emission probabilities are defined as:
Transition among hidden states, i.e., $a_{ij} : P(\omega_j(\theta+1)|\omega_i(\theta))$
Emission of a visible state, i.e., $b_{jk} : P(v_k(\theta)|\omega_j(\theta))$
with limiting conditions as, $\sum_j a_{ij} = 1$ for all i and $\sum_k b_{jk} = 1$ for all j.

Projected problem is a learning problem of HMM. The problem states that given a set observed states sequence V^{Θ} and any hidden state is given by $\omega(\theta)$, the task is to determine the probabilities a_{ij} and b_{jk} using forward backward algorithm. We start with the above defined initial arbitrary values of a_{ij} and b_{jk} and find more accurate values of a_{ij} and b_{jk} at the end of Baum–Welch or forward–backward algorithm as illustrated below.

The probability that the model produces a sequence V^{Θ} of visible states is

$$P(V^{\Theta}) = \sum_{r=1}^{r_{max}} P(V^{\Theta}|\omega_r^{\Theta})P(\omega_r^{\Theta}) \tag{6.1}$$

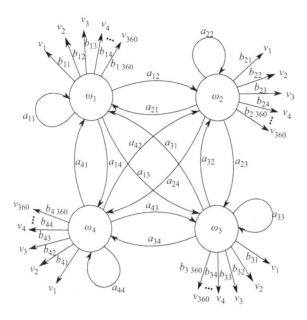

Figure 6.2   State transition diagram of the proposed Hidden Markov Model

where each r indexes a particular sequence $\omega_r^{\Theta} = \{w(1), w(2), \ldots, w(\theta)\}$ of Θ hidden states. In the general case of c hidden states, there will be $r_{max} = c^{\Theta}$ possible terms in the sum of (6.1).

As we are dealing with first-order Markov process, the factors in (6.1) can be written as (6.2) and (6.3):

$$P(\omega_r^{\Theta}) = \prod_{\theta=1}^{\Theta} P(\omega(\theta)|\omega(\theta-1)) \tag{6.2}$$

$$P(V^{\Theta}|\omega_r^{\Theta}) = \prod_{\theta=1}^{\Theta} P(v(\theta)|\omega(\theta)) \tag{6.3}$$

Combining results of (6.2) and (6.3), previously described (6.1) can be rewritten as (6.4):

$$P(V^{\Theta}) = \sum_{r=1}^{r_{max}} \prod_{\theta=1}^{\Theta} P(v(\theta)|\omega(\theta))P(\omega(\theta)|\omega(\theta-1)) \tag{6.4}$$

We denote our model – the a's and b's – by ϕ and using Bayes formula, probability of the model given observed sequence is given by (6.5):

$$P(\phi \mid V^{\Theta}) = \frac{P(V^{\Theta} \mid \phi) P(\phi)}{P(V^{\Theta})} \tag{6.5}$$

Now, $\alpha_j(\theta)$ and $\beta_i(\theta)$ can be defined as shown in (6.6) and (6.7):

$$\alpha_j(\theta) = \begin{cases} 0 & t = 0 \text{ and } j \neq \text{ initial state} \\ 1 & t = 0 \text{ and } j = \text{ initial state} \\ \left[\sum_i \alpha_i(\theta - 1)a_{ij} \right] b_{jk}v(\theta) & \text{otherwise} \end{cases} \qquad (6.6)$$

$$\beta_i(\theta) = \begin{cases} 0 & \omega_i(\theta) \neq \omega_0 \text{ and } \theta = \Theta \\ 1 & \omega_i(\theta) = \omega_0 \text{ and } \theta = \Theta \\ \sum_j \beta_j(\theta + 1)a_{ij}b_{jk}v(\theta + 1) & \text{otherwise} \end{cases} \qquad (6.7)$$

However, this way of determining $\alpha_j(\theta)$ and $\beta_i(\theta)$ is mere estimates of their true values, as we do not know the actual values of a_{ij} and b_{jk} in (6.6) and (6.7). We can calculate improved values of $\alpha_j(\theta)$ and $\beta_i(\theta)$ by defining $\gamma_{ij}(\theta)$ (shown in (6.8)), which is the probability of transition between $\omega_i(\theta - 1)$ and $\omega_j(\theta)$, given the model generated the entire training sequence V^Θ by any path:

$$\gamma_{ij}(\theta) = \frac{\alpha_i(\theta - 1)a_{ij}b_{jk}\beta_j(\theta)}{P(V^\Theta \mid \phi)} \qquad (6.8)$$

Hence, we find an improved estimation of a_{ij} and b_{jk} as $\hat{a}_{ij}$ and $\hat{b}_{jk}$ through (6.9) and (6.10), respectively:

$$\hat{a}_{ij} = \frac{\sum_{\theta=1}^{\Theta} \gamma_{ij}(\theta)}{\sum_{\theta=1}^{\Theta} \sum_k \gamma_{ik}(\theta)} \qquad (6.9)$$

$$\hat{b}_{jk} = \frac{\sum_{\theta=1, v(\theta)=v_k}^{\Theta} \sum_l \gamma_{jl}(\theta)}{\sum_{\theta=1}^{\Theta} \sum -l\gamma_{jl}(\theta)} \qquad (6.10)$$

If we denote the a's and b's of our model by ϕ and use Baye's formula, probability of the model given observed sequence is

$$P(\phi \mid O_1^\theta) = \frac{P(O_1^\theta \mid \phi)P(\phi)}{P(O_1^\theta)} \qquad (6.11)$$

6.3.2 Training, testing, and inferences through HMM

The proposed training through HMM comprises two phases, that is, learning phase and lip-based-authentication phase as shown in Figure 6.3. In the learning phase, sample images undergo various steps. Region of interest, that is, the lip portion is segmented from the given image. Then, its texture-and shape-based features are extracted and an aggregated feature vector is formed that uniquely defines respective classes. Hence, in different training conducted for each class, different HMMs are formed representing

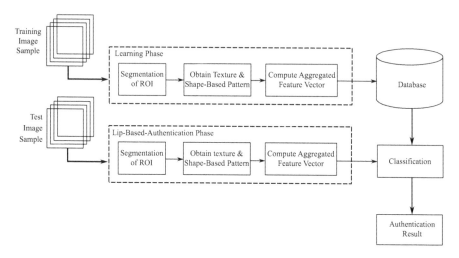

Figure 6.3   Training and testing phases of lip-based authentication

a unique user. Thus, a total of $(n + 1)$ classes are formed, where n is the number of different user and an extra class representing all non-authenticated (unintended) users. Related experiments are discussed in the next section.

6.4   Experimental verification and results

Experiments are conducted to justify the proposed work. The proposed method of authentication of user for handheld device needs dataset, where lip images are captured through front camera of handheld device. Two suitable databases are created for the same purpose, which is described in detail in Section 6.4.2. The experiment of authentication is carried 3 times with 2, 4, and 9 numbers of authenticated users, making a total of 3, 5, and 10 different classes to be classified into, where an extra class for each case is the class for all non-authenticated users. The detailed description about the experiment that includes experimental constraints, assumptions, about used databases, parameters of evaluation, phases of training and testing, and results are discussed in detail in this section.

6.4.1   Assumptions and constraints in the experiment

The experiments of proposed methodology are conducted on partially constrained situations. The NITRLip datasets, which are used for testing, are created using static cameras. The datasets are created in indoor lighting conditions with varied illumination, and do not consider very high change in illumination. The images are consciously captured by the subjects. Users do not wear any lip paints, and do not have facial expressions. In practical cases, the presence of these may degrade the performance of the proposed method. However, the experiments are performed over greyscale

Table 6.1   Detail of capturing device and mechanism in testing databases

Database	Capturing device	Exposure details	Capturing spectrum	Capturing condition	Illumination condition	Image format
NITRLipV1	Canon PowerShot A1100IS Camera	Aperture: F2.7, Shutter speed: 1/60 s 1/25 s	Visible Spectrum (VS)	Constrained	Indoor, Varied	JPEG
NITRLipV2	Front Camera of Lenovo K3 Note	Aperture: F2.4, Shutter speed: 1/20 s 1/12 s	Visible Spectrum (VS)	Constrained	Indoor, Varied	JPEG

Table 6.2   Detail of subjects in testing databases

Database	Number of subjects	Number of images	Age group	Genders	Ethnicity	Expression
NITRLipV1	15	109	(20,40) years	{male,female}	Indian	Nil
NITRLipV2	20	100	(20,30) years	{male,female}	Indian	Nil

images. We have purposefully discarded the colour information to verify the strength of this approach. Adding colour information will help achieving better performance.

6.4.2   Databases used

NITRLipV1 has been captured by Canon PowerShot A1100IS with aperture F2.7 and shutter speed varying from 1/60 s to 1/25 s. The images are indoor and illumination conditions are varied intentionally. The capturing condition is purely constrained as the subject cooperates the acquisition. The resulting images are saved in JPEG format. Fifteen subjects with age group 20–40 have participated in the database. A total of 109 images have been captured. All the subjects are Indian. Both male and female subjects have participated in the database.

NITRLipV2 has been captured by front camera of Lenovo K3 Note with aperture F2.4 and shutter speed varying from 1/20 s to 1/12 s. The subjects are asked to hold the mobile phone in the usual posture they use it, and they have captured their face images with front camera of the mobile. The images are indoor and illumination conditions are varied intentionally. The resulting images are saved in JPEG format. Twenty subjects with age group 20–30 have participated in the database. A total of 100 images have been captured. All the subjects are Indian. Both male and female subjects have participated in the database.

A detail of the parameters of NITRLipV1 and NITRLipV2 can be found in Tables 6.1 and 6.2.

6.4.3   *Parameters of evaluation*

- **Recall**: Sensitivity or true positive rate represents the proportion of positive cases that were correctly identified. In multi-class classification, as in the proposed scenario, average recall gives the corresponding information. It gives the average of all positive cases that were correctly identified.
- **Precision**: Precision is the predicted positive cases that were correct. In the proposed multi-class classification, precision of classification is defined as average of precision of all the classes.
- **Balanced accuracy**: Accuracy is the proportion of total number of correct predictions. In multi-class classification, balanced accuracy is given by the arithmetic mean of class specific accuracies.
- **Error rate**: Error rate is given by $(1 - Accuracy)$. In the proposed multi-class classification, error rate is given by the average of misclassification for each individual classes.
- **Confusion matrix**: In the presented confusion matrix, it is depicted 'how many times class M is identified by class N?' The diagonal elements of the matrix represent cases when $M = N$, that is, true positives. Non-diagonal elements indicate erroneous matches.

6.4.4   *Results and analysis*

In each experiment conducted, we choose $n + 1$ random subjects from the database, where first n chosen subjects are considered to be intended users, and the data of last chosen subject are used to observe the behaviour of the system when an intruder (or imposter) attempts to access the system. In our available NITRLipV1 and NITRLipV2 databases, we have 4–10 images/subject. If a subject is randomly chosen, half of the images of every subject are used for training and residue is subjected to testing. We vary n as 3, 5, and 10 only, as we consider the assumption that number of intended users of a handheld device will not be very high.

A total of nine types of experiments are conducted as described below. Due to random selection, different combination of subjects are chosen each time. Hence, we record different performance in each instance of running the experiment. Finally, we report the consolidated performance quantified with the parameters of evaluations mentioned previously.

Experiment Type #1:

- Database used: NITRLipV1
- Subjects randomly chosen : 3 (2 intended users + 1 intruder)
- Applied training mechanism: HMM with 3 classes

Possibility of such random selection: in $\binom{15}{3} = 455$ ways
Number of times Experiment 1 is conducted: 46 times

Experiment Type #2:

- Database used: NITRLipV2
- Subjects randomly chosen : 3 (2 intended users + 1 intruder)
- Applied training mechanism: HMM with 3 classes

Possibility of such random selection: in $\binom{20}{3}$ = 1,140 ways
Number of times Experiment 2 is conducted: 114 times

Experiment Type #3:

- Database used: NITRLipV1
- Subjects randomly chosen : 5 (4 intended users + 1 intruder)
- Applied training mechanism: HMM with 5 classes

Possibility of such random selection: in $\binom{15}{5}$ = 3,003 ways
Number of times Experiment 3 is conducted: 301 times

Experiment Type #4:

- Database used: NITRLipV2
- Subjects randomly chosen : 5 (4 intended users + 1 intruder)
- Applied training mechanism: HMM with 5 classes

Possibility of such random selection: in $\binom{20}{5}$ = 15,504 ways
Number of times Experiment 4 is conducted: 1,551 times

Experiment Type #5:

- Database used: NITRLipV1
- Subjects randomly chosen : 10 (9 intended users + 1 intruder)
- Applied training mechanism: HMM with 10 classes

Possibility of such random selection: in $\binom{15}{10}$ = 3,003 ways
Number of times Experiment 5 is conducted: 301 times

Experiment Type #6:

- Database used: NITRLipV2
- Subjects randomly chosen : 10 (9 intended users + 1 intruder)
- Applied training mechanism: HMM with 10 classes

Possibility of such random selection: in $\binom{20}{10}$ = 184,756 ways
Number of times Experiment 6 is conducted: 1,847 times

Experiment Type #7:

- Database used: merged NITRLipV1 and NITRLipV2
- Subjects randomly chosen : 3 (2 intended users + 1 intruder)
- Applied training mechanism: HMM with 3 classes

Possibility of such random selection: in $\binom{35}{3} = 6{,}545$ ways
Number of times Experiment 7 is conducted: 66 times

Experiment Type #8:

- Database used: merged NITRLipV1 and NITRLipV2
- Subjects randomly chosen : 5 (4 intended users + 1 intruder)
- Applied training mechanism: HMM with 5 classes

Possibility of such random selection: in $\binom{35}{5} = 324{,}632$ ways
Number of times Experiment 8 is conducted: 325 times

Experiment Type #9:

- Database used: merged NITRLipV1 and NITRLipV2
- Subjects randomly chosen : 10 (9 intended users + 1 intruder)
- Applied training mechanism: HMM with 10 classes

Possibility of such random selection: in $\binom{35}{10} = 183{,}579{,}396$ ways
Number of times Experiment 9 is conducted: 184 times

We here illustrate one particular random instance of Experiment Type #8 that may help readers to understand the experimental process. In Experiment Type #8, five random subjects are chosen from combined pool of NITRLipV1 and NITRLipV2. Let us assume that randomly the subjects $1{,}013M$ (5 images), $1{,}014M$ (8 images), $2{,}005M$ (6 images), $2{,}006F$ (7 images), and $2{,}014F$ (4 images) are chosen. Hence we further randomly choose 3, 4, 3, 4, and 2 images, respectively, from these subjects (classes) for training through HMM. Remaining 2, 4, 3, 3, and 2 images of respective classes are used for testing purpose. In this case, the users of the handheld device are $U_1 \equiv 1{,}013M$, $U_2 \equiv 1{,}014M$, $U_3 \equiv 2{,}005M$, $U_4 \equiv 2{,}006F$, and $\overline{U_{(1-4)}} \equiv 2{,}014F$ and the performances are recorded with respect to U_1, U_2, U_3, U_4, and $\overline{U_{(1-4)}}$. In some other iteration of this random experimentation, the users are mapped to some other subject in the database and performance is recorded.

Figures 6.4, 6.5, and 6.6 are presented to illustrate convergence of HMM training for three sample experiments with 3, 5, and 10 iterations respectively.

From 46 executions of Experiment Type #1, we find the confusion matrix for NITRLipV1 database with two users as described in Table 6.3. From 114 executions of Experiment Type #2, we find the confusion matrix for NITRLipV2 database with

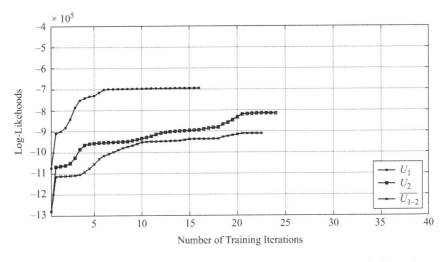

Figure 6.4   A sample plot showing convergence of HMM training with three classes

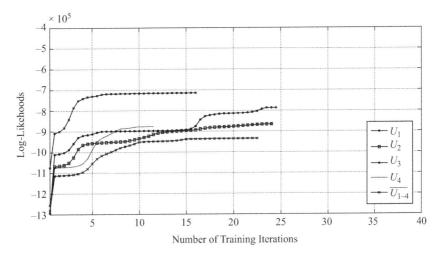

Figure 6.5   A sample plot showing convergence of HMM training with five classes

two users as described in Table 6.4. From 301 executions of Experiment Type #3, we find the confusion matrix for NITRLipV1 database with four users as described in Table 6.5. From 1,551 executions of Experiment Type #4, we find the confusion matrix for NITRLipV2 database with four users as described in Table 6.6. From 301 executions of Experiment Type #5, we find the confusion matrix for NITRLipV1 database with nine users as described in Table 6.7. From 1,847 executions of Experiment Type #6, we find the confusion matrix for NITRLipV2 database with nine users as described in Table 6.8. From 66 executions of Experiment Type #7, we find

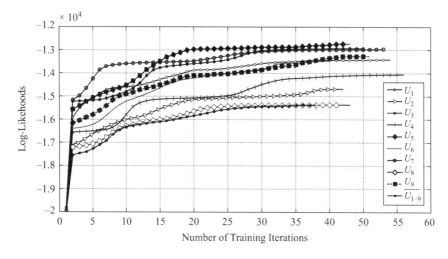

Figure 6.6   A sample plot showing convergence of HMM training with ten classes

Table 6.3   Confusion matrix for NITRLipV1 database with two users

Users	U_1	U_2	$\overline{U_{(1-2)}}$	Accuracy(%) ↓
U_1	89	1	2	96.74
U_2	2	91	2	95.79
$\overline{U_{(1-2)}}$	2	0	87	97.75
	Balanced accuracy →			96.76

Table 6.4   Confusion matrix for NITRLipV2 database with two users

Users	U_1	U_2	$\overline{U_{(1-2)}}$	Accuracy(%) ↓
U_1	340	3	1	98.84
U_2	1	338	2	99.12
$\overline{U_{(1-2)}}$	1	4	335	98.53
	Balanced accuracy →			98.83

the confusion matrix for merged NITRLipV1 and NITRLipV2 database with two users as described in Table 6.9. From 325 executions of Experiment Type #7, we find the confusion matrix for merged NITRLipV1 and NITRLipV2 database with four users as described in Table 6.10. From 184 executions of Experiment Type #7, we find the confusion matrix for merged NITRLipV1 and NITRLipV2 database with nine users as described in Table 6.11.

Table 6.5   Confusion matrix for NITRLipV1 database with four users

Users	U_1	U_2	U_3	U_4	$\overline{U}_{(1-4)}$	Accuracy(%) ↓
U_1	660	9	7	4	4	96.49
U_2	8	670	11	6	5	95.71
U_3	5	12	654	8	7	95.34
U_4	9	10	6	674	3	96.01
$\overline{U}_{(1-4)}$	8	4	12	6	648	95.58
		Balanced accuracy →				95.83

Table 6.6   Confusion matrix for NITRLipV2 database with four users

Users	U_1	U_2	U_3	U_4	$\overline{U}_{(1-4)}$	Accuracy(%) ↓
U_1	4,544	65	81	45	86	94.25
U_2	103	4,654	114	54	87	92.86
U_3	74	93	4,612	48	102	93.57
U_4	82	89	77	4,570	38	92.40
$\overline{U}_{(1-4)}$	83	57	96	64	4,621	92.22
		Balanced accuracy →				93.06

Table 6.7   Confusion matrix for NITRLipV1 database with nine users

Users	U_1	U_2	U_3	U_4	U_5	U_6	U_7	U_8	U_9	$\overline{U}_{(1-9)}$	Accuracy(%) ↓
U_1	641	9	8	4	6	8	12	8	4	8	90.54
U_2	8	643	6	6	8	10	6	4	11	7	90.69
U_3	11	7	637	5	7	7	6	7	5	4	91.52
U_4	12	5	11	656	8	7	8	7	8	3	90.48
U_5	6	9	5	10	644	6	8	7	8	8	90.58
U_6	8	8	1	9	6	621	8	9	7	2	91.46
U_7	9	11	6	7	4	5	626	10	4	5	91.12
U_8	4	8	9	6	7	7	6	630	8	7	91.04
U_9	12	7	8	8	6	3	6	11	663	10	90.33
$\overline{U}_{(1-9)}$	5	7	11	8	2	7	9	7	4	633	91.34
				Balanced accuracy →							90.91

Table 6.12 presents the consolidated results achieved on these two databases executing the nine types of experiments. It can be observed from the table that precision and recall values are very close, indicating the stability of the proposed method.

Table 6.8   Confusion matrix for NITRLipV2 database with nine users

Users	U_1	U_2	U_3	U_4	U_5	U_6	U_7	U_8	U_9	$\overline{U_{(1-9)}}$	Accuracy(%) ↓
U_1	4,921	46	68	56	71	57	45	81	47	85	89.85
U_2	51	4,887	58	73	82	90	47	36	40	62	90.05
U_3	58	72	4,902	71	52	66	48	50	63	74	89.85
U_4	66	43	68	4,976	82	75	58	40	42	71	90.13
U_5	51	56	64	70	4,798	73	80	47	40	56	89.93
U_6	63	61	56	54	49	4,833	47	52	92	61	90.03
U_7	75	83	47	61	75	44	4,954	32	84	67	89.71
U_8	73	38	47	43	74	53	70	4,516	66	58	89.64
U_9	63	62	55	71	41	65	66	59	4,765	76	89.52
$\overline{U_{(1-9)}}$	44	68	66	78	51	48	71	44	73	4,861	89.95
					Balanced accuracy →						89.87

Table 6.9   Confusion matrix for merged NITRLipV1 and NITRLipV2 database with two users

Users	U_1	U_2	$\overline{U_{(1-2)}}$	Accuracy(%) ↓
U_1	201	3	1	98.05
U_2	0	199	4	98.03
$\overline{U_{(1-2)}}$	2	3	203	97.60
	Balanced accuracy →			97.89

Table 6.10   Confusion matrix for merged NITRLipV1 and NITRLipV2 database with four users

Users	U_1	U_2	U_3	U_4	$\overline{U_{(1-4)}}$	Accuracy(%) ↓
U_1	971	11	11	15	12	95.20
U_2	16	964	12	9	10	95.35
U_3	13	10	960	13	12	95.24
U_4	17	14	8	978	10	95.23
$\overline{U_{(1-4)}}$	14	16	10	9	982	95.25
	Balanced accuracy →					95.25

6.5   Conclusions

The article presented an HMM-based mm lip recognition for limited users of a hand-held device. Tests are made on two small databases. The balanced accuracy in case of three, five, and ten classes are observed. While accuracy in case of three classes

Table 6.11   Confusion matrix for merged NITRLipV1 and NITRLipV2 database with nine users

Users	U_1	U_2	U_3	U_4	U_5	U_6	U_7	U_8	U_9	$\overline{U_{(1-9)}}$	Accuracy(%) ↓
U_1	531	8	10	4	7	7	4	8	6	8	89.54
U_2	7	527	5	5	6	8	7	7	8	7	89.78
U_3	7	7	527	6	6	5	7	7	8	8	89.63
U_4	8	10	7	538	4	5	8	7	8	8	89.22
U_5	7	8	6	5	529	8	6	9	7	4	89.81
U_6	8	8	9	7	2	531	7	8	5	5	90.00
U_7	3	6	5	8	7	9	545	9	9	8	89.49
U_8	5	8	10	7	6	5	7	542	8	7	89.59
U_9	9	8	8	7	4	3	9	6	540	9	89.55
$\overline{U_{(1-9)}}$	6	7	7	8	7	6	8	3	6	526	90.07
					Balanced accuracy →						89.67

Table 6.12   Summary of results achieved on NITRLipV1 and NITRLipV2 databases

Classes ↓	Database ↓	Balanced accuracy (%)	Precision (%)	Recall (%)	Error rate (%)
Three	NITRLipV1	96.76	96.74	96.76	3.24
	NITRLipV2	98.83	98.83	98.83	1.17
	NITRLipV1+NITRLipV2	97.89	97.89	97.89	2.11
Five	NITRLipV1	95.83	95.56	95.83	4.17
	NITRLipV2	93.06	93.73	93.06	6.94
	NITRLipV1+NITRLipV2	95.25	95.26	95.25	4.75
Ten	NITRLipV1	90.91	91.03	90.91	9.09
	NITRLipV2	89.87	89.87	89.87	10.13
	NITRLipV1+NITRLipV2	89.67	89.67	89.67	10.33

(two users) is approximately 99%, it falls to approximately 90% when ten classes (nine users) are considered. From the confusion matrices, it is evident that this fall of accuracy of 10% is due to the increase in number of classes. As this research focuses on use of handheld device by limited number of users, this limitation of scalability is not an issue. This approach can satisfactorily produce performance in considered situation. For practically using this methodology, any template replacement algorithm can be embedded into the biometric system to overcome slight challenges faced due to seasonal change of lip.

References

[1]   Jain AK, Flynn P, and Ross A. Handbook of biometrics. Secaucus, NJ, USA: Springer-Verlag New York, Inc.; 2008. ISBN: 978-0-387-71040-2.

[2]   Tsuchihashi Y. Studies on personal identification by means of lip prints. Forensic Science. 1974;3:233–248. DOI: 10.1016/0300-9432(74)90034-X.

[3]   Suzuki K, and Tsuchihashi Y. Personal identification by means of lip prints. Journal of Forensic Medicine. 1970;17:52–57.

[4]   Coward RC. The stability of lip pattern characteristics over time. Journal of Forensic Odontostomatology. 2007;25(2):40–56.

[5]   Gondivkar SM, Indurkar A, Degwekar S, and Bhowate R. Cheiloscopy for sex determination. Journal of Forensic Dental Sciences. 2009;1(2):56–60. DOI: 10.4103/0974-2948.60374.

[6]   Kasprzak J, and Leczynska B. Cheiloscopy–human identification on the basis of lip prints. Warsaw, Poland: CLK KGP Press; 2001.

[7]   Lievin M, and Luthon F. Lip features automatic extraction. In: Proceedings of International Conference on Image Processing; 1998. p. 168–172. DOI: 10.1109/ICIP.1998.727160.

[8]   Leung Sh, Wang SL, and Lau WH. Lip image segmentation using fuzzy clustering incorporating an elliptic shape function. IEEE Transactions on Image Processing. 2004;13(1):51–62. DOI: 10.1109/TIP.2003.818116.

[9]   Hsu CY, Yang CH, Chen YC, and Tsai Mc. A PSO-SVM lips recognition method based on active basis model. In: Proceedings of the 2010 Fourth International Conference on Genetic and Evolutionary Computing; 2010. p. 743–747. DOI: 10.1109/ICGEC.2010.188.

[10]   Wang SL, and Liew AWC. Physiological and behavioural lip biometrics: a comprehensive study of their discriminative power. Pattern Recognition. 2012;45(9):3328–3335. DOI: 10.1016/j.patcog.2012.02.016.

[11]   Bakshi S, Raman R, and Sa PK. Lip pattern recognition based on local feature extraction. In: Annual IEEE India Conference; 2011. p. 1–4. DOI: 10.1109/INDCON.2011.6139357.

[12]   Lowe DG. Distinctive image features from scale-invariant keypoints. International Journal of Computer Vision. 2004;60(2):91–110. DOI: 10.1023/B:VISI.0000029664.99615.94.

[13]   Bay H, Ess A, Tuytelaars T, and Gool LV. Speeded-up robust features (SURF). Computer Vision and Image Understanding. 2008;110(3):346–359. DOI: 10.1016/j.cviu.2007.09.014.

[14]   Choras M. The lip as a biometric. Pattern Analysis and Applications. 2010;13(1):105–112. DOI: 10.1007/s10044-008-0144-8.

[15]   Choras M. Emerging methods of biometrics human identification. In: Proceedings of Second International Conference on Innovative Computing, Information, and Control; 2007. p. 365–365. DOI: 10.1109/ICICIC.2007.283.

[16]   Smacki L, Porwik P, Tomaszycki K, and Kwarcinska S. The lip print recognition using Hough transform. Journal of Medical Informatics & Technologies. 2010;14:31–38.

[17]   Smacki L, and Wrobel K. Lip print recognition based on mean differences similarity measure. In: Burduk R, Kurzynski M, Wozniak M, Zolnierek A, editors. Computer Recognition Systems 4. Vol 95 of Advances in Intelligent and Soft Computing; 2011. p. 41–49. DOI: 10.1007/978-3-642-20320-6_5.

[18] Kim JO, Lee W, Hwang J, Baik KS, and Chung CH. Lip print recognition for security systems by multi-resolution architecture. Future Generation Computer System. 2004;20(2):295–301. DOI: 10.1016/S0167-739X(03)00145-6.

[19] Wrobel K, and Doroz R. Method for identification of fragments of lip prints images on the basis of the generalized Hough transform. Journal of Medical Informatics & Technologies. 2013;22:189–194.

[20] Wrobel K, Doroz R, and Palys M. A method of lip print recognition based on sections comparison. In: Proceedings of the 2013 International Conference on Biometrics and Kansei Engineering; 2013. p. 47–52. DOI: 10.1109/ICBAKE.2013.10.

[21] Jain AK, and Hong L, Y K. A multimodal biometric system using fingerprint, face and speech. In: Proceedings of the Second International Conference on Audio and Video-based Biometric Person Authentication; 1999. p. 182–187.

[22] Ichino M, Sakano H, and Komatsu N. Multimodal biometrics of lip movements and voice using kernel Fisher discriminant analysis. In: Proceedings of 9th International Conference on Control, Automation, Robotics and Vision; 2006. p. 1–6. DOI: 10.1109/ICARCV.2006.345473.

[23] Lievin M, Delmas P, Coulon PY, Luthon F, and Fristol V. Automatic lip tracking: Bayesian segmentation and active contours in a cooperative scheme. In: Proceedings of IEEE International Conference on Multimedia Computing and Systems; 1999. p. 691–696. DOI: 10.1109/MMCS.1999.779283.

[24] Ooi WC, Jeon C, Kim K, Han DK, and Ko H. Effective lip localization and tracking for achieving multimodal speech recognition. In: Proceedings of IEEE International Conference on Multisensor Fusion and Integration for Intelligent Systems; 2008. p. 90–93. DOI: 10.1109/MFI.2008.4648114.

[25] Pocovnicu A. Biometric security for cell phones. Informatica Economica. 2009;13(1):57–63.

[26] Saevanee H, and Bhatarakosol P. User authentication using combination of behavioural biometrics over the touchpad acting like touch screen of mobile device. In: Proceedings of International Conference on Computer and Electrical Engineering; 2008. p. 82–86. DOI: 10.1109/ICCEE.2008.157.

[27] Franzgrote M, Borg C, Tobias Ries BJ, *et al.* Palmprint verification on mobile phones using accelerated competitive code. In: Proceedings of International Conference on Hand-Based Biometrics; 2011. p. 1–6. DOI: 10.1109/ICHB.2011.6094309.

[28] Choras M, and Kozik R. Contactless palmprint and knuckle biometrics for mobile devices. Pattern Analysis and Applications. 2012;15(1):73–85. DOI: 10.1007/s10044-011-0248-4.

[29] Shabeer HA, and Suganthi P. Mobile phones security using biometrics. In: International Conference on Computational Intelligence and Multimedia Applications; 2007. p. 270–274. DOI: 10.1109/ICCIMA.2007.182.

[30] Tresadern P, Cootes TF, Poh N, *et al.* Mobile biometrics: combined face and voice verification for a mobile platform. IEEE Pervasive Computing. 2013;12(1):79–87. DOI: 10.1109/MPRV.2012.54.

[31] Choras RS. Lip-prints feature extraction and recognition. In: Choras RS, editor. Image Processing and Communications Challenges 3. Vol 102 of Advances in Intelligent and Soft Computing; 2011. p. 33–42. DOI: 10.1007/978-3-642-23154-4_4.

[32] Choras M. Lips recognition for biometrics. In: Tistarelli M, Nixon MS, editors. Advances in Biometrics. Vol 5558 of Lecture Notes in Computer Science; 2009. p. 1260–1269. DOI: 10.1007/978-3-642-01793-3_127.

[33] Gomez E, Travieso CM, Briceno JC, and Ferrer MA. Biometric identification system by lip shape. In: Proceedings of 36th Annual International Carnahan Conference on Security Technology; 2002. p. 39–42. DOI: 10.1109/CCST.2002.1049223.

[34] Lai JY, Wang SL, Shi XJ, and Liew AWC. Sparse coding based lip texture representation for visual speaker identification. In: Proceedings of 19th International Conference on Digital Signal Processing; 2014. p. 607–610. DOI: 10.1109/ICDSP.2014.6900736.

[35] Wang SL, and Liew A. ICA-based lip feature representation for speaker authentication. In: Proceedings of Third International IEEE Conference on Signal-Image Technologies and Internet-Based System; 2007. p. 763–767. DOI: 10.1109/SITIS.2007.37.

[36] Chan CH, Goswami B, Kittler J, and Christmas W. Local ordinal contrast pattern histograms for spatiotemporal, lip-based speaker authentication. IEEE Transactions on Information Forensics and Security. 2012;7(2):602–612. DOI: 10.1109/TIFS.2011.2175920.

[37] Goswami B, Chan CH, Kittler J, and Christmas B. Local ordinal contrast pattern histograms for spatiotemporal, lip-based speaker authentication. In: Fourth IEEE International Conference on Biometrics: Theory Applications and Systems; 2010. p. 1–6. DOI: 10.1109/BTAS.2010.5634469.

[38] Wang SL, Leung SH, and Lau WH. Lip segmentation by fuzzy clustering incorporating with shape function. In: Proceedings of IEEE International Conference on Acoustics, Speech, and Signal Processing. Vol. 1; 2002. p. I-1077–I-1080. DOI: 10.1109/ICASSP.2002.5743982.

[39] Bakshi S, Raman R, and Sa PK. NITRLipV1: a constrained lip database captured in visible spectrum. ACM SIGBioinformatics Record. 2016;6(1):2–2. DOI: 10.1145/2921555.2921557.

[40] Raman R, Sa PK, Majhi B, and Bakshi S. Acquisition and corpus description of a constrained lip database captured from handheld devices: NITR-LipV2 (MobioLip). ACM SIGBioinformatics Record. 2017;7(1):2–2. DOI: 10.1145/3056351.3056353.

Chapter 7

Mobile device usage data as behavioral biometrics

*Tempestt J. Neal[1], Damon L. Woodard[2],
and Aaron D. Striegel[3]*

7.1 Introduction

Mobile devices house extremely private and sensitive information. Most, if not all, modern smartphones are designed to mimic the functionality of computers while remaining mobile, small, and computationally fast. These devices are often used for banking, scheduling, social networking, education, and, of course, many forms of communication. Recent statistics[4] suggest a growing dependency on mobile technology. For instance, 64 percent of adults own a smartphone, and 7 percent use their devices for online activities due to no at-home broadband access. More than 55 percent of smartphone owners use their device to research a health condition, for mobile banking, sharing community events, or driving navigation [1]. As these devices become increasingly complex, the monetary value of these devices also increases, further encouraging theft of the device or the information it holds. It is reported that malware increased by 58 percent from 2011 to 2012, 32 percent of which was used for obtaining information about the owner [2]. It is, therefore, important to secure these devices to ensure that this information is not accessed by individuals with malicious intentions.

Common security measures on mobile devices include personal identification numbers (PINs), alphanumeric passwords, and/or graphical passwords. Unfortunately, users may forget the password, use it for multiple accounts, write it down, or have to contact customer service to recover or reset a password. Additionally, shoulder surfing attacks (i.e., an intruder retrieving the password through direct observation) are possible, along with the inconvenience of entering the password multiple times [3]. Due to these disadvantages, researchers are now exploring biometrics, or the

[1]Department of Computer and Information Science and Engineering, University of Florida, USA
[2]Department of Electrical and Computer Engineering, University of Florida, USA
[3]Department of Computer Science and Engineering, University of Notre Dame, USA
[4]Based on information obtained from a U.S. population through telephone surveys and the American Trends Panel in 2014.

authentication of individuals through physiological or behavioral traits, as a more reliable and convenient means to securing mobile devices.

Modern mobile devices include several refined sensors capable of capturing high-quality data samples that can be fed into embedded or offline biometric systems. Hence, manufacturers are now incorporating face and fingerprint recognition technologies as biometric security solutions [4,5]. There are pitfalls in these deployments, however, particularly in usability. In a recent study [6], fingerprint recognition was voted as faster and more accurate compared to the face unlocking service, even though wet fingers caused problems in fingerprint recognition and face recognition was time-consuming and hindered in low-light conditions. Both methods are susceptible to spoofing via photographs and fingerprint molds, but the study also revealed that most users were unaware of possible security breaches. Beyond that, researchers must properly manage additional hardware, account for maintenance costs, support high computational processes in resource-limited devices, deal with the vulnerability of traits to injury and natural development, and cope with the possibility of not being able to offer such services to users with physical disabilities [3].

Given the shortcomings of passwords and physical biometrics [7], researchers are now exploring behavioral biometrics. Behavioral biometrics identify individuals based on patterns of behavior and generally do not require additional hardware. Behavioral systems allow for continuous verification (i.e., active authentication), whereas physiological systems can only authenticate users at the point-of-entry. In other words, in a physiological-based biometric system, a user must initially identify himself by presenting a physical trait to access a banking application. Once the user's identity is validated, he can access his financial records, and consequently the device as a whole. The device is now susceptible to attack, even though the user has been verified as the owner of the device. In contrast, a behavioral-based biometric system could continuously monitor the actions of the user as he interacts with the device beyond the point-of-entry, and take the appropriate action once an action (or a series of actions) is taken out of the norm. As a result, behavioral systems do not interrupt the natural flow of use of the device due to no required explicit use of the device's sensors or the need to input a password. Therefore, users may find a behavioral scheme more intuitive as they are able to interact with the device as they normally would.

Behavioral biometrics relies on consistency in user behavior as he or she interacts with a mobile device. Prominent trends in the device's usage, such as repetitive morning routines, could be extracted as biometric features, as they are the most consistent and unique to the user. The features effectively build a behavioral profile for the user, which is used to verify that the same user has possession of the device in later sessions. Profiles are created through the use of very common data types available in modern mobile devices, such as application usage or WiFi connectivity. Hence, a user's *usage data*, or the data gathered from the use of the mobile device, can be used for verification. Termed behavioral profiling, this method allows for continuous, unobtrusive identity verification based purely on behavioral data.

This chapter provides a broad overview of the advancements made on mobile platforms in regards to behavioral biometrics via usage data. Section 7.2 introduces the modules necessary for a complete biometric system. Sections 7.3 and 7.4 further elaborate on the data collection and feature extraction modules, respectively, as they relate to mobile device usage data. Section 7.5 provides an overview of related research literature. Finally, Section 7.6 discusses several research challenges in this area and Section 7.7 summarizes the chapter.

7.2 Biometric system modules

A typical biometric system consists of data collection, feature extraction, matching, and database modules. The data collection module involves the process of collecting the raw biometric trait from which the features will be extracted during the feature extraction module. On a mobile platform, data collection could incorporate the sensors manufactured into the device, such as cameras and microphones. In the context of utilizing usage data, the raw data would consists of activities carried out by the user, such as making a phone call or sending a text message. For example, Trung *et al.* detail the data collection process for capturing a large gait dataset through accelerometer sensors [8]. Formally, we define *usage data* as any trackable activity carried out in a normal fashion (thus excluding any direct and intentional use of hardware for verification purposes) on mobile devices. Usage data encompasses the output of the available sensors present in modern mobile devices or the response of the device as the user provides input. We loosely define mobile devices as the family of mobile, hand-held computing devices equipped with an operating system and mobile device specific protocols. In this chapter, cellular phones (e.g., smartphones) are the primary focus.

Feature extraction is the process of retaining the raw data values that will be used for identity verification. The data collection module could result in a user log similar to that in Table 7.1, which requires conversion into a format that can be quantitatively processed by a computer in later modules. Additionally, features are usually selected based on the results of a raw data analysis, which identifies prominent patterns in the data. Hence, the feature extraction module is responsible for properly converting the raw data into a quantitative format, and then retaining the values most distinct to the user. Once the desired features are extracted, the raw data can usually be discarded. The features create a behavioral profile for the device owner, which ultimately summarizes habitual actions.

Table 7.1   Sample user log

Application name	Date	Time
Scientific Calc	January 1, 2000	8:00
Car Race	January 1, 2000	8:39
How To	January 1, 2000	10:11
ChatTalk	January 1, 2000	12:35
CookGlutenFree	January 1, 2000	13:24
Car Race	January 1, 2000	14:01

The matching module typically consists of three different forms of classifiers that either confirm or deny that features were obtained from the same individual. A Type I classifier outputs a single class as the label for a new feature set. A nearest-neighbor classifier is a Type I classifier, as it assigns the label associated with the nearest training sample. Type II classifiers assign a score that corresponds with a confidence measure, or probability, of a class being the label of the new feature set. For instance, Bayes classification rule assigns a conditional probability to each class, and chooses the label of the class with the highest probability. A Type III classifier assigns a ranking, $1, 2, \ldots, N$, to each class. This classifier could simply order the output of a Type II classifier according to the most likely class to the least, and replace the score with the rank. Refer to [9] for further elaboration on classifier types in biometric applications.

7.3   Data collection

Data collection represents the process for gathering or more aptly instrumenting the device for the purpose of enabling feature extraction. As noted earlier, there exists a wide variety of sensors available for data collection on the smartphone ranging from *environmental* to *movement* to *location* to *interactive* (see Table 7.2 for sensor information). In the case of *environmental* sensing, a variety of sensors can extract information about the local environment, including aspects such as temperature, pressure, audio, or video, via the on-board smartphone sensors. For the case of *positional* sensing, on-board sensors capture either local signals with regards to movement (accelerometer for movement, magnetometer for direction, gyroscope for orientation) or wider range location signals. Location sensors attempt to determine either absolute (Global Positioning System (GPS), cellular/WiFi triangulation) or relative (Bluetooth beacons) locations of the smartphone user. Finally, there exists a fourth class of sensing extracted from how the user utilizes the smartphone and its respective applications. Interaction data can include coarse data, such as the phone or screen being turned on/off, applications being invoked, or the data usage of said applications. Finer-grained interaction data can include individual messages (SMS, iMessage) to interactive sessions (Skype) to even security interactions (permission granting, login attempts) and application usage (screens navigated or intra-application usage).

A further complication arises with regards to the vast array of sensors from the variety of timescales by which a sensor or instrument gathers data or can be configured to gather data. The sensors themselves might provide data *periodically* across a variety of configurable timescales or *aperiodically* via callbacks when important events occur. Instrumentation may possess OS-specific application programming interfaces (APIs) or one may need to periodically poll such data on the smartphone. For example, nearly all movement-based sensors may be configured across a variety of sampling rates ranging from tens of times per second (e.g., 60 Hz for an accelerometer) to only once per second (e.g., 1 Hz for the magnetometer for direction). Similarly, environmental sensors may range in terms of their periodicity, such as the temperature sensor versus the microphone for audio capture. Other data may need to be constructed

Table 7.2   Sensors found in modern mobile devices

Sensor	Description
Accelerometer	Positional sensor that measures velocity changes. Determines device orientation in three dimensions. Commonly used for motion detection.
Gyroscope	Positional sensor for obtaining orientation information at a higher precision in comparison to the accelerometer. Commonly used for rotation detection.
Magnetometer	Positional sensor that measures the strength of the magnetic field in three physical axes. Commonly used in compasses and metal detection applications.
Barometer	Environmental sensor that measures atmospheric pressure to determine how high the device is above sea level. Improves GPS accuracy.
Proximity	Positional sensor that measures the user's physical position in relation to the screen. Normally used during calls to determine if the device is held to the ear for automatic screen shut-off. Uses invisible infrared LED reflections off nearby objects.
Light Sensor	Environmental sensor that measures the ambient light level for automatic adjustments to the display's brightness.
Touchscreen	Input device layered on the display of the device to allow user input and control of the device via touch gestures.
Camera	Front and back cameras capture high-quality images and video.
Near Field Communication	A set of protocols that allow communication between two devices that are within near proximity of each other. Used in commerce, gaming, and social networking.
Cellular	Positional sensors via tower triangulation and proximity. Efficiently divides the finite radio frequency spectrum among multiple users.
Bluetooth	Positional wireless technology standard for exchanging data over short distances from fixed and mobile devices, connecting several devices at a time.
WiFi	Provides positional information via a local area wireless computer networking technology that allows connection to a network resource (e.g., the Internet) via a wireless network access point.
GPS	Positional navigation device that calculates geographical location via GPS satellites. Used in maps and turn-by-turn navigation.
Thermometer	Environmental sensor responsible for monitoring the internal temperature of the device. Prevents overheating of the device.
Pedometer	Uses the accelerometer to count the steps of the device user.
Microphone	Used to capture high-quality sound, such as voice during calls.

periodically, such as periodically scanning for nearby Bluetooth devices or periodically recording the amount of data as sent by WiFi or cellular. Finally, significant data streams with biometric information are often aperiodic and find their roots in OS-level callbacks. Examples of such callbacks include notifications, screen activation or de-activation, location changes, or wireless roaming.

From the perspective of a prospective biometric system, the key challenge after determining what data are viable to capture is balancing the granularity of said data with the energy costs associated with the data for both gathering and analysis.

For nearly all consumer devices, developers are largely constrained to work within the public API afforded by the operating system as dominant platforms now require at least a cursory application analysis before the application can be published for widespread availability. While approaches, such as MDM (Mobile Device Management) or MAM (Mobile Application Management), such as AirWatch, afford the distribution of non-public applications that can utilize private APIs, backward compatibility and even inter-patch compatibility may not be guaranteed. One alternative for a researcher is to adapt a platform-style approach, whereby the entire mobile device ROM is replaced with a customized deployment, such as espoused by the NetSense [10] and PhoneLab [11] studies. Most likely though, many applications tend to operate only with the public API through a user-level application, including the studies by the Massachusetts Institute of Technology (MIT) Reality Lab [12], the Copenhagen Networks study [13], and the NetHealth study.

Furthermore, while certain portions of the data are relatively robust and stable, much of the data as gathered by the instrumentation process will be exceptionally noisy. For instance, location data can vary dramatically in terms of precision ranging from accuracy at times on the order of meters (e.g., rich WiFi triangulation and Bluetooth beacons) to the order of hundreds of meters (limited GPS, only cellular triangulation). Moreover, the sampling rate as selected for various movement sensing may need to compromise in order to achieve energy balance but yet sacrifice the fidelity needed for accurate features. Hence, while smartphones possess a rich variety of sensors capable of gathering a wide sampling of data, the appropriate tuning and evaluation of said sensors is a decidedly non-trivial task.

7.4   Feature extraction

7.4.1   Name-based features

Name-based features are nominal values that typically include processes opened, closed, and controlled via interaction with the touchscreen. For instance, application downloads and installations, web browsing activity, and text input are likely name-based features as the corresponding nominal value reflects the name of the visited entity, such as a website's URL. However, a property of a suitable biometric trait is its ability to allow for quantifiable measurements. Hence, these features may require conversion into a numeric representation that, unfortunately, may have no sense of order and cannot be used in mathematical formulas (e.g., computing sums and averages) (refer to [14] for distance measures using nominal values). Consequently, researchers have resorted to categories, frequencies, and sequences to represent name-based features.

Categories Categorical representation groups a large variety of nominal values. Consider a feature vector of 10 mobile applications, labeled A through J. Applications A and D may be mapped to category phone; likewise, the remaining applications could belong to category messaging. Hence, the number of nominal variables is reduced to a broader categorical representation. Both Bassu *et al.* [15] and

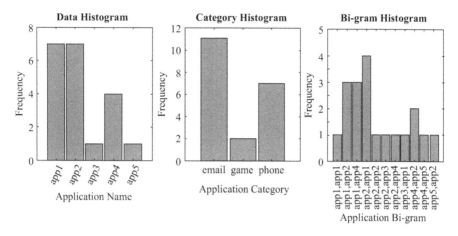

Figure 7.1   Graphical representation of nominal usage data features

Branscomb [16] use a similar categorical representation for application traffic. There is the possibility of information loss in this approach, however. For instance, exact values may provide more discriminating information about the user in comparison to concluding that the user's action(s) occurs in larger groups.

Frequencies Frequency counts provide insight into how often a certain task is performed. Through counting the data elements, percentages can be computed for various analytic conclusions. For example, a certain application could account for 50 percent of the data for half of the users in a dataset. Chi-square statistical tests also incorporate the use of frequencies, and could be useful in analyzing raw usage data for observing prominent trends and patterns. Likewise, examining frequencies during certain times or days could be useful. Consider the case that users are most active on mobile devices during work hours; counting the number of actions during this time frame could confirm or deny this hypothesized usage pattern. Neal *et al.* [17] employ a frequency model for application, Bluetooth, and WiFi traffic by counting the most visited entities overall and during certain time frames. Likewise, Fridman *et al.* [18] use the frequencies of application and web browsing traffic as features.

   Frequencies are also easy to visualize (refer to Figure 7.1). Visualization of the data is beneficial for the researcher's audience. Graphs are sometimes easier to understand compared to textual descriptions, and they provide support for the interpretation of experimental results.

Sequences The order in which data values appear could be extracted if it is suspected that a user performs certain actions in a particular order. N-grams (the representation of *n* consecutive terms in a sequence) are a common representation of text in stylometry. For instance, Brocardo *et al.* [19] study short messages as a biometric by evaluating 3-gram, 4-gram, and 5-gram features and Fridman *et al.* [18] represent text features as *n*-grams. Researchers could use *n*-grams to determine the chances of observing one value immediately following or preceding another. Dependencies among the features are more obvious using *n*-grams, and prominent consecutive actions could be further evaluated for positive contribution to discriminating users.

7.4.2   Positional features

Positional features rely on the assumption that the user is moving, and therefore, attempt to model that movement. Positional data will likely serve complimentary to usage data, as some may argue that these measurements are not strictly usage data; the user does not have to be interacting with the device at all, but can instead only be carrying the device at the time of data capture. However, positional features define the user's behavior in terms of location or motion, and therefore, fit well into an efficient behavioral profile.

Location features are derived from the output of positional sensors, such as GPS, that provide information on the geographical location of the user. For instance, a large majority of users may have a consistent travel pattern going to and leaving from work. Deviations from these travel patterns could indicate an intrusion. On the other hand, motion features aim to model a user's bodily movements to detect human activities. Walking, running, and even arm swing are motion activities that are trackable via the built-in family of sensors. Researchers may discover correlations between a user's activity and the way in which the device is handled. For instance, some users may pace back and forth while on phone calls. Some users may always carry their phone in a backpack, while others prefer a holster. These are all situations that the sensors can detect. In the event that an activity is completed out of the norm, a biometric implementation could take further action to properly secure the device. In summary, while location features aim to determine the *whereabouts* of the user, motion features aim to determine the *what and how* of the user's activities. An answerable question by accelerometer measurements could be simply is the user walking and if so, how fast. The biometric service's task is then to determine if the user's current walking pattern aligns with the usual walking pattern.

Figure 7.2 models three groups of location solutions (satellites, radio frequency (RF) signals, and sensors) on mobile platforms [20]. Satellite technologies are typically responsible for outdoor positioning due to wide coverage and high accuracy. Specifically, GPS has been extended over the years to include multiple satellite systems for positioning. Additionally, recent smartphones now have two satellite technologies available, GPS and GLONASS (GLObal NAvigation Satellite System), increasing the performance of satellite solutions. In some devices, assisted GPS (A-GPS) also boosts GPS performance via additional resources. RF solutions are now standard mobile device (particularly, smartphones) features. Service providers offer connectivity to network resources via cellular networks (GSM/CDMA cells), and WiFi technologies provide access to online resources. RF technologies may be more useful in determining location indoors when satellite solutions are unavailable. Lastly, embedded sensors, such as accelerometers, provide additional location awareness when the former technologies are weak. For example, accelerometers can

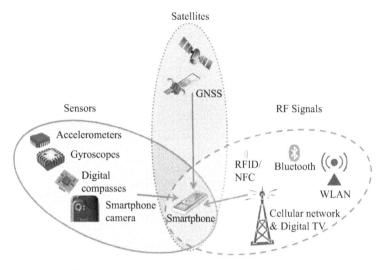

Figure 7.2   Mobile device location solutions. [Reproduced by permission of [20]]

calculate the device's speed. However, sensor technologies provide a lesser sense of geographical location, but provide more insight into the user's movement, such as walking and running. There are advantages to each of the three methods; hence, there have been hybrid approaches, as explored in [20].

7.4.2.1   Location features

The most common location solutions on mobile platforms studied throughout the research literature for biometric purposes are GPS, GSM/CDMA, and WiFi.

GPS   Fridman *et al.* [18] use longitude and latitude pairs as GPS features. According to the authors, GPS features provide location data at higher resolutions in time and space compared to GSM/CDMA features. However, during data collection, GPS sampling is reduced to prolong the battery life of the device, implying excessive battery drain caused by GPS technologies. Pei *et al.* [20] evaluate a hybrid approach of location detection that includes GPS features and also notice the issue of battery drain when employing GPS, as shown in Figure 7.3.

GSM/CDMA Cell   An approach introduced by Shi *et al.* [21] addresses the issue of battery drain by capturing data from multiple sensors in an adaptive manner. Initially, only low power sensors are utilized, and if necessary, more power-consuming sensors are included to reach the desired performance level. The authors use cell IDs to obtain location information. Unique cell ID (four-part data entry, specified as mobile country code, mobile network code, local area code, and cell ID) sequence patterns learned over a sliding window are extracted as features. Compared to WiFi and GPS, cell IDs are available at no extra cost, offer coarse location information, and are more energy efficient. Additionally, cell ID sequences are consistent on a daily basis and are more discriminating than single cell IDs.

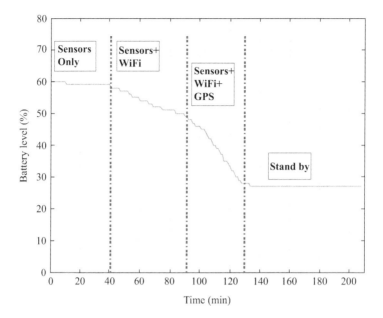

Figure 7.3   Battery drain on a smartphone starting with 60 percent of battery power. Significant battery drain occurs once GPS is turned on. [Reproduced by permission of [20]]

WiFi WiFi features have been extracted similarly to name-based features. Neal *et al.* [17] extracted the names of the 10 most visited WiFi access points as features. Fridman *et al.* [18] also use WiFi as a location feature when users are indoors (GPS features are used outdoors), but specific WiFi features are not discussed. However, the authors note that location features encompass both GPS and WiFi; therefore, we assume that GPS and WiFi features are the same (i.e., latitude/longitude pairs).

7.4.2.2   Motion features

Gait Gait recognition, known to reveal human characteristics, such as age, gender, and physical disabilities, is a common transparent approach to motion detection on mobile platforms via accelerometer measurements. Particularly, the accelerometer measures acceleration in three spatial dimensions, x, y, and z. Researchers are able to use this information to track and process an individual's gait information as he or she walks with an attached device. Figure 7.4 shows accelerometer readings from a medical study [22] that attempts to distinguish healthy patients from those diagnosed with Parkinson's disease, where each line corresponds with a dimension. Gait, however, is influenced by several physical and environmental conditions, such as fatigue, intoxication, clothing, ground conditions, and location of the device on the user.

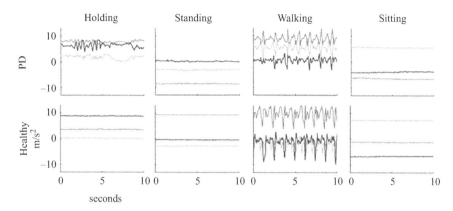

Figure 7.4   Accelerometer readings for a healthy patient versus a patient diagnosed with Parkinson's disease. [Reproduced by permission of [22]]

There are typically four phases in a gait-based behavioral biometric system: data collection, preprocessing, cycle detection, and analysis. During data collection, users perform an action while carrying the device in a natural manner, such as on a belt, inside of a holster, or in a pocket, as accelerometer measurements are recorded. During the preprocessing phase, linear interpolation creates equal intervals between samples, and filtering reduces any additional noise due to misplacement of the device, screen taps, or automatic vibrations. A cycle is the time between two steps, which are identified by local maximums and minimums during cycle detection. Finally, the analysis phase processes the cycles in both time and frequency domains (it is claimed in [23] that a user's typical walking frequency is 2 Hz). The statistics of the cycles, such as average cycle length and frequency, are studied. Additionally, acceleration moments are evaluated to reveal information about the distribution of the samples. The cyclic rotation metric (each cycle of a probe set is repetitively matched against an enrolled set to provide a distance metric between the two) determines if two gait patterns belong to the same individual according to an established threshold [24]. Hence, cycle components are extracted as features.

However, noncycle-based approaches have been studied as well. Machine learning approaches are also considered and have resulted in lower error rates [25]. Kwapisz *et al.* [26] define the following as features extracted from accelerometer output:

- average acceleration,
- standard deviation,
- average absolute difference between the readings,
- average resultant acceleration, defined as $\sqrt{x_i^2 + y_i^2 + z_i^2}$, where x, y, and z are the dimensions of the accelerometer readings,
- time between sinusoidal wave peaks, and
- binned distribution of each axis.

Nickel, Wirtl, and Busch [27] extract similar features and compute the minimum, maximum, mel-frequency cepstral coefficients (MFCCs), and bark-frequency cepstral coefficients (BFCCs) as additional features (refer to Reference [28] for more information on MFCCs and BFCCs). The authors find that combining the BFCC feature from all axes with a magnitude vector yielded the best results when users walked normally in a test environment. Alternatively, Vildjiounaite *et al.* [29] employ a simpler method by normalizing and averaging right-foot and left-foot steps separately to create a feature template that models the shape of each signal.

Activity recognition Accelerometers, magnetometers, and gyroscopes have been combined to recognize certain activities, such as standing versus descending stairs. In this case, feature extraction depends on the identified activity. Hence, researchers have included activity recognition prior to feature extraction. For instance, Shi *et al.* [21] use the JigSaw pipeline [30] that can identify the user as stationary, while walking, cycling, running, or in a vehicle. Once the activity is identified, the usual accelerometer features, such as mean, standard deviation, and spectral peaks, are extracted. Similarly, Shoaib *et al.* [31] attempt to identify walking, running, sitting, standing, and climbing up and down stairs; however, the authors take into account accelerometer, gyroscope, and magnetometer measurements. After adding the magnitude of the axes as a fourth dimension to each sensor, the authors simply extract the mean and standard deviation from each sensor as features. Pei *et al.* [20] also evaluated all three sensors to recognize sitting, walking normally and quickly, standing, and sharp and gradient turns. Features included the mean, variance, median, interquartile range, skewness, kurtosis, and calculations among the dominant frequencies, such as amplitude.

7.4.3 Touch features

Touch features are extracted from data values that are a result of touch gestures on the device's screen. Xu *et al.* [32] identify four common touch gestures:

1. *Keystroke*: Finger tap on the screen, usually via soft keyboard.
2. *Slide*: Finger moves across the screen while maintaining contact.
3. *Pinch*: Two-finger gesture, usually used in zooming.
4. *Handwriting*: Alternative input method, sometimes done with a stylus.

Several touch gesture features can be captured as the user presses the screen [33]. Basic features include:

- *x/y-coordinates* of the finger position.
- *Pressure* of the object (i.e., finger) as it presses the screen.
- *Size* of the object pressing the screen.
- *Time* of event occurrence.
- *Acceleration*, computed by squaring the time between an action up and down event.
- *Dwell/Hold time*, or the duration of the keystroke (time from pressing to releasing a key) [32,34].

- *Flight time*, or the time interval between keystrokes [32].
- Handwriting features have reflected forensic applications that typically use the arrangement, slant, alignment, and design of characters [32].

Some researchers have adopted more complex approaches for feature extraction. Xu, Zhou, and Lyu [32] extract trajectory features of each finger, along with features that define the correlation of the two fingers as pinch gesture features. Similarly, trajectory features that represent the directional information of a slide gesture are extracted as slide gesture features. Shi *et al.* [21] extract the least square linear gradient, the angle between two gradient values, and the distance between fingers as multi-touch features. During experimentation, the authors notice that no two users shared the exact spread and pinch touch gestures.

Both Cai *et al.* and Gascon *et al.* employ motion sensors for obtaining touch gesture data as the soft keyboard of mobile devices uses these sensors for acceleration, torque, and device position measurements. Gascon *et al.* [35] use a time-based feature extraction method, wherein a data point is computed every T seconds while typing. Accelerometer, gyroscope, and orientation sensors are each measured in three dimensions, where orientation is usually derived from accelerometer data. The nine values are normalized by subtracting the mean, linear, and 3-degree spline. Finally, 2,376 features are extracted from the normalized values and are clustered into 5 groups:

1. Simple Statistics (root mean square, mean, standard deviation, variance, maximum and minimum)
2. Spline Coefficients (5-degree smoothing spline coefficients)
3. Spline Simple Coefficients (5-degree smoothing spline simple statistics)
4. iFFT Spline Features (5-degree spline fit curve inverse DFT (discrete Fourier transform))
5. iFFT Signal Features (sensor signal inverse DFT)

Similarly, Cai and Chen [36] hypothesize that device vibrations as a result of keystrokes are strongly correlated with the typed key. Features related to the pitch angle (associated with rotation about the x-axis) are extracted from motion signals, where a motion signal consists of the time of capture and the normalized pitch and roll (associated with rotation about the y-axis) angles. Specifically, the angles between the direction shown above and below the x-axis and width of the pitch signals are extracted.

7.4.4   Voice features

Voice features aim to detect and model the human voice by filtering out environmental noise. In mobile devices, voice data are guaranteed primarily via phone calls, but also through voice-to-text and video chat applications. However, according to Jain *et al.* [37], the communication channel usually degrades the voice signal. Furthermore, changes in an individual's age and emotional and physical states lead to inconsistencies in voice signals. Hence, Vildjiounaite *et al.* [29] describe voice as a weak biometric and claims, for said reasons, that speech signals should be a part of a multimodal biometric system. Researchers have explored this multimodal approach, choosing the following components of the voice signal as features [21]: microphone input is divided into W frames and the energy entropy, signal energy, zero crossing rate, spectral rolloff, spectral centroid, and spectral flux are calculated as features. Statistical features are derived from these features, including the standard deviation and mean of energy entropy and the standard deviation by mean ratio of signal energy and spectral flux. Refer to reference [38] for further mathematical explanation.

7.5   Research approaches

Usage data offers many data types (e.g., application traffic, GPS coordinates, and text input). Reports from various news sources support the argument that each data type provides an exponential amount of information available to researchers for verification purposes. For instance, according to one news source, mobile users have access to over 1.4 billion and 1.2 billion applications in Google's Play Store and Apple's App Store, respectively [39]. Another source predicted that 47.7 million WiFi access points would be available by the close of 2014 [40]. Moreover, a 2015 article reported that the United States led in daily phone usage, spending an average of 4.7 hours per day on mobile devices and consuming 20 gigabytes of data per month [41]. Such alarming statistics suggest a growing attachment to mobile technologies; as the use of these devices become a part of everyday routines, the information available from how these devices are used increases. As a consequence, many researchers are studying behavioral biometrics on mobile platforms. This section describes a few of these approaches.

7.5.1   *Application traffic*

Evaluation of application traffic as a behavioral biometric trait is a logical approach; there is virtually a mobile application for every purpose, ranging from weight loss to shopping. Applications often reflect user interests and needs; hence, the hypothesis is that applications are distinct to the user that installs them. Researchers should be aware that like most behavioral traits, application usage is less likely distinct among a large user set; however, the differences between users may be more noticeable within a smaller set.

Bassu, Cochinwala, and Jain [15] model application traffic according to the application's name, device location (i.e., home, office, or away), time of access, mobility of the user, and device orientation after determining that application traffic follows a power-law distribution, and as a consequence, may be a poor stand-alone mobile biometric. Applications are categorized according to marketplace themes, time values are discretized to reflect the time of the day (i.e., morning, afternoon, and evening), and location values are categorized as static, slow, and fast. A Bayesian classifier is used for matching, which makes an independence assumption among features, with the exception of interdependent time and location values. Therefore, given feature vector $(X_1, \ldots, X_n, T, L)$, where X_i, T, and L, represent application

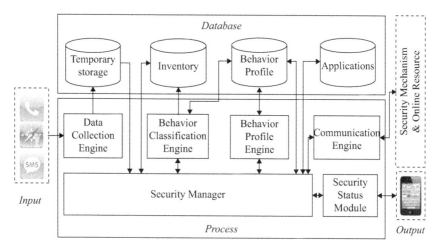

Figure 7.5 A novel behavior profiling framework [42]

usage, time, and location features, respectively, and prior probability, $P(U)$, the joint probability is computed as

$$P(X_1, X_2, \ldots, X_n, T, L, U) = P(U) \times P(T, L \mid U) \times \prod_{i=1}^{N} P(X_i \mid T, L, U).$$

Additionally, Laplace smoothing and random forests are used to ensure a non-zero Bayesian probability and stable estimates. The authors propose a background service to record and extract the features from mobile devices. The service is responsible for modeling the user's data, maintaining an updated model of the user's data, and verifying the user as the owner of the device. Once a reliable user model is established on the back-end, it is pushed back to the user's device via the biometric service, which subsequently performs active authentication. Finally, because the service periodically updates the user's model parameters, there is no need to frequently uninstall and install the service. Additionally, any user model can be pushed to any device. Therefore, for performance evaluation, a model can be pushed to a user's device that did not create the model. With this flexibility, the user is now acting as an imposter, and the performance of the system can be explored. Hence, a self-updating mobile biometric service that records and quantizes application traffic is developed, making it useful in a continuous authentication scheme.

Similarly, Li *et al.* suggest a mobile biometric framework with several engines (refer to Figure 7.5[5]) that work interdependently for user verification [42]. Prior to

---

[5] *International Journal of Information Security*, Active authentication for mobile devices utilizing behavior profiling, vol. 13, 2014, pp. 229–244, Fudong Li, Nathan Clarke, Maria Papadaki, Paul Dowland, © Springer-Verlag Berlin Heidelberg 2013. With permission of Springer.

implementation and testing, however, the authors analyze application traffic (standard, telephone, and SMS applications) to determine which features provide the most discriminating contribution. Using the MIT Reality Mining Project dataset (refer to Table 7.3 for database descriptions[6]) and a similar application representation as proposed by Bassu *et al.*, the authors discover that each user shared only two phonebook application location values with other users on average. Likewise, each user shared one telephone number with another user. For standard applications, when the application names largely overlapped among users, the location value was useful. The duration of calls and the time of calls, however, had no discriminating properties. Once the most significant features were found, the authors determined that a rule-based classification approach provided the lowest error rates, given that this approach accounts for large intra-class variations (i.e., large changes in individual behavioral profiles). Specifically, an alarm is raised if

$$1 - \frac{\sum_{i=1}^{N}\left(\frac{Occurrence\ of\ Feature_{ix}}{\sum_{x=1}^{M} Occurrence\ of\ Feature_{ix}}\right)}{N} \geq threshold,$$

where i is the feature, x is the value of *Feature$_i$*, M is the total number of values for *Feature$_i$*, and N is the total number of features. Similar to [15], a dynamic behavioral profile is considered to account for behavioral changes from the valid user, and a smoothing function is introduced to make matching decisions based on several combined events. Finally, all information obtained from the preliminary steps help to develop a multi-engine architecture. The data collection engine captures the application traffic and extracts the relevant features. The behavior profile engine computes the user models based on historical data. The behavior classification engine handles the verification process, and the communication engine allows communication with the online back-end.

This multi-engine framework is subsequently tested in [43] through the development of background service, Sentinel. Sentinel collects the time of access, application name, and access location and then implements the rule-based classifier. Sentinel is optimized to account for the limited resources available in mobile devices, having a small memory requirement of 6 kilobytes. Sentinel's user acceptance is also examined; through a nine-question survey, the authors find that only 56.4 percent of the 55 participants used point-of-entry security methods (e.g., PIN and passwords). Perhaps, if an alternative, more convenient method is available, such as behavioral profiling, users will be more inclined to secure their devices. Interestingly, this hypothesis is strengthened, as the authors also find that 71 percent of the participants are in favor of the Sentinel approach.

Branscomb [16] proposes Mouflon, a service similar to Sentinel, that collects various application-related data, including access date and time, name, and category

---

[6]Table 7.3 summarizes the databases used in the discussed approaches from various researchers. Several data collection environments and protocols are seen, reflecting the lack of an evaluation standard in this area. Section 7.6 elaborates on this further.

Table 7.3   Database descriptions

Provider	Number of users	Data types	Description
Massachusetts Institute of Technology [42]	106	Applications, phone, SMS	Data were collected from September 2004 through June 2005. Noted as publicly available.
Drexel University [18]	200	Text, applications, websites, location	A tracking application was installed for at least 30 days and up to 5 months. Data were collected with a 1-s resolution. All users were students.
University of Notre Dame [17]	200	Applications, Bluetooth, WiFi	Data were collected from August 2011 through February 2013. Application usage records include user ID, timestamp, name of application, and bytes received/transferred. Bluetooth and WiFi connectivity records include user ID, timestamp, device name, MAC address, and RSSI strength. All users were students.
Enron [19]	150	E-mails	Contains over 200,000 e-mails. Two hundred average words per e-mail. E-mails are written in plain text and are both business-related and personal.
Norwegian Information Security Lab [44]	30	Arm swing	Twenty-three males and seven females, ages 19–47. Users walked normally for 20 meters. Sensor is attached to lower right arm during first two sessions and lower left arm during last two sessions. One hundred and twenty total samples are collected.
Fordham University [26]	36	Gait	Users were instructed to walk, jog, and climb up and down stairs while carrying a phone in front pants leg pocket. Accelerometer data were collected using a sampling frequency of 50 ms (20 samples per second).
Hochschule Darmstadt (CASED) [27]	36	Gait	Twenty-nine males and seven females, ages 20–49. Two sessions were approximately 24 days apart for each user. Users were instructed to walk on a flat hallway with a phone in a pouch on the right hip. Twenty-four normal walks and 16 fast walks are recorded, equaling to 30 min of walking data per user.
The Institute of Scientific and Industrial Research [8]	736	Gait	Users (382 males and 354 females, ages 2–78) walked twice within 9 m long straight lines with a sensor (included accelerometer, gyroscope, and compass into single device) centered on the back waist. 6D signal sequences were captured.

(Continues)

Table 7.3   (Continued)

Provider	Number of users	Data types	Description
Shenzhen Research Institute [32]	32	Touch gestures	Java-based application collected four touch gesture types over 15 min. Two hundred touch gesture sequences were collected from each user. Three users were selected for long-term (1 month) data acquisition.
Pervasive Systems [31]	4	Human physical activities	Fifty samples per second are collected from an accelerometer, magnetometer, and gyroscope while users walk, run, sit, stand, and ascend and descend stairs for 3–5 min indoors. Phones were placed in the right jean pocket, belt, right arm, and right wrist. Male users, ages 25–30. Noted as publicly available.
University of California, Davis [36]	1	Keystrokes	Three datasets (each have multiple sessions of 4–25 consecutive keystrokes) collected on a number-only soft keyboard in landscape mode on different days as the user interacted naturally with the device.
Computer Security Group [35]	315	Touch gestures	Users wrote short (160 characters), predefined passages on a soft keyboard in English. Three hundred and three users entered text once, 12 users entered text 10 times.

(audio, messaging, browser, settings, etc.). Like Sentinel's multiple interconnected engines, Mouflon has multiple components that work together. The MouflonRecorder extracts and uploads usage data from the device every 30 minutes. A WeeklyChecker module ensures that data are uploaded once a week. An upload service encrypts and delivers the data to the researchers. Once the researchers receive and decrypt the data, classifier, IBk, is used to authenticate six participants. IBk is considered a lazy classifier, operating mainly as a nearest-neighbor classifier. As new instances are presented to the system, the class of the nearest training instance is returned as the class of the new instance. For five of the six users, authorized and unauthorized users are accurately identified at least half of the time. However, Branscomb notes that it is important that the ratio of user instances to intruder instances in the training set is close to 1:1 to allow for sufficient modeling of both the user and intruder for efficient performance.

There are major similarities between these approaches. For instance, application traffic is mostly represented as a tuple of the application's name, time of access, and location of access. Including application categories is also common practice. However, there are a few weaknesses among these approaches. First, there is an assumption that a user's behavior is consistent. Second, the proposed services are trained on a single user; if multiple users use the device, the services would not be able to discriminate between multiple valid users and an intruder. Lastly, when analyzing performance, an intruder is represented by the other users in the database. Therefore, a true intruder model is needed; specifically, a phone thief (individual with intentions to steal the device) and information thief (individual with intentions to steal the information contained in the device) should both be evaluated.

7.5.2 Text

It is reported that the U.S. population sent more than 2 trillion text messages in 2011 [45]. Several third-party applications also allow SMS-like (e.g., chatting) communication capabilities. Considering how often a mobile device user inputs text, it is reasonable to consider it as a behavioral biometric. Related closely to stylometry, the core assumption is that writers have a distinct linguistic style that can be extracted for verification purposes. Common stylistic traits are word length, function words (e.g., context-free words), vocabulary distributions, and the results of content analyses. Stylometric features are typically characterized as character-level (alphabet, digits, white-space, punctuation, *n*-grams), lexical (vocabulary richness, misspellings), syntactic (parts of speech, sentence structure), or semantic (synonyms, semantic dependencies). The research literature is widely focused on instances that include a small set of authors and large documents. However, a 2008 study revealed that messages sent on mobiles by female participants were an average of 58 characters. Similarly, messages sent by males averaged 47 characters [46]. Text and chat messages are also typically poorly structured grammatically. Therefore, traditional features are likely unavailable in these messages and researchers are tasked with deriving alternate stylometric features.

Brocardo, Traore, Saad, and Woungang [19] attempt to address the issue of authorship verification with short messages. Although the authors do not extract actual text from mobile devices, they break down the Enron e-mail dataset into 250 and 500 character blocks and separately evaluate 3-gram, 4-gram, and 5-gram features. The proposed algorithm involves extracting a user's *n*-grams from a training set and classifying a new text block based on a predefined threshold. Specifically:

- Training
  1. Each users' training data is divided into two sets, namely training set and block set.
  2. All unique *n*-grams are found in the training set and the block set is broken into equal-character blocks.
  3. Unique *n*-grams are found in each block in the block set.
- Testing
  1. Given two sets of data, determine the percentage of unique *n*-grams shared by each block in the block set of one user and the training set of the other user.

2. Conclude that the data are from the same user if each percentage is greater than or equal to a varying threshold (threshold varies to achieve an equal error rate (EER) closest to zero percent).

Performance decreases as the number of authors increases (87, 92, and 107 authors were tested) and with the smaller block size, indicating the need for further studies primarily on mobile platforms. A 500-character block is not exactly representative of the message sizes on mobile devices. Additionally, only one feature type is evaluated, and alternative features could improve performance. However, this work is a positive step towards evaluating short messages for verification purposes.

7.5.3 Movement

Nickel, Wirtl, and Busch [27] study gait with respect to the machine learning algorithm, k-nearest neighbor (kNN) with Euclidean distance. Gait data were collected from 36 users that were instructed to walk normally and quickly on a flat carpet with the device carried on the users' right hips. The authors discover that it takes approximately 7 s for the data collection, feature extraction, and matching modules to complete on the mobile platform. They also note that features are dependent on the walking velocities, resulting in the need for activity recognition for identifying the velocity.

However, Kwapisz *et al.* [26] counter Nickel's notion by claiming that it is unnecessary to know the user's activity in order to verify the user. In essence, the authors extend Nickel's work by investigating several walking activities, and therefore, walking velocities. Thirty-six participants were instructed to carry phones in their pockets and perform four tasks: walking, jogging, and ascending and descending stairs. The authors choose decision trees and neural networks as the classifiers for the matching module and also investigate identification and authentication performance. Identification and authentication are often used interchangeably, but in some contexts these terms have different definitions. In an identification task, the biometric system attempts to associate an individual with an identity already known to the system. An authentication task attempts to verify that an individual is who he or she claims. As it relates to this chapter, we focus on authentication methods as we would like to explore actively, or continuously, authenticating a mobile device user. Nonetheless, though Kwapisz's implementation requires three additional seconds (10 s total) to authenticate users compared to Nickel's implementation, walking and jogging provided the best results. Walking is considered the ideal activity, given that some users are not physically able to jog.

Kwapisz's approach extends previous efforts that focused only on walking as an activity. However, like in other approaches, the devices were carried in set locations. In Kwapisz's implementation, the devices were carried in a pocket, and Nickel's data collection required the attachment of the device to the right hip. Exploration of movement with unconstrained device positions under a more demanding data collection protocol that includes various clothing and flooring remains as a complicated task.

Lastly, an unconventional approach is taken that is strongly related to gait. Specifically, arm swing while walking aids in balance and energy efficiency; researchers Gafurov and Snekkenes [44] investigate arm swing as a weak behavioral biometric. The authors attach a motion recording sensor to the lower arm to measure the three accelerations available from the three-axis accelerometer. Even though this is not a realistic setup, it closely resembles a user carrying the device in their hand, which is sometimes done in real-world scenarios. The maximum amplitudes in the frequency domain are retained as features, and authentication is determine based on the Euclidean distance between the enrolled and probe feature sets. Namely, given feature vectors $V = (v_1, \ldots, v_n)$ and $W = (w_1, \ldots, w_n)$, the similarity score S is represented as

$$S(V, W) = \sqrt{\sum_{i=1}^{n} (v_i - w_i)^2}. \tag{7.1}$$

The authors acknowledge that this is not a stand-alone approach for active authentication, but suggest it as an accompanying security measure to a stronger approach. However, based on a pool of 30 users, a Rank-1 verification probability of 71.7 percent is achieved. Unfortunately, arm swing is affected by several factors that should be further tested, including carrying objects in the hands, walking speeds, and walking with one or both hands in a pocket.

7.5.4 Touch

A 2011 report by a technology market intelligence company predicted that 97 percent of all smartphones would have touchscreens by 2016 [47]. As a result, biometric researchers have become increasingly interested in extracting touch gesture behavior for the purpose of active authentication.

Transparent background service, SilentSense, is introduced by Bo *et al.* [23] to determine if the current user of the device is the owner or a guest through touch gestures and the reactions from the device as a result of these gestures. Specifically, pressure, area, duration, and position of the gesture and acceleration and rotation of the device as a reaction of the gesture are investigated. The authors initially claim that tap, scroll, fling, and multi-touch gestures will result in different amplitude vibrations. However, after further investigation, it is observed that the device's reaction to touch gestures is indistinguishable. Subsequently, the authors incorporate walking features in combination with touch gesture features. After evaluating the results of 100 users, a guest could always be accurately identified after 12 steps. Similarly, the actual device owner is accurately identified after seven steps.

Zheng *et al.* [48] also monitor acceleration, pressure, size, and time of tapping behavior. Data are fed into a one-class classifier to verify users with an average EER of 3.65 percent. Five PINs (3244, 1111, 5555, 12597384, 12598416) are tested among 80 participants. They observe that each user demonstrates consistent

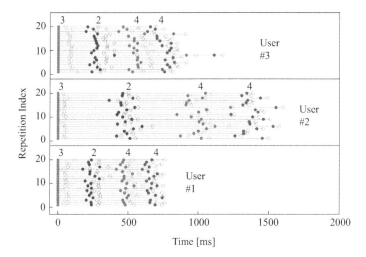

Figure 7.6   Timing of tapping on the smartphone from three different users, shown in three vertical panels. Each user typed 20 times of the number string "3244". The solid dots represent key-press time, and the open dots are key-release time. Different colors represent the timestamps of different digits [48]

and distinct tapping behavior when entering the specified PINs (refer to Figure 7.6[7]). The authors also observe that acceleration is proportional to tap force and touch size is related to finger size and tap force. Valid and invalid class eight-digit PIN distributions were also completely isolated, whereas there was small overlap in the four-digit case, indicating that longer PINs are more unique to individuals than shorter PINs. Mimic attacks, such as through shoulder surfing, were also quantitatively measured by having two imposters closely observe another as the latter used the device. One imposter has the same gender and similar finger size as the legitimate user. The other imposter has a different gender and larger fingers. Experimental results indicated that shoulder surfing had no significant bearing on reproducing the actions of the legitimate user. The authors believe that this is due to three aspects: (1) the chosen features are independent of each other, and mimicking multiple features is difficult, (2) timing is measured in milliseconds, which is higher than human perception, and (3) physical differences (e.g., larger hands) hinder reproduction of actions. In another work presented by Alariki and Manaf [33], it is also acknowledged that reproducing touch gestures through shoulder surfing is quite difficult, given that user-specific characteristics, such as force, speed, and pressure, are hard to mimic.

Xu, Zhou, and Lyu [32] attempt to understand the distinctiveness and permanence of keystroke, slide, and pinch gestures. This approach is able to classify with

---

an accuracy of over 80 percent, reflecting reasonable distinctiveness. Moreover, when allowing three to five consecutive slide or pinch gestures, error rates approach zero percent. However, even though pinch and slide operations are most permanent, permanence is not maintained without using an adaptive approach (i.e., adjusting the enrolled behavioral profile according to new samples). The authors also observe that touch gestures require a large training set for authentication purposes.

7.5.5 Multimodal approaches

It is well-known in the biometric community that performance often improves through the use of multiple modalities. This is also explored in mobile devices as Saevanee, Clarke, Furnell, and Biscione [34] investigate a multimodal system using linguistic profiling, keystroke dynamics, and behavioral profiling. Linguistic profiling, a common approach to stylometry, attempts to verify individuals based on linguistic morphology. Like the multi-engine application monitoring approach developed in [42], the authors devise a multimodal framework with four processing engines: data collection, biometric profile, authentication, and communication. The authors evaluate a variety of experimental setups. Specifically, data types are analyzed first individually and later combined via simple sum and matcher weighting. Additionally, three classifiers, namely kNN radial basis function, and feed-forward multi-layered perceptron neural networks, are examined. Finally, three usage levels, infrequent, moderate, and frequent, are also investigated with the hypothesis that usage levels will have an impact on performance because usage levels correlate with the amount of usage data available. Based upon experimental results, the authors determine that usage levels have minimum impact on classification performance and the matcher weighting technique provides the best results. Additionally, results further confirm that multimodal approaches increase classification performance.

Neal, Woodard, and Striegel [17] combine application traffic with Bluetooth and WiFi data to evaluate the distinctiveness and permanence of the three individually and combined via feature-level fusion. Using one of the largest databases in this area (200 users), the authors extract 70 frequency-based features to model user behavior. Features were examined in various time frames, namely morning, lunch, work, and night hours. Seven feature vectors are created according to data type and the time frame of feature extraction. A simple nearest-neighbor classifier similar to the Overlap method presented in [14] is used for matching. Namely, a matching score S is defined as

$$S = M - \sum_{k=1}^{M} Q_k, \quad \text{where}$$

$$Q_k = \begin{cases} \left| N_k^G \cap N_k^P \right| \div \left| N_k^G \cup N_k^P \right|, & \text{if } \left| N_k^G \right| \text{ and } \left| N_k^P \right| \geq 1 \\ 0, \text{otherwise} \end{cases}$$

and M represents the number of feature vectors, and sets N_k^G and N_k^P represent the kth feature vector in the gallery and probe, respectively. $| N_k^G |$ and $| N_k^P |$ can be zero if a user is inactive. Like Saevanee, the authors discover that a feature vector that includes all three data types offers the best classification performance. The authors also find that features representing a user's week of mobile device usage provides better results than those representing a day, concluding that the distinctiveness of the features is most maintained over a week's span. Results also suggest that WiFi features are the most permanent. In fact, the authors report a 13 percent performance decrease when evaluating permanence in application and Bluetooth features. This suggests that users WiFi activity is most stable over time, and updating a user's application and Bluetooth behavioral profiles as time progresses should be considered. Overall, the valid user of the device is accurately recognized on average 85 percent of the time. A weakness to this approach, however, is how the researchers determined the time frames for feature extraction. Instead of analyzing the raw data for efficient behavior profiling, the authors choose time frames based on intuition, assuming that users interact differently with their devices during different parts of the day.

Fridman, Weber, Greenstadt, and Kam [18] also consider combining application traffic with other modalities, specifically text (i.e., stylometry), web browsing activity, and the physical location of the device. As seen in [17], a rather large database is employed (200 users); however, the time span of recorded data is a fraction of the length present in the database used in [17]. Additionally, this work reports the results of fusing all four modality via the Chair-Varshney decision-level fusion rule, whereas in [17], the authors consider the modalities separately and combined. Nonetheless, Fridman *et al.* report promising results with the chosen behavioral traits, achieving an EER of 0.05 and 0.01 after 1 and 30 min of device interaction, respectively. To achieve these values, the authors create a tracking application to record user usage data. For each modality, a maximum likelihood classifier is used, fusing probabilities of events, x_t, across time, Ω:

$$H^* = \arg\max_{i \in 0,1} \prod_{x_t \in \Omega}^{b} P(x_t \mid H_i). \tag{7.2}$$

The authors report that the location-based classifier contributed the most to the fused biometric system. This is also seen in the work presented by Neal *et al.*, as WiFi data, which is highly correlated with location, is the most distinctive and permanent. Selecting suitable data was challenging; however, mobile platform development is largely centered around user convenience and the reliability of the device. When implementing a mobile biometric system, issues related to limited resources, speed, and battery drain must be addressed. The authors note that battery drain was the biggest problem of their implementation, and opted to choose modalities which reduced this issue.

Finally, Vildjiounaite *et al.* [29] attempt to join voice and gait data via weighted sum fusion, achieving an EER range of 9.14–11.8 percent. Correlation and Fast Fourier Transform (FFT) scores are used for matching gait information. Correlation scores provide insight into the shape of the gait signal and FFT scores describe the frequency component distributions. The Munich Automatic Speaker Verification (MASV) environment is used for voice recognition via Gaussian Mixture Models. To properly analyze their approach, the authors ask 31 participants to walk in a normal and hurried fashion with the phone in the hip pocket, chest pocket, and in hand. Likewise, speech samples were contaminated with three different signal-to-noise ratio (SNR) conditions, namely 20, 10, and 0 dB. When considering the modalities separately, the EER of voice recognition exceeded 40 percent in high-noise conditions and the lowest EER (13.5 percent) of gait recognition is achieved when the phone is carried in the chest pocket. Their experimental results suggest that combining the two modalities significantly improves performance.

Multimodal approaches help to overcome the issue of an unavailable trait in the event that a user is in inactive. Additionally, the discussed approaches support the notion that use of multiple traits increases the performance on mobile platforms. Another benefit of this approach is it indirectly works around the assumption of consistent user behavior. Consider an implementation that authenticates in an adaptive manner, i.e., applying additional sensors as needed. If the user is behaving uncharacteristically in one trait, but normally in the others, the system could use a majority voting scheme to determine that the user is indeed the actual owner of the device. However, the disadvantage is the limitation of resources available on mobile devices. Fridman *et al.* ran into the issue of battery drain due to incorporating multiple data types. In the future, researchers must be conscious of choosing data types that do not computationally overwhelm the device. There is also the issue of determining which traits to combine. For instance, Vildjiounaite *et al.* combine voice and gait due to the weaknesses of each trait with the intentions of making an overall accurate system. However, if the user rarely moves with the device, the implementation becomes essentially a single modality voice system, which then suffers dramatically alone. Finally, further work should determine an optimal fusion level. For instance, Neal *et al.* use feature-level fusion, while Fridman *et al.* employ decision-level fusion. Feature-level fusion may cause performance to suffer if a certain trait performs poorly, especially if all traits are given equal weight. Decision-level fusion, however could be more convenient, given that the system would not require retraining if a new trait is introduced. Overall, researchers are left with the tasks of managing device resources and choosing appropriate combination schemes in multimodal systems.

7.6 Research challenges

Future research should attempt to address the following issues: lack of consistency in user behavior, preserving the privacy of usage data, lack of a standard evaluation platform, and engineering limitations.

Lack of consistency in user behavior There is an assumption that user behavior is maintained throughout the use of the device [49]. However, this is often not the case. As mobile devices continue to expand in functionality, size, and cost, users will adjust accordingly. Lifestyle changes, such as starting a family or a new job, could have an effect on how the device is used. Additionally, if a single device is used by multiple legitimate users, the service may only be able to authenticate a single user and inaccurately decide that the others are intruders due to changes of behavior between users. Therefore, it is critical to further examine methods that will keep behavioral profiles updated in a non-intrusive way. Some researchers have incorporated an adaptive method of keeping a userâŁ™s profile current. Alternatively, Zheng *et al.* [48] suggest contacting the service provider to disable the authentication process to allow for re-training. However, once the authentication service is turned off, the device is susceptible to attack. This approach would likely require a second security measure, most likely point-of-entry, removing the advantages of transparency. Hence, Abdulwahid *et al.* [7] note that some frameworks are not fully behavioral, and therefore, not fully transparent. Hence, at least three tasks remain as a result of inconsistent user behavior:

1. Deriving an optimal adaptive method to keep user profiles current, thereby accounting for sudden changes in behavior without frequent inaccurate authentications.
2. Creating a fully transparent system despite behavior fluctuations and/or,
3. Effectively combining behavioral and physiological biometrics when either is unavailable or unreliable for maintaining accuracy.

Preserving the privacy of usage data   Behavioral biometrics not only identify an individual, but they can also reveal habits and interests, which could result in the inference of certain characteristics about the user. Therefore, a user may be hesitant to allow a biometric service to continuously monitor his or her actions. Hence, as the data gathered will be used for the purpose of biometrics, the data as stored in an intermediate form presents important security and privacy concerns. For instance, should the data be encrypted (as seen in [16]) or anonymized before storage and what impact would said processing have on energy? Similarly, what type of data should be analyzed and would the anonymization process remove important data attributes? Furthermore, to what extent should the instrumentation process peek into the user data? Will there be sufficient information available via system logs? Moreover, does the user truly understand to what extent said information is being analyzed for the benefit of improved security?

Lack of a standard evaluation platform   Meng *et al.* [49] emphasize the need for a larger user study on a standard evaluation platform for full evaluation of the various implementations. Currently, each proposed approach uses a different set of users (usually small) and there is no one widely accepted approach. There lacks a standard database that encompasses all usage data from a large set of mobile device users captured under realistic conditions. Instead, there are multiple databases for the same trait; it becomes difficult to decipher if experimental results reflect any bias or tuning on the created database. Furthermore, there lacks a strong notion of baseline comparisons considering that databases are collected under dissimilar conditions. It becomes harder to compare approaches given these circumstances.

Engineering limitations   Complications arise as a result of the design and engineering of modern mobile devices. First, usually all running processes can be terminated by the user, posing a problem for a biometric service that could be potentially rendered inactive if closed accidentally by the legitimate user or intentionally by an intruder. Additionally, researchers are tasked with determining how to continuously monitor the device once it is placed in sleep or power-saving states [21].

Second, Branscomb [16] notes that log permissions in some dominant mobile OS platforms have been modified to be only accessible to system applications. System logs are typically used for debugging, and therefore include valuable information regarding how the device is used. As a result of this change, biometric services may require root permissions, and implementation may become more difficult under these restrictions.

Finally, biometric services on these devices should not pose an inconvenience due to excessive resource drain. Implementations should efficiently manage power and memory resources, and quickly learn user models. Additionally, there is not a natural extension of behavioral biometrics from desktop computers to mobile devices. Modalities commonly associated with computers, such as mouse movements, are not available in mobile devices. Even though mobile devices are becoming functionally similar to computers, researchers should not expect similar biometric implementations or results.

7.7 Summary

Rapid advancements in mobile technologies have resulted in a strong dependency on mobile devices. These devices are now being used for a multitude of tasks, including banking and social networking. Passwords and physical biometrics are most commonly used to secure the sensitive information contained in mobile devices. These methods, however, are frequently inefficient due to various malicious attacks, such as shoulder surfing and spoofing. As a result, behavioral biometrics have been explored for convenient, continuous, and non-intrusive authentication.

In such systems, data collection encapsulates the process of properly instrumenting the device for environmental, movement, location, and interactive data capture via on-board sensors. During this phase, researchers should carefully consider appropriate tuning and evaluation of sensors, along with efficient implementation within the constraints imposed by dominant development platforms.

Features extracted from raw usage data depend on the output of the sensors employed. Name-based, positional, touch, and voice features are the most common evaluated throughout the literature. Name-based features are nominal values that typically represent the name of a visited entity, such as a website's URL. Positional features are extracted from location and motion sensors, such as GSM/CDMA cells and accelerometers, and model a user's geographical location or physical activity. Touch gestures, such as keystrokes, are represented by a variety of touch features, such as the duration of a keystroke. Finally, voice data captured via the device's microphone is sometimes considered a weak biometric given it is highly affected by environmental noise. Nonetheless, features, such as signal energy and zero crossing rate, are extracted from voice data.

Additionally, various matching techniques have been implemented in this area. Bayesian classifiers and a modified version of the Overlap categorical classifier are investigated for application traffic, where data are typically modeled as a tuple of the application's name and time and location of access. Text input via the touchscreen's soft keyboard has little exploration; however, 500-character text blocks and *n*-grams have been analyzed with reasonable results. Gait data are traditionally processed with a cycle-based approach, where a cycle represents a user's step pattern. However, some authors have resorted to machine learning methods with improved results. Among touch gestures, it is observed that longer input sequences are more distinguishable than shorter sequences. Additionally, researchers have found that it is difficult for an intruder to mimic touch gesture behavior via shoulder surfing, given that characteristics, such as force and pressure, are hard to reproduce. Finally, these methods have been combined into multimodal systems that have shown improvement in user verification, and helped to cope with unavailable or weak modalities.

Finally, several research challenges remain in this area, such as properly dealing with inconsistent user behavior and engineering limitations, preserving the privacy of usage data, and deriving a standard evaluation platform. Consequently, there are significant opportunities for researchers to make contributions to this developing field of research. Furthermore, recent advances in this area have provided evidence in support of a continuous, intuitive, and transparent approach to mobile device security.

References

[1]   Smith A. *U.S. Smartphone Use in 2015* [online]. *Internet, Science and Tech: Pew Research Center*; 2015. Available from http://www.pewinternet.org/2015/04/01/us-smartphone-use-in-2015/ [Accessed 15 Jan 2016]

[2]   Feng, T., DeSalvo, N., Xu, L., Zhao, X., Wang, X., and Shi, W. "Secure session on mobile: An exploration on combining biometric, trustzone, and user behavior". *6th International Conference on Mobile Computing, Applications and Services (MobiCASE)*. IEEE; 2014, pp. 206–215

[3]   Amin, R., Gaber, T., ElTaweel, G., and Hassanien, A. "Biometric and traditional mobile authentication techniques: Overviews and open issues". *Bio-inspiring Cyber Security and Cloud Services: Trends and Innovations*. Springer; 2014. pp. 423–446

[4]   *Ice Cream Sandwich* [online]. Available from http://developer.android.com/about/versions/android-4.0-highlights.html [Accessed 07 Jan 2016]

[5]   *About Touch ID Security on iPhone and iPad* [online]. Available from https://support.apple.com/en-us/HT204587 [Accessed 07 Jan 2016]

[6]   Bhagavatula, C., Iacovino, K., Kywe, S., Savvides, M., Ur, B., and Cranor, L. "Biometric authentication on iPhone and Android: Usability, perceptions, and influences on adoption". *USEC, Internet Society*. 2015

[7]   Abdulwahid A., Clarke, N., Stengel, I., Furnell, S., and Reich, C. "Continuous and transparent multimodal authentication: Reviewing the state of the art". *Cluster Computing*; 2016, vol. 19(1), pp. 455–474. DOI: http://dx.doi.org/10.1007/s10586-015-0510-4

[8]   Trung, N., Makihara, Y., Nagahara, H., Mukaigawa, Y., and Yagi, Y. "Performance evaluation of gait recognition using the largest inertial sensor-based gait database". *5th IAPR International Conference on Biometrics (ICB)*. IEEE; 2012, pp. 360–366

[9] Tulyakov, S., and Govindaraju, V. "Classifier combination types for biometric applications". *Computer Vision and Pattern Recognition Workshops*; IEEE; 2006, p. 58

[10] Striegel, A., Liu, S., Meng, L., Poellabauer, C., Hachen, D., and Lizardo, O. "Lessons learned from the netsense smartphone study". *ACM SIGCOMM Computer Communication Review*. ACM; 2013, vol. 43(4), pp. 51–56

[11] Nandugudi, A., Maiti, A., and Ki, T. "Phonelab: A large programmable smartphone testbed". *Proceedings of First International Workshop on Sensing and Big Data Mining*. ACM; 2013, pp. 1–6

[12] Eagle, N., and Pentland, A. "Reality mining: Sensing complex social systems". *Personal and Ubiquitous Computing*. Springer-Verlag; 2006, vol. 10(4), pp. 255–268.

[13] Stopczynski, A., Sekara, V., Sapiezynski, P., *et al.* "Measuring large-scale social networks with high resolution". *PloS One*. Public Library of Science; 2014, vol. 9(4), pp. 1–24.

[14] Boriah, S., Chandola, V., and Kumar, V. "Similarity measures for categorical data: A comparative evaluation". SIAM; 2008

[15] Bassu, D., Cochinwala, M., Jain, A. "A new mobile biometric based upon usage context". *IEEE International Conference on Technologies for Homeland Security (HST)*; 2013, pp. 441–446

[16] Branscomb, A. *Behaviorally Identifying Smartphone Users*. Masters thesis, 2013

[17] Neal, T., Woodard, D., and Striegel, A. "Mobile device application, Bluetooth, and Wi-Fi usage data as behavioral biometric traits". *IEEE 7th International Conference on Biometrics Theory, Applications and Systems (BTAS)*. IEEE; 2015, pp. 1–6

[18] Fridman, L., Weber, S., Greenstadt, R., and Kam, M. "Active authentication on mobile devices via stylometry, application usage, Web browsing, and GPS location". *IEEE Systems Journal*, vol. PP(99), pp. 1–9. doi: 10.1109/JSYST.2015.2472579

[19] Brocardo, M., Traore, I., Saad, S., and Woungang, I. "Authorship verification for short messages using stylometry". *International Conference on Computer, Information and Telecommunication Systems (CITS)*. IEEE; 2013, pp. 1–6

[20] Pei, L., Guinness, R., Chen, R., *et al.* "Human behavior cognition using smartphone sensors". *Sensors*. Multidisciplinary Digital Publishing Institute; 2013, vol. 13(2), pp. 1402–1424

[21] Shi, W., Yang, F., Jiang, Y., Yang, F., and Xiong, Y. "Senguard: Passive user identification on smartphones using multiple sensors". *IEEE 7th International Conference on Wireless and Mobile Computing, Networking and Communications (WiMob)*. IEEE; 2011, pp. 141–148

[22] Albert, M., Toledo, S., Shapiro, M., and Kording, K. "Using mobile phones for activity recognition in Parkinson's patients". *Frontiers in Neurology*. 2012, vol. 3, pp. 1–7

[23] Bo, C., Zhang, L., Li, X., Huang, Q., and Wang, Y. "Silentsense: Silent user identification via touch and movement behavioral biometrics". *Proceedings*

of the 19th Annual International Conference on Mobile Computing and Networking. ACM; 2013, pp. 187–190

[24] Ferrero, R., Gandino, F., Montrucchio, B., Rebaudengo, M., Velasco, A., and Benkhelifa, I. "On gait recognition with smartphone accelerometer". *4th Mediterranean Conference on Embedded Computing (MECO)*; IEEE; 2015, pp. 368–373

[25] Nickel, C., and Busch, C. "Classifying accelerometer data via Hidden Markov Models to authenticate people by the way they walk". *IEEE International Carnahan Conference on Security Technology (ICCST)*. IEEE; Oct 2011, pp. 1–6

[26] Kwapisz, J., Weiss, G., and Moore, S. "Cell phone-based biometric identification". *Fourth IEEE International Conference on Biometrics: Theory Applications and Systems (BTAS)*; IEEE; 2010, pp. 1–7

[27] Nickel, C., Wirtl, T., and Busch, C. "Authentication of smartphone users based on the way they walk using k-NN algorithm". *Eighth International Conference on Intelligent Information Hiding and Multimedia Signal Processing (IIH-MSP)*. IEEE; 2012, pp. 16–20

[28] Gulzar, T., Singh, A., and Sharma, S. "Comparative analysis of LPCC, MFCC and BFCC for the recognition of Hindi words using artificial neural networks". *International Journal of Computer Applications*. 2014, vol. 101(12), pp. 22–27

[29] Vildjiounaite, E., Mäkelä, S., Lindholm, M. *et al.* "Unobtrusive multimodal biometrics for ensuring privacy and information security with personal devices". *Pervasive Computing.* Springer; 2006. pp. 187–201

[30] Lu, H., Yang, J., Liu, Z., Lane, N., Choudhury, T., and Campbell, A. "The Jigsaw continuous sensing engine for mobile phone applications". *Proceedings of the 8th ACM Conference on Embedded Networked Sensor Systems.* ACM; 2010, pp. 71–84

[31] Shoaib, M., Scholten, H., and Havinga, P. "Towards physical activity recognition using smartphone sensors". *IEEE 10th International Conference on Ubiquitous Intelligence and Computing and 10th International Conference on Autonomic and Trusted Computing.* IEEE; 2013, pp. 80–87

[32] Xu, H., Zhou, Y., and Lyu, M. "Towards continuous and passive authentication via touch biometrics: An experimental study on smartphones". *Symposium On Usable Privacy and Security*; Menlo Park, CA, 2014. pp. 187–198

[33] Alariki, A., and Manaf, A. "Touch gesture authentication framework for touchscreen mobile devices". *Journal of Theoretical and Applied Information Technology.* 2014, vol. 62(2), pp. 493–498

[34] Saevanee, H., Clarke, N., Furnell, S., and Biscione, V. "Text-based active authentication for mobile devices". *ICT Systems Security and Privacy Protection.* Springer; 2014. pp. 99–112

[35] Gascon, H., Uellenbeck, S., Wolf, C., and Rieck, K. "Continuous authentication on mobile devices by analysis of typing motion behavior". *Sicherheit*; 2014. pp. 1–12

[36] Cai, L., and Chen, H. "TouchLogger: Inferring keystrokes on touchscreen from smartphone motion". *HotSec.* 2011, vol. 11, p. 9

[37]   Jain A., Ross A., and Nandakumar K. *Introduction to Biometrics.* New York: Springer Science & Business Media; 2011. p. 33

[38]   Giannakopoulos, T., Kosmopoulos, D., Aristidou, A., and Theodoridis, S. "Violence content classification using audio features". *Advances in Artificial Intelligence.* Springer; 2006. pp. 502–507

[39]   Smith D. *Google Play Has More Apps Than Apple Now, But It's Still Behind in One Key Area* [online]. *Tech Insider: Business Insider Inc*; 2015. Available from http://www.businessinsider.com/google-play-vs-apple-app-store-2015-2 [Accessed 11 Jan 2016]

[40]   Wakefield J. *One WiFi Hotspot for Every 150 People, Says Study* [online]. *Technology: BBC News*; 2014. Available from http://www.bbc.com/news/ technology-29726632 [Accessed 11 Jan 2016]

[41]   Harper J. *Hello? Americans Now Spend Five Hours a Day – On Their Phones* [online]. *The Washington Times*; 2015. Available from http://www. washingtontimes.com/news/2015/feb/10/smart-phone-nation-americans-now-spend-five-hours-/ [Accessed 11 Jan 2016]

[42]   Li, F., Clarke, N., Papadaki, M., and Dowland, P. "Active authentication for mobile devices utilising behaviour profiling". *International Journal of Information Security.* Springer; 2014, vol. 13(3), pp. 229–244

[43]   Li, F., Wheeler, R., and Clarke, N. "An evaluation of behavioral profiling on mobile devices". *Human Aspects of Information Security, Privacy, and Trust.* Springer; 2014. pp. 330–339

[44]   Gafurov, D., and Snekkenes, E. "Arm swing as a weak biometric for unobtrusive user authentication". *IIHMSP'08 International Conference on Intelligent Information Hiding and Multimedia Signal Processing*; IEEE; 2008, pp. 1080–1087

[45]   O'Grady M. *SMS Usage Remains Strong in the US: 6 Billion SMS Messages are Sent Each Day* [online]. *Forrester Research, Inc*; 2012. Available from http://blogs.forrester.com/michael_ogrady/12-06-19-sms_usage_remains_ strong_in_the_us_6_billion_sms_messages_are_sent_each_day [Accessed 13 Jan 2016]

[46]   Battestini, A., Setlur, V., and Sohn, T. "A large scale study of text-messaging use". *Proceedings of the 12th International Conference on Human Computer Interaction with Mobile Devices and Services*; ACM; 2010, pp. 229–238

[47]   *97% of All Smartphones Will Have Touchscreens by 2016* [online]. *ABI Research*; 2011. Available from https://www.abiresearch.com/press/97-of-all-smartphones-will-have-touchscreens-by-20/ [Accessed 16 Jan 2016]

[48]   Zheng, N., Bai, K., Huang, H., and Wang, H. "You are how you touch: User verification on smartphones via tapping behaviors". *IEEE 22nd International Conference on Network Protocols (ICNP)*; 2014, pp. 221–232

[49]   Meng, W., Wong, D., Furnell, S., and Zhou, J. "Surveying the development of biometric user authentication on mobile phones". *Communications Surveys Tutorials.* IEEE; 2015, vol. 17(3), pp. 1268–1293

Chapter 8

Continuous mobile authentication using user–phone interaction

Xi Zhao[1], Tao Feng[2], Xiaoni Lu[1], Weidong Shi[2], and
Ioannis A. Kakadiaris[2]

8.1 Introduction

According to a recent market analysis [1], 2.56 billion people across the world will use smartphones by 2018. Mobile devices have become increasingly ubiquitous compared to traditional desktop PCs. As well as being personal communication and entertainment devices, many of them are distributed by companies to their employees with customer relationship management, enterprise resource planning apps installed. Pervasive usage and multifunctionality, especially critical organization data access and/or process control, make device security extremely important. However, these devices frequently suffer from security threats by intruders.

It is assumed that genuine users only attempt to use their own devices; intruders, however, attempt to bypass the authentication measures in others' phones. The threats can be categorized into two types.

The Prelogin Phase threat: To protect the system from external attackers, the current 4-digit numeric PINs used in most phones only have a theoretical potential entropy of $\log_2 (10^4) = 13.3$ bits. Furthermore, practical entropy for 4-digit PINs is likely to be much lower, as is the case with passwords [2]. It is clear that this kind of authentication method is not secure enough to protect increasingly sensitive data and information due to its very low information entropy.

The Postlogin Phase threat: This is an important concept that differentiates continuous user authentication from point-based authentication. The threat is based on the consideration that a mobile device could be accessed by an intruder when it is in a postlogin state (i.e., a user left his postlogin mobile device unattended or a device has been compromised by the intruder(s) during login).

Current point-based mobile user authentication technology can only provide limited protection during prelogin phase or payment. Most devices only provide screen password security or fingerprint authentication during login. For usability reasons, some apps opt to store user credentials to avoid explicit access authentication. This increases the risk of critical data/process theft. As a result, there is a strong need

[1]School of Management, Xi'an Jiaotong University, P. R. China
[2]Computer Science Department, University of Houston, USA

for a complementary continuous authentication to ensure the owner's identity after login stage. Explicit authentication mechanisms are not appropriate for continuous authentication scenarios due to usability reasons. The most desirable user authentication method should be able to implicitly and continuously perform user identification in the background without disrupting natural user–phone interaction [3].

Security vs. usability: Modeling the trade-off between usability and security achieved by a continuous authentication solution, we use false accept rate (FAR) and false reject rate (FRR) metrics. FAR is the percentage of authentication decisions that allow access to an unauthorized user. FRR is the percentage of authentication decisions, where an authorized user is denied access.

A solution exhibiting a low FAR and a high FRR is more secure but not user friendly, because genuine users tend to be rejected more by their devices. A solution with a low FRR and a high FAR is more user friendly but less secure, because intruders can be accepted with a high probability. The goal for continuous user authentication is to minimize both metrics.

This chapter aims to introduce and discuss continuous user authentication during natural user–phone interaction. Two types of user authentication are considered, including touch gesture-based and keystroke-based methods.

8.2   Previous works

Mobile user authentication can be categorized into two groups: physiological biometrics, which rely on static physical characteristics (e.g., fingerprints [4] and facial features [5]), and behavioral biometrics, which employ identity-invariant features of human behavior to identify or verify people during their daily activities [6].

8.2.1   Touch gesture-based mobile authentication

Previous research on behavioral biometrics has shown that biometric features can be extracted when users carry out some activities such as arm gestures [7,8], strokes, or signatures using a pen or stylus [9,10]. Shahzad *et al.* [11] authenticated users based on how they input the gestures. Sherman *et al.* [12] employed the features of the users' free-form gestures to verify them. Yamazaki *et al.* [13] identified individuals with features such as stroke shape, writing pressure, and pen inclination. Cao *et al.* [14] discussed a quantitative human performance model of making single-stroke pen gestures. Ferrer *et al.* [15] segmented and extracted identity characteristics from signatures using local patterns (e.g., the local binary pattern and the local directional pattern). Bailador *et al.* [16] proposed authenticating a user by his 3D signature, which is captured by a 3D accelerometer sensor available on modern mobile devices. Impedovo *et al.* [17] presented the state-of-the-art in automatic signature verification. However, identifying users with pen strokes or signatures cannot be pervasively applied in mobile user authentication because only a few types of mobile devices provide a pen or stylus. Also, the user may not want to draw his signature to unlock the phone because of privacy. Furthermore, most users opt to interact directly with their phone using their fingers for convenience.

The studies of touch gestures for biometric identification on mobile devices are quite recent [18]. Li *et al.* [19] employed features from touch gestures such as the position and area of first touch, duration of slide, and average curvature of slide to build a continuous authentication scheme for smartphones. Hui *et al.* [20] generated a continuous authentication scheme on smartphones using different touch operations such as keystroke, slide, pinch, and handwriting. They found that from a classification accuracy perspective, the slide gesture is the best to classify users uniquely, while handwriting performs the worst. Some other continuous authentication schemes include SilentSense [21], which extracts features from finger movements and user motion patterns and uses Support Vector Machine to achieve high accuracy, and LatentGesture [22], which uses Support Vector Machine and random forest classifier. Feng *et al.* [23] extracted finger motion speed and acceleration of touch gesture as features for identification. Luca *et al.* [24] directly computed the distance between traces using the DTW algorithm. Sae-Bae *et al.* [25] used different combinations of 5 fingers to design 22 touch feature certification. They computed the DTW distance and Frechet distance between multi-touch traces. However, users must perform predetermined touch gestures for authentication, which can be opaque and compulsive to users. Frank *et al.* [26] studied the correlation between 30 analytic features of touch traces and used the method of K-nearest-neighbors and support vector machine to classify these analytic features. Zhao *et al.* [27] extracted graphic touch gesture feature (GTGF) from touch traces so that the 2D movement dynamics and pressure value can be represented explicitly by plotting bar shape and intensity values. Zhao *et al.* [28] adopted the pressure dynamics and represented them as the profiles of statistical touch dynamics image (STDI) instances. STDI learns the intrapersonal variations from many traces of a user so that scores are computed between pairs of gallery models and probe traces, which significantly reduces the computational time.

8.2.2 *Keystroke-based mobile authentication*

The method of authenticating users through analyzing keystroke and mouse movement was first used in the desktop environment. In consideration of weak authentication mechanisms on mobile handsets, Clarke *et al.* [29] started to authenticate users based on their typing characteristics. In their work, they used features such as the key hold time, which is the time period between pressing two keys, and the error rate. Later, Zahid *et al.* [30] also extracted a viable features' set from keystroke dynamics for authentication. And Campisi *et al.* [31] proposed to authenticate users using keystroke dynamics when users perform the input on a mobile phone keypad. Contrary to these methods, which extract features via the hardware keyboard, the virtual keyboard can provide new, useful features such as pressure data, finger size, and virtual key typing location. Trojahn *et al.* [32] first authenticated users using the extra information provided by the virtual keyboard. Saevanee *et al.* [33] made use of features such as key hold time, interkey and finger pressure to verify users. Maiorana *et al.* [34] demonstrated the feasibility of using keystroke dynamics to authenticate users on mobile phones. Hwang *et al.* [35] built up a keystroke dynamics-based authentication (KDA) system on mobile devices, which adds artificial rhythms and gets better performance. Chang *et al.* [36] proposed a new graphical-based password KDA system, which can improve KDA effectiveness in touch screen handheld mobile devices.

8.3   Touch gesture features

The rationale of touch gesture-based biometrics is that due to the dependency on the user's hand geometry and muscle behavior, gesture sample series from the same identity should have limited variations in dynamics of finger movements and tactile pressure. These intraclass variations should be smaller than interclass variations of gestures performed by other identities in specific usage scenarios.

Touch gestures are normally represented as a series of time stamped, x–y coordinates of contacted fingertips, contact size, and/or pressure. Most touch gesture types performed by users when interacting with phones include click, double click, hold, sliding up, sliding down, flicking right, flicking left, pinching, and spreading. Some common features are computed from touch gestures as basic representation in user authentication methods, as described below.

Contact Size is the contact surface area between user's finger and the touch screen surface. The contact size value can be affected by how hard the user touches the screen, and therefore, it is also used as an approximation of touch pressure on pressure-insensitive touch screens.

Pressure is the force or thrust on a touch screen applied by users in contact, normally expressed as force per unit area. On devices with pressure-insensitive touch screens, it can be directly measured.

Touch Location indicates the swipe location preference. For instance, when performing a vertical swipe gesture, some users like to do it on the left part of the touch screen, while others may prefer the right part of the touch screen.

Swipe/Zoom Speed reflects how fast a user performs a swipe/zoom gesture. Although this feature might be affected by user's current emotional state or environment, it is usually determined by the user's finger and hand muscles. This feature can be calculated by collecting the touch time and location.

Swipe/Zoom Length represents the length of the swipe or zoom gestures. This feature is application dependent. For example, during a left-to-right screen scroll operation in the launcher application, some users may swipe all the way across the touch screen while others may only swipe a short distance.

Swipe/Zoom Curvature is another useful feature, which represents the slope of a user's swipe or zoom gestures. The consistency of this feature can be seen from the swipe/zoom gestures shown in Figure 8.1.

Click Gap represents the time difference between two clicks. This feature is especially useful when dealing with virtual typing gestures. Its value is the difference of time stamp data between two clicks.

Context represents the contextual information when users perform touch gestures. It can be represented by Apps, locked screens, or assistive screens. As shown in Figure 8.1, the running application context is extremely important for identity

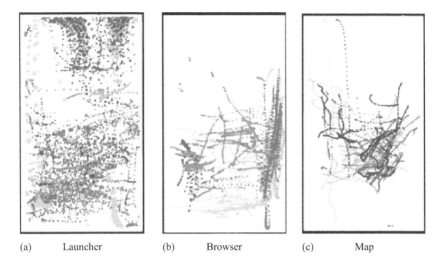

(a)   Launcher   (b)   Browser   (c)   Map

Figure 8.1   One user's touch data in different applications. (Courtesy of Tao et al. [37].)

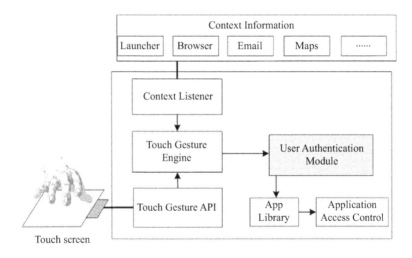

Figure 8.2   User authentication architecture overview

recognition. A user's touch gestures in the launcher application are significantly different from the same user's touch gestures in Email, Browser, or Map applications.

8.4   User authentication schema overview

Figure 8.2 demonstrates the architecture overview for mobile user authentication. It collects touch gesture inputs and context information from the *Touch Gesture API* and *Context Listener*, respectively. The collected raw data are then transferred to the *Touch Gesture Engine* for data preprocessing and feature extraction and then evaluated in the *User Authentication Module*, which computes the score metrics and outputs the authentication results. When a user tries to access an App, the system uses the *App Library* to check whether the App allows unauthorized users. If it does not, the *Application Access Control* prompts a password dialogue to request an explicit password. Otherwise, no action is performed unless there is a need to update the trained templates/models due to a misclassification (adaptive component).

8.5 Dynamic time warping-based method

8.5.1 One nearest neighbor–dynamic time warping

To perform user identity recognition, a one nearest neighbor (1NN) classifier and dynamic time warping (DTW) are combined. This allows us to capture the variety of user's touch screen data by maintaining different gesture templates per application and adapt them over time and user behavior.

DTW is considered an efficient way to measure the similarity between two time series. It works by computing the Euclidean distance between any two input sequences of feature vectors and finds the optimal sequence alignment using dynamic programming.

1NN classifier is a nonparametric method for classifying objects based on the closest training example in the feature space. When applying it, the DTW distance is calculated between an incoming touch gesture and all candidate gesture templates. The label assigned to the incoming gesture is that of the closest gesture template according to the DTW distance.

8.5.2 Sequential recognition

Using the single newest incoming gesture would not capture the temporal correlation of consecutive gesture inputs, thus it is suggested to use sequential user identity recognition to observe γ number of consecutive gesture examples and accumulate their individual DTW distances (resulting from each pair of gesture comparison). We call γ the *authentication length* and use it as a metric to define the number of most recent gestures used before providing an identity recognition result.

Gestures within the authentication length will be normalized and aggregated. Then, the 1NN classifier is employed using the aggregated value to recognize the identity of the new touch input sequence.

8.5.3 Multistage filtering with dynamic template adaptation

Comparing an incoming gesture with all the available template gestures requires a high and unacceptable computational cost. Stored templates must be regularly updated with new training gestures to compensate for the user's gestural variations over time. To allow user adaptation without increasing the storage or recognition delay, a multistage

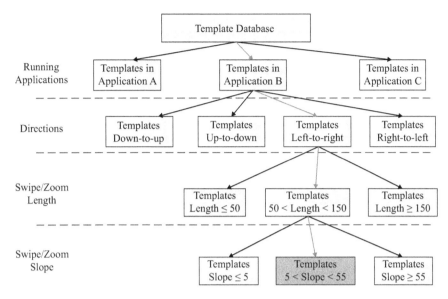

Figure 8.3   Illustration of the process of multistage filtering. (Courtesy of Tao et al. [37].)

filtering technique is employed and combined with dynamic template adaptation to reduce the computational complexity while maintaining good performance.

To reduce the DTW distance calculation in each touch gesture recognition process, a multistage filtering hierarchy over swipe/zoom gestures and click gestures is adopted. For swipe/zoom gestures, the hierarchy consists of the following four levels: (1) running applications, (2) direction, (3) swipe/zoom length, and (4) swipe/zoom curvature or slope. Changes of order in the hierarchy would not result in a different sample selection result. Figure 8.3 shows an example of this hierarchy for a swipe input gesture in application B, with a direction of left-to-right, a swipe length of 100, and a slope of 30. For click gestures, the hierarchy consists of just two levels: (1) running application and (2) click location.

The running application layer ensures that only gestures belonging to the same application are always compared. The direction layer further divides swipe/zoom templates into four classes: left-to-right, right-to-left, up-to-down, and down-to-up, while the click location further divides click templates into nine grid areas. The swipe/zoom length and swipe/zoom curvature layers are employed to reduce the variation caused by usage behavior following the direction layer. These two layers guarantee that only swipe/zoom gestures with similar length and curvature are compared. The above process helps reduce the number of templates a new incoming gesture needs to be compared against, thus improving recognition delay.

To prevent the template storage from growing unbounded as a user's touch screen usage behavior evolves over time, we set a threshold on the number of training examples for each application to limit the size of the template sets. When a new template adaptation request is detected (misclassification), it is verified if the size of the library exceeds the threshold. If not, the new template will be added to the set directly. Otherwise, the oldest template will be replaced by the new template to maintain a constant computational complexity. Because if a touch gesture is recognized as an unauthorized input, the service will ask for a password when users try to access a sensitive or customized application. A misclassification can be detected if the user instantly inputs the correct password and accesses the application.

8.5.4 Experimental results

This method has been implemented as an Android background service that implicitly collects touch screen data and authenticates user identity continuously. The App was installed by 23 smartphone users (14 males and 9 females), and 100 guest users were recruited to play with a subset of phones (13 phones). The performance of real-time user authentication was measured.

The experiment consisted of two main phases:

- *Off-device Simulation Phase:* We built an off-device touch screen data analysis using MATLAB®. Data were collected from 23 phones for three weeks. Each data sequence includes timestamp, raw touch data, and the underlying running application. Twenty-three users' first week of touch data was used as training templates, and the subsequent two weeks of data were employed as testing data.
- *On-device Testing Phase:* We incorporated both online training and testing modules into a background service. The on-device training session took one week and collected about 2,000 touch gestures for each user. The user can customize the mode (training or notification) and the authentication length parameter. If the notification mode is selected, the authentication result will be shown in real time in the middle of the screen, and the device will be locked if an unauthorized user is detected.

Tables 8.1 and 8.2 demonstrate the simulation results. Table 8.1 presents the true positive results under four different template database size settings. For instance,

Table 8.1 Off-device simulation performance—True positive (courtesy of Tao et al. [37])

Percentage	Authentication length									
	1	2	3	4	5	6	7	8	9	10
All templates	58	64	70	75	79	82	89	90	92	93
89% templates	56	57	63	67	74	78	82	84	85	88
67% templates	52	57	58	64	69	72	78	79	80	82
56% templates	48	52	57	60	68	71	72	78	78	81

"*All Templates*" indicates results are being calculated using all templates collected in a week, whereas "*56% Templates*" means 44% of the templates are eliminated by merging them with similar templates. An interesting finding that supports our hypotheses on pattern representation is: as template size decreases, the accuracy also decreases. This degradation is caused by the loss of information while reducing the number of templates. Significant accuracy improvements can be found as authentication length increased (for all cases with different template database sizes). When the authentication length is 8, the true positive and true negative for "*All Templates*" already exceed 90%. Therefore, we set the authentication length to 8 in the on-device testing phase. We observe a similar performance trend for the metric of true negative, as shown in Table 8.2.

We implemented the complete service (including both online training and notification modes) on Android phones for on-device practical test. After implicitly logging the classification results for natural touch screen usage data over one week for 13 mobile device users, we computed the true positive rate. By requesting guest users to test these mobile devices, we computed the true negative rate.

Table 8.3 demonstrates the true positive and true negative rates from the implemented background authentication service. For all users, a true positive rate of 91% or more and a true negative rate of 93% or more were achieved under uncontrolled environments in real time. Accuracy performance is evaluated based on touch sequences

Table 8.2 Off-device simulation performance—True negative (courtesy of Tao et al. [37])

Percentage	Authentication length									
	1	2	3	4	5	6	7	8	9	10
All templates	76	78	81	83	84	86	88	90	91	92
89% templates	60	62	66	71	72	73	75	76	78	89
67% templates	52	54	60	61	62	68	70	72	73	86
56% templates	46	48	53	54	56	60	62	71	71	85

Table 8.3 On-device testing accuracy (courtesy of Tao et al. [37])

Percentage	User												
	U1	U2	U3	U4	U5	U6	U7	U8	U9	U10	U11	U12	U13
True positive	96	91	90	91	92	95	93	91	90	90	91	90	89
True negative	99	96	95	94	92	92	96	91	95	91	91	92	91

that combine eight touch gestures. Concurrently, we also measured power consumption. From the energy usage data collected, we found that the power consumption has an average value of 88 mW, and does not exceed 6.2% battery usage.

8.6   Graphic touch gesture-based method

8.6.1   Feature extraction

The six types of touch gestures in Table 8.4 are collected, where generality and usability of GTGF-based methods are ensured. To exclude gesture-type variations, the captured traces are prefiltered into one of these six types based on gestures' start and end point. The Euclidean distances between two fingers at the start and end are used to judge the multifingertip gesture types.

Cubic interpolation is used to resample the traces in terms of their x–y coordinates, time, and pressure series. Traces with different numbers of points and time intervals can be normalized and registered. Thus, registered single-touch tip trace $\hat{S}^0$ (i.e., UD, DU, LR, RL) consist of c samples and registered multiple touch tip traces $\hat{S}^0, \hat{S}^1$ (i.e., ZI, ZO) consist of $2c$ samples.

We further extract GTGF from the normalized trace $\hat{S}$. A zero-valued image template $\mathscr{T}$ is first created with resolution set to $H \times W$, as the resolution of the image depicted in Figure 8.4. Then, a block c_n is used which has a width of three columns to represent a sample $\hat{s}_n$. The block is evenly divided into an upper subblock c_n^p and a lower subblock c_n^d. In general, the upper subblock describes the movement of the trace along the X direction exclusively, and the lower subblock describes the movement along the Y direction exclusively.

The height H_n^p and the intensity of the upper and lower subblocks (I_n^p, I_n^d), respectively, are the three important properties computed from the trace samples. The height H_n^p defines the profile of the GTGF image representing the tactile pressure of the nth sample, and is proportional to the pressure value p_n (Equation (8.3)). The intensity I_n^p defines the brightness of the upper part of the image and represents the movement dynamics along the X direction. It depends on the direction and absolute value of the $\hat{x}_n$ gradient (Equation (8.1)). Similarly, the intensity I_n^b defines the brightness of the

Table 8.4   Gesture descriptions

	Gesture	Fingertips	Touch type
DU	Slide up	Thumb or index finger	Single
UD	Slide down	Thumb or index finger	Single
LR	Flick right	Thumb or index finger	Single
RL	Flick left	Thumb or index finger	Single
ZI	Pinch	Thumb and index finger	Multiple
ZO	Spread	Thumb and index finger	Multiple

lower part of the image and represents the movement dynamics along the Y direction. It depends on the direction and absolute value of the $\hat{y}_n$ gradient (Equation (8.2)). For traces $\hat{S}^0$ and $\hat{S}^1$ in a multiple fingertip gesture (i.e., Pinch, Spread), we extract two GTGFs T^0 and T^1, respectively:

$$I_n^p = \left\lceil 128 * \frac{U_x - \Delta\hat{x}_n}{U_x} \right\rceil \tag{8.1}$$

$$\Delta\hat{x}_n = \hat{x}_n - \hat{x}_{n-1}$$

$$I_n^b = \left\lceil 128 * \frac{U_y - \Delta\hat{y}_n}{U_y} \right\rceil \tag{8.2}$$

$$\Delta\hat{y}_n = \hat{y}_n - \hat{y}_{n-1}$$

$$H_n^p = \left\lceil \frac{H}{2} * \frac{\hat{p}_n}{L_p} \right\rceil \tag{8.3}$$

where $\lceil\rceil$ is the ceiling function.

The construction of GTGF has multiple advantages. First, original traces have different spatial topology and temporal duration. GTGF solves the difficulty in registering these traces into a canonical 2D template $\mathcal{T}$. Second, the dynamics are considered to be an important factor in other pattern recognition problems (e.g., facial expression recognition [38] and speech recognition [39]). However, they have

Figure 8.4   Examples of the GTGF extraction. The first row includes GTGF extracted from a user while the second row includes GTGF extracted from another user. The movement and pressure dynamics can be explicitly represented in GTGF

not been commonly considered in the touch gesture in the mobile authentication literature. The GTGF is able to represent the gesture dynamics in terms of movement and pressure intuitively and explicitly. Third, due to the inhomogeneity between movement and pressure data, it is difficult to combine their discriminative power at the feature level. However, GTGF takes both features into consideration and combines their discriminative power.

8.6.2   Statistical touch dynamics images

As mentioned earlier, comparing incoming touch gesture with stored gestures exhaustively is costly in terms of computation, time, and battery power. Instead of building a touch gesture selection hierarchy to assign a subset of directly comparable templates for the incoming probe gesture, we build STDIs to achieve the similar purpose. To capture intrapersonal variations while keeping the method computationally efficient, subject-specific STDIs are learned from the GTGF feature space as depicted in Figure 8.5. The expense of the score computation is then proportional only to the number of subjects rather than the number of features in the gallery.

STDI learns the intraclass variation modes from the training set and synthesizes the new instances using the learned base feature and affine combinations of these learned modes. In the testing phase, STDI expresses the probe trace in the new basis as synthesized instances. The distance between the synthesized probe instance and the original probe instance is computed for verification. This distance tends to be greater when the probe is from other users than genuine users on whose touch gestures STDI model is trained. Because STDI is learned with specific knowledge of the user, its representation power decreases for others. Thus, less similarity exists between the synthesized and original instances extracted from others' traces.

The training of an STDI is presented in Algorithm 1. It takes as input a set of touch traces per gesture type from the user and outputs the trained STDI. This training process runs repeatedly for all types of touch gestures (i.e., LR, RL, DU, UD, ZI, ZO) to finish the training process. Thus, six STDIs are trained for the six gestures, respectively, for each user.

Algorithm 1 Training Algorithm (Per Gesture Type)

Input: The training set of touch gesture (type t) traces $\mathbb{S}_u^t$ from the user u.
Output: The user-specific STDI D_u^t.

1. For each trace $S_i \in \mathbb{S}_u^t$:
    1. Extract the GTGF T_i (Section 8.6.1),
    2. Reshape the 2D matrix T_i into a vector t_i,
2. Learn the base vector $\bar{v}$ and basis feature vectors v_i by applying Statistical Analysis on $\{t_i\}$ (Equation (8.4)).

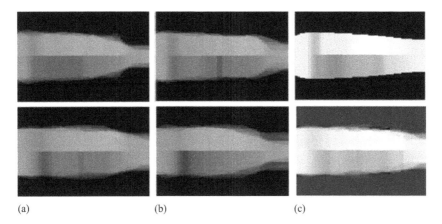

(a)   (b)   (c)

Figure 8.5   *A demonstration of the STDI in (8.4) and its instance synthesis. (a) The*
appearance of the STDI with the coefficient b_1 set to $3\sigma_1$ (top) and
$-3\sigma_1$ (bottom) with other coefficients fixed to 0; (b) the appearance of
the STDI with the coefficient b_2 set to $3\sigma_2$ (top) and $-3\sigma_2$ (bottom) with
other coefficients fixed to 0; (c) the probe feature (top) and the
synthesized instance (bottom), where the test instance and the STDI are
from the same user. It can be observed that the synthesized instances
tend to be more similar with their original features when the probe
instances originate from the same user as the STDIs. (Courtesy of Zhao
et al. [28].)

Because the gesture traces have been registered and scaled in feature extraction,
the GTGF can be directly used for training without further normalization or alignment.
For every $T \in \mathbb{R}^{H \times W}$ in the training set, it is first reshaped into $t \in \mathbb{R}^{HW}$. The lost
2D shape information due to reshaping can be easily restored by reverse reshaping
during score computation. As in (8.4), a vectorized feature v can be expressed as a
base $\bar{v}$ plus a linear combination of n basis feature vectors v_i. The standard approach
to computing the STDI is to apply principal component analysis (PCA) to the training
vectors [40]:

$$v = \bar{v}_0 + \sum_{i=1}^{n} b_i v_i = \bar{v}_0 + Vb \tag{8.4}$$

where the coefficients b_i are the feature parameters and the base $\bar{v}_0$ is the mean
of training vectors. The vectors v_i are n eigenvectors corresponding to the n largest
eigenvalues. The number n is determined based on the sum of the n largest eigenvalues
which is greater than a portion (κ) of the sum of all the eigenvalues. We can assume
that the vectors v_i are orthonormal after PCA and the parameters b_i are assumed to
follow the Gaussian distributions $\mathcal{N}(0, \sigma_i^2)$. We can further express the term $\sum_{i=1}^{n} b_i v_i$
as Vb.

The testing algorithm is presented in Algorithm 2. It takes a probe trace S_p and a trained model D as input, where the model D and the trace S_p share the same gesture type. The output is a score d representing the distance between the trace S_p and the user u represented by the trained model D_u'.

Algorithm 2 Testing Algorithm

Input: A probe trace S_p' and the trained model D_u'.
Output: The score d.

1.   Extract the GTGF T_p' from the probe trace S_p',
2.   Reshape T_p' into v_p',
3.   Synthesize the instance $\tilde{v}_p'$ using the parameters $\hat{b}^e$ (Equations (8.5)–(8.8)),
4.   Reversely reshape the vector $\tilde{v}_p'$ to $\tilde{T}_p'$
5.   Compute the score d between T_p' and its responding instance $\tilde{T}_p'$ (Equation (8.9)).

The trained STDI has the capability to synthesize different instances by varying the parameters b. Given an input v_p' for testing, its responding instances $\tilde{v}_p'$ can be synthesized from (8.4) using the estimated parameters b_i^x. The estimation is performed using (8.5):

$$b^e = V^T(v_p' - \bar{v}_0). \tag{8.5}$$

Instead of directly applying b^e to (8.4), we limit the range of these parameters by applying the function f to increase the separability between classes. The function f limits the distribution of the synthesized instances close to the training class in the feature space:

$$\hat{b}^e = f(b^e). \tag{8.6}$$

$$\text{for } b_i \in b^e, \; f(b_i) = \begin{cases} b_i & \text{if } abs(b_i) \le \rho\sigma_i \\ \rho\sigma_i \, \text{sgn}\,(b_i) & \text{otherwise} \end{cases} \tag{8.7}$$

Then, the corresponding instance $\tilde{v}_p'$ can be synthesized as follows:

$$\tilde{v}_p' = \bar{v}_0 + V\hat{b}^e. \tag{8.8}$$

The L_1 distance is computed between the input instance T_p' and its corresponding one $\tilde{T}_p'$ given the STDI D_u' as the score metrics.

$$L_1(\tilde{T}_p', T_p') = |T_p' - \tilde{T}_p'|. \tag{8.9}$$

8.6.3   *User authentication algorithms*

Single User Case: The user verification algorithm is designed for a single user. In this case, the six touch gestures from the genuine owner are collected. A set of user-specific

STDIs are trained for each of these gesture types. During the online testing stage, the type of the probe trace is retrieved first and the corresponding STDI is selected. Then, a distance is computed between the probe trace and the selected STDI, and is further compared with the threshold for verification.

Algorithm 3 The User Verification Algorithm

Input: Six user-specific STDIs D, the probe trace S_p, the threshold ξ.
Output: A boolean decision B.

1. Extract the GTGF T_p from S_p as in Section 8.6.1,
2. Select the STDI D^t from D that shares the same gesture type as the probe S_p,
3. Compute the score d given D^t and T_p, as in Algorithm 2.
4. Compute and output the B:
    1. if $d < \xi$, $B = true$; otherwise, $B = false$

Multiple User Case: The open set user recognition algorithm is designed for the case of multiple users, where user identity management (UIM) is needed. Mobile devices with UIM can reject unauthorized users and change the personalized settings according to the authorized user identity. Touch traces from all authorized users are collected and a set of six STDIs are trained for each user. During the online stage, the type of the probe trace is retrieved first and compared with all STDIs in the gallery from the same type. Distances are computed given the probe trace and the STDIs. Note that the distance computation only needs to be repeated for all subjects, instead of all traces in the gallery. The minimum distance is then selected and compared with the threshold for authentication. Only if it is lower than a personalized threshold, its corresponding identity is returned as the result.

Algorithm 4 The User Identification Algorithm

Input: Sets of STDI D_u for the user group, the probe trace S_p, the threshold ξ.
Output: The user identity $\mathscr{I}$.

1. Extract the GTGF T_p from S_p as in Section 8.6.1,
2. For each subject u:
    1. Select the STDI D_u^t from D_u which shares the same gesture type as the probe S_p,
    2. Compute the score d_u given D_u^t and T_p, as in Algorithm 2,
3. find the smallest d_m among all d_us,
4. if $d_m \geqslant \xi_m$,
    then $\mathscr{I} = Null$,
    else $\mathscr{I}$ is the identity corresponding to the d_m.

Table 8.5   A comparison of EER and RR for different fusion schemes and methods

	GTGF-S	GTGF-M	STDI-S	STDI-M	GTGF-A	STDI-A
EER	6.3%	10.5%	4.7%	9.6%	4.6%	4.1%
RR	84.1%	79.2%	91.5%	80.7%	86.9%	88.4%

8.6.4   Experimental results

Evaluation of GTGF and STDI performance is conducted with a dataset including 78 subjects. The data acquisition contains six sessions collected over several months. The first session served as a pilot session, where only 30 subjects were recruited. All 78 subjects participated in all the following sessions from the second to the sixth. The explanation and practice procedure were repeated for the newly recruited subjects in the second session.

Performance of STDI-based user authentication was evaluated by using different combinations of touch gestures (combining single-touch tip gestures, combining multiple touch tip gestures, and combining all six gestures) and comparing the method with the literature. After score computation, we obtained one score matrix for each gesture. Then, we fused the score matrix of multiple gestures using the sum rule, which is proved by Kitller *et al.* [41] to be superior in comparison to other rules (i.e., product, min, max, median rule). Table 8.5 depicts a comparison of the EER and RR for different fusion schemes and methods: (i) GTGF-S: fusion of the score matrices of four single-touch tip gestures computed from the aforementioned GTGF method [27]; (ii) GTGF-M: fusion of the score matrices of the two multiple touch tip gestures computed from the GTGF method [27]; (iii) STDI-S: fusion of the score matrices of four single touch tip gestures computed from the STDI method; (iv) STDI-M: fusion of the score matrices of the two multiple touch tip gestures computed from the STDI method; (v) GTGF-A: fusion of the score matrices of the six gestures computed from the aforementioned GTGF method [27]; (vi) STDI-A: fusion of the score matrices of the six gestures computed from the STDI method. As in most score-level fusion schemes, combining more scores from different channels increases the overall performance. This trend can be observed from the decrease in EER and increase in RR from the STDI-M (including only two multitouch gestures) to the STDI-S (including four single-touch gestures) and further to the STDI-A schemes (including all the six gestures).

To test the performance and usability of STDI in a real-world scenario, we embedded the method as the user authentication module into the aforementioned user authentication schema on Android devices. Twenty subjects' data were acquired. Each subject used the device (a Samsung Galaxy S3 phone with Android version 4.1.1) for three days as their daily phone and their touch traces in the context of Android Launcher were recorded. Open set user recognition is excluded from this test because the Android system (version 4.1.x) does not currently support multiple user

Table 8.6   Average computational time (ms) per probing trace

Gallery size	30	50	80	100
GTGF	32.22	46.26	67.32	81.36
STDI	12.83	12.90	12.87	12.92

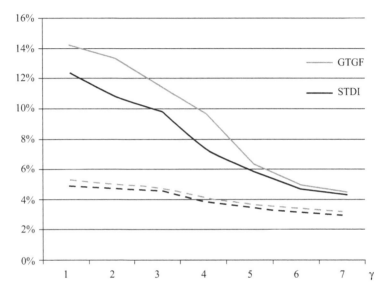

Figure 8.6   A comparison of the FRR and FAR results for continuous authentication for the GTGF and the STDI methods. The Y axis is the percentage of the FAR and FRR; the X axis is the number of consecutive traces (γ) used in continuous authentication. The solid lines represent the FRR results and the dashed lines represent the FAR results

account management. The distance from the probe trace and the distances from the following $\gamma - 1$ consecutive probe traces are averaged to obtain the score for authentication. If it is above the threshold, the user is judged as an unauthorized intruder and invokes explicit user authentication. Table 8.6 depicts the average computational time per probe trace used with different gallery size at the online testing stage. As the gallery size increased from 30 traces per gesture to 100 traces, the computational time for GTGF also increased significantly from 32.22 to 81.36 ms. However, the computational time for STDI remains at around 12.90 ms.

Figure 8.6 compares the continuous authentication results for GTGF and STDI in terms of the average FRR (solid line) and average FAR (dash line). The X axis is the number of consecutive traces used in decision making. FARs for all methods reduced in a small but steady rate as the γ increased. Meanwhile, the FRRs of all methods decreased significantly as the number of traces increased. For the GTGF and STDI methods, the decreasing rate reduced after $\gamma > 5$ and almost converged when $\gamma = 7$. When $\gamma > 5$, the delay caused by heavier computational burden became noticeable for GTGF. However, no obvious delay has been observed for the STDI method, even when $\gamma = 7$.

8.7 Virtual key typing-based method

8.7.1 Feature extraction

The data demonstrated in Table 8.7 depict different typing habits of two users when typing the characters "H" and "E". User2 tends to take more time than User1 to type "E" after typing an "H". In Table 8.8, the result demonstrates that when typing the character "R", typically User3 will press much longer than User4. And Table 8.9 demonstrates the typing pressure difference between two users when typing the character "E", which is one of the most frequently used characters. User5 normally presses the virtual keyboard harder than User6. The aforementioned variations are due to many factors, such as finger size, holding habit, typing habit, finger muscles, etc.

Thus, features from the raw typing data are extracted to represent user identity. First, letter combinations are extracted in the login stage. The 40 most frequently used key combinations are recorded, chosen based on the Statistical Distributions

Table 8.7   Two users' time interval data when typing characters "H" and "E" (courtesy of Tao et al. [42])

Time Interval of "H" and "E"	Sample numbers									
	1	2	3	4	5	6	7	8	9	10
User1	1	2	2.2	2.1	2.3	5.1	2.1	2.2	2.2	2.2
User2	4.8	9	6.1	7.8	6.1	5.4	5	5.1	4.5	6.1

Table 8.8   Two users' press time data when typing character "R" (courtesy of Tao et al. [42])

holding time of "R"	Sample numbers									
	1	2	3	4	5	6	7	8	9	10
User3	20	18	11	12	14	11	10	8	14	10
User4	6	6	6	6	6	7	6	7	7	6

of English Text Research result [43] in the postlogin stage. This is because to calculate all the possible virtual key combinations may introduce an overhead for the smartphone's computational ability. Then, because users often have unique typing behaviors regarding how long a virtual key is held during virtual typing, the holding time of each key is extracted into our feature set. Additionally, the tactile pressure provided by the capacitive touch screen is recorded.

In total, 40 features are extracted from virtual key combinations, 41 from key holding times (for the characters used in our virtual keyboard) for the data without pressure information, and 41 more pressure features from the data with pressure information.

8.7.2 User authentication

Three types of classifiers have been adopted, including Random Forest, Decision Tree, and Bayes Net. The classification results are used to improve smartphone security in the following scenarios. First, during login, the method explicitly asks the user to input a password and subsequently extracts features of the virtual key typing inputs. Second, in the postlogin phase, the method applies virtual typing-based user recognition whenever a smartphone user enters inputs using the virtual keyboard.

Different standards and goals are pursued for different authentication scenarios in terms of FAR and FRR. For login control, a low FAR is the primary objective—during login security is more important than usability. Postlogin, using virtual typing verification, a low FRR is the primary objective—during normal user–smartphone interactions, usability is more important. This is because the frequency of the interactions ensures a rapid detection of intruders even for larger FAR values.

During the postlogin phase, authenticity of the mobile user is continuously monitored in a user-transparent fashion. The method achieves this by intercepting virtual typing inputs, and strives to achieve a low FRR. However, a user's virtual keyboard typing inputs may vary in time. Solutions that rely on just a single input instance of virtual keyboard typing are unlikely to be reliable and accurate. Thus, sequence aggregation is adopted. To control the quality of the aggregated user verification performance, we use two metrics: the (i) the *sliding window length* (SWL), the length of touch input sequences, and (ii) the *authentication threshold* (AT), for aggregating results. The AT metric is used as a reference for setting the lower bound on the SWL: If the number of accepted touch inputs during one sequence is below the threshold,

Table 8.9 Two users' pressure data when typing character "E" (courtesy of Tao et al. [42])

Pressure of "E"	Sample numbers									
	1	2	3	4	5	6	7	8	9	10
User5	0.66	0.67	0.64	0.64	0.68	0.64	0.68	0.66	0.66	0.68
User6	0.6	0.59	0.59	0.57	0.58	0.59	0.6	0.58	0.6	0.59

Table 8.10   Result of login stage

Algorithm	Only Time Feature		Time and Pressure Feature		Time and Pressure under Haptics	
	FAR	FRR	FAR	FRR	FAR	FRR
J48	19.3%	55%	27%	22.2%	21.5%	17.3%
RF	25.2%	25%	22.7%	11.1%	19.2%	11%
BN	17.8%	60%	10.4%	11.1%	16.9%	10.2%

the method considers that the current user is unauthorized and invokes an explicit authentication process.

8.7.3   Experiment results

Our virtual typing authentication aims to authenticate user identity both in the login stage and the background in the postlogin stage. Thus, different strategies need to be implemented in the two different scenarios. For the login stage, security is more important, so security should not be compromised for usability. The online evaluation involved 40 subjects and lasted for about 3 weeks.

8.7.3.1   Login stage result

The performance result of the login stage authentication is presented in Table 8.10. J48, RF, and BN denote J48 decision tree, random forest, and Bayes Net algorithms, respectively. Time Feature, Time & Pressure Feature, and Time & Pressure under Haptics stand for authentication using only time features, including pressing time and time interval, authentication using both time and pressure features, and authentication using both time and pressure features with haptics feedback, respectively. The authentication performance employing only time features is not good enough to maintain both security and usability. The authentication performances employing both pressure and time features with or without the haptics feedback are better and roughly at a same level due to the adoption of the pressure information. Both of them decrease the FRR to around 11.0% with a FAR, respective, of 10.4% and 16.9% using Bayes Net.

8.7.3.2   Postlogin stage result

In the postlogin stage, achieving a low FRR is vital for its usability, while a reasonable FAR is sufficient to provide good security levels. In order not to disturb an authorized user, the method leverages virtual typing input sequences rather than single or short virtual typing inputs. If none of the input units within a sliding window is recognized as inputs from an authorized user, the user is locked out of the system.

Figure 8.7 depicts the verification results with haptics feedback. The best result with 3.8 FAR and 0.7 FRR is achieved when the input length is 40 characters. It can be observed that when typing with haptics feedback, the input character length can be further reduced from 40 to 15 to reach a sufficient authentication performance with

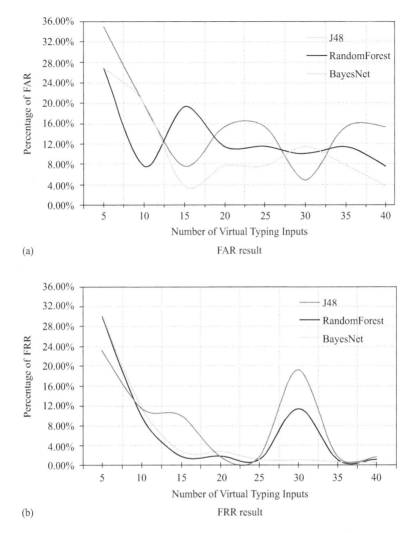

(a)   FAR result

(b)   FRR result

Figure 8.7   *Virtual key typing FAR and FRR performance results with pressure data*
under different classification algorithms and typing sequence lengths
with haptics feedback. (Courtesy of Feng et al. [42].)

3.8% FAR and 2.8% FRR. This greatly improves security because the input length is
shortened over 62.5% ((40 − 15)/40).

8.8   Conclusion

We have described the threats encountered by the pervasive mobile devices, and that
the goal of continuous user authentication is to balance security and usability during

natural user–phone interaction. Literature on two types of user authentication has been reviewed, including touch gesture-based and keystroke-based authentication methods. General touch gesture features have also been elaborated as the feature basis for both types of methods.

Three methods have been described: the first one using dynamic time warping, the second one using GTGF and statistical modelling, and the last one using virtual key typing features. The first two methods are good examples of continuous user authentication, which extract user identity from touch gestures throughout the usage sessions. The last method is a good example of user authentication, which takes effects when users apply virtual key typing. We have evaluated the effectiveness of the methods using collected databases as well as online testing user studies.

Acknowledgments

Partially supported by the National Nature Science Foundation of China, under Award No. 61303121, No. 61472313, No. 11401464, and the Department of Homeland Security (DHS) under Award Number N66001-13-C-3002. We thank the following individuals for help in this work: Guodong Guo, Xiaoming Liu, Arun Ross, Kevin Bowyer, Patrick Flynn, Zhimin Gao, and Yang Lu.

References

[1]   Worldwide Smartphone Markets: 2011 to 2015—Analysis, Data, Insight and Forecasts;. Available online at http://www.researchandmarkets.com/reports/1871240/worldwide_smartphone_markets_2011_to_2015.

[2]   Weir M, Aggarwal S, Collins M, and Stern H. Testing metrics for password creation policies by attacking large sets of revealed passwords. In: Proc. 17th ACM Conference on Computer and Communications Security; 2010. p. 162–175.

[3]   Jakobsson M, Shi E, Golle P, and Chow R. Implicit authentication for mobile devices. In: Proc. Fourth USENIX Conference on Hot Topics in Security; 2009. p. 1–9.

[4]   Li S, and Kot AC. Fingerprint combination for privacy protection. IEEE Transactions on Information Forensics and Security. 2013;8(2):350–360.

[5]   Zhao X, Shah SK, and Kakadiaris IA. Illumination normalization using self-lighting ratios for 3D2D face recognition. In: Proc. 12th European Conference on Computer Vision, Workshops and Demonstrations. Firenze, Italy; 2012. p. 220–229.

[6]   Messerman A, Mustafic T, Camtepe SA, and Albayrak S. A generic framework and runtime environment for development and evaluation of behavioral biometrics solutions. In: Proc. 10th International Conference on Intelligent Systems Design and Applications, Cairo, Egypt; 2010. p. 136–141.

[7]   Gafurov D, and Snekkenes E. Arm swing as a weak biometric for unobtrusive user authentication. In: Proc. IEEE International Conference on Intelligent Information Hiding and Multimedia Signal Processing, Harbin, China; 2008. p. 1080–1087.

[8]   Feng T, Zhao X, and Shi W. Investigating mobile device picking-up motion as a novel biometric modality. In: Proc. IEEE Sixth International Conference on Biometrics: Theory, Applications and Systems. Washington, DC; 2013. p. 1–6.

[9]   Faundez-Zanuy M. On-line signature recognition based on VQ-DTW. Pattern Recognition. 2007;40(3):981–992.

[10] Kholmatov Ar, and Yanikoglu B. Biometric authentication using online signatures. In: Computer and Information Sciences 2. Springer; 2004. p. 373–380.

[11] Shahzad M, Liu A X, and Samuel A. Secure unlocking of mobile touch screen devices by simple gestures: you can see it but you can not do it. In: Proc. 19th ACM Annual International Conference on Mobile Computing & Networking, Miami, FL, USA; 2013. p. 39–50.

[12] Sherman M, Clark G, Yang Y, *et al.* User-generated free-form gestures for authentication: Security and memorability. In: Proc. 12th ACM Annual International Conference on Mobile Systems, Applications, and Services, Bretton Woods, NH, USA; 2014. p. 176–189.

[13] Yamazaki Y, Mizutani Y, and Komatsu N. Extraction of personal features from stroke shape, writing pressure and pen inclination in ordinary characters. In: Proc. Fifth IEEE International Conference on Document Analysis and Recognition; 1999. p. 426–429.

[14] Cao X, and Zhai S. Modeling human performance of pen stroke gestures. In: Proc. SIGCHI Conference on Human Factors in Computing Systems. San Jose, CA: ACM; 2007. p. 1495–1504.

[15] Ferrer MA, Morales A, and Vargas J. Off-line signature verification using local patterns. In: Proc. Second IEEE National Conference on Telecommunications, Catolica San Pablo, Arequipa, Peru; 2011. p. 1–6.

[16] Bailador G, Sanchez-Avila C, Guerra-Casanova J, and de Santos Sierra A. Analysis of pattern recognition techniques for in-air signature biometrics. Pattern Recognition. 2011;44(10):2468–2478.

[17] Impedovo D, and Pirlo G. Automatic signature verification: the state of the art. IEEE Transactions on Systems, Man, and Cybernetics, Part C: Applications and Reviews. 2008;38(5):609–635.

[18] Shi W, Yang F, Jiang Y, Yang F, and Xiong Y. Senguard: Passive user identification on smartphones using multiple sensors. In: Proc. Seventh IEEE International Conference on Wireless and Mobile Computing, Networking and Communications; 2011. p. 141–148.

[19] Li L, Zhao X, and Xue G. Unobservable Re-authentication for Smartphones. In: Proc. 20th Annual Network & Distributed System Security Symposium. San Diego, CA; 2013. (Available online at http://internetsociety.org/doc/unobservable-re-authentication-smartphones).

[20] Xu H, Zhou Y, and Lyu MR. Towards continuous and passive authentication via touch biometrics: an experimental study on smartphones. In: Symposium on Usable Privacy and Security; 2014. p. 187–198.

[21] Bo C, Zhang L, Li XY, Huang Q, and Wang Y. Silentsense: silent user identification via touch and movement behavioral biometrics. In: Proc. 19th

ACM Annual International Conference on Mobile Computing & Networking; 2013. p. 187–190.

[22]   Saravanan P, Clarke S, Chau DHP, and Zha H. LatentGesture: active user authentication through background touch analysis. In: Proc. Second International Symposium of Chinese CHI. ACM; 2014. p. 110–113.

[23]   Feng T, Liu Z, Kwon KA *et al.* Continuous mobile authentication using touchscreen gestures. In: Proc. IEEE Conference on Technologies for Homeland Security. Waltham, MA; 2012. p. 451–456.

[24]   De Luca A, Hang A, Brudy F, Lindner C, and Hussmann H. Touch me once and i know it's you!: implicit authentication based on touch screen patterns. In: Proc. SIGCHI Conference on Human Factors in Computing Systems. ACM; 2012. p. 987–996.

[25]   Sae-Bae N, Memon N, and Isbister K. Investigating multi-touch gestures as a novel biometric modality. In: Proc. IEEE Fifth International Conference on Biometrics: Theory, Applications and Systems. Washington, DC; 2012. p. 156–161.

[26]   Frank M, Biedert R, Ma ED, Martinovic I, and Song D. Touchalytics: On the applicability of touchscreen input as a behavioral biometric for continuous authentication. Proc IEEE Transactions on Information Forensics and Security. 2013;8(1):136–148.

[27]   Zhao X, Feng T, and Shi W. Continuous mobile authentication using a novel graphic touch gesture feature. In: Proc. IEEE Sixth International Conference on Biometrics: Theory, Applications and Systems. Washington, DC; 2013. p. 1–6.

[28]   Zhao X, Feng T, Shi W, and Kakadiaris IA. Mobile user authentication using statistical touch dynamics images. Proc IEEE Transactions on Information Forensics and Security. 2014;9(11):1780–1789.

[29]   Clarke NL, and Furnell S. Authenticating mobile phone users using keystroke analysis. International Journal of Information Security. 2007;6(1):1–14.

[30]   Zahid S, Shahzad M, Khayam SA, and Farooq M. Keystroke-based user identification on smart phones. In: Proc. 12th International Symposium on Recent Advances in Intrusion Detection. Saint-Malo, France; 2009. p. 224–243.

[31]   Campisi P, Maiorana E, Lo Bosco M, and Neri A. User authentication using keystroke dynamics for cellular phones. Signal Processing, IET. 2009;3(4):333–341.

[32]   Trojahn M, and Ortmeier F. Biometric authentication through a virtual keyboard for smartphones. International Journal of Computer Science & Information Technology. 2012;4(5):1–12.

[33]   Saevanee H, and Bhattarakosol P. Authenticating user using keystroke dynamics and finger pressure. In: Proc. Sixth IEEE Consumer Communications and Networking Conference. Las Vegas, NV; 2009. p. 1–2.

[34]   Maiorana E, Campisi P, González-Carballo N, and Neri A. Keystroke dynamics authentication for mobile phones. In: Proc. 26th ACM Symposium on Applied Computing. Taichung, Taiwan; 2011. p. 21–26.

[35]   Hwang Ss, Cho S, and Park S. Keystroke dynamics-based authentication for mobile devices. Computers & Security. 2009;28(1):85–93.

[36] Chang TY, Tsai CJ, and Lin JH. A graphical-based password keystroke dynamic authentication system for touch screen handheld mobile devices. Journal of Systems and Software. 2012;85(5):1157–1165.

[37] Feng T, Yang J, Yan Z, Tapia EM, and Shi W. TIPS: Context-aware implicit user identification using touch screen in uncontrolled environments. In: Proc. of the 15th ACM Workshop on Mobile Computing Systems and Applications. Santa Barbara, CA; 2014. p. 9.

[38] Zeng Z, Pantic M, Roisman GI, and Huang TS. A survey of affect recognition methods: audio, visual, and spontaneous expressions. IEEE Transactions on Pattern Analysis and Machine Intelligence. 2009;31(1):39–58.

[39] Hasan T, and Hansen JH. A study on universal background model training in speaker verification. IEEE Transactions on Audio, Speech, and Language Processing,. 2011;19(7):1890–1899.

[40] Zhao X, Dellandréa E, Zou J, and Chen L. A unified probabilistic framework for automatic 3D facial expression analysis based on a Bayesian belief inference and statistical feature models. Image and Vision Computing. 2013;31(3):231–245.

[41] Kittler J, Hatef M, Duin RP, and Matas J. On combining classifiers. IEEE Transactions on Pattern Analysis and Machine Intelligence. 1998;20(3):226–239.

[42] Feng T, Zhao X, Carbunar B, and Shi W. Continuous mobile authentication using virtual key typing biometrics. In: Proc. 12th IEEE International Conference on Trust, Security and Privacy in Computing and Communications. Melbourne, Australia; 2013. p. 1547–1552.

[43] Data-compression. http://www.data-compression.com/english.

Chapter 9
Smartwatch-based gait biometrics
Andrew Johnston[1]

9.1 Introduction

Smartwatches are the newest trend in mobile technology, with over 216 million devices expected to be sold globally by the conclusion of 2016 [16]. These devices offer much of the same technology, including sensors, antennae (Bluetooth and Wi-Fi), and operating systems as their smartphone counterparts. Consequently, smartwatches create a new dimension for mobile gait biometrics.

Significant research has been conducted on smartphone-based gait biometric systems. Although the performance of these systems has increased steadily over the years, as of writing, there are no popular commercial implementations of gait biometric systems. This can be attributed to the "orientation problem" and the "off-body carry problem." The orientation problem lies in the fact that many of these systems assume that a person will carry a phone in a specific orientation, namely, in a front pocket, screen upwards (i.e., top of the screen is closest to the person's waist) and screen facing out (i.e., away from the leg). Although such an assumption is fine for a research scenario, it is immediately apparent that such an assumption will fail in any realistic implementation. Further, as smartphones continue to grow in size (with flagship phones such as the Nexus 6P, Galaxy Note 5, and iPhone 6S having screens in excess of 5.5 in.), many people are opting to carry the phone in "off-body" positions, such as in coats, backpacks, or handbags. These positions are unlikely to guarantee the device will be close to the body and is therefore less suitable for gait biometric purposes. Certain strategies (such as orientation-invariant features [17]) have been proposed, but they are both complicated to implement and still make unrealistic assumptions to compensate (e.g., orientation-invariant studies assumed consistent placement of the device).

Smartwatches are an attractive option because they solve many of the problems that have implicitly plagued traditional smartphone-based systems. Smartwatches are by definition watches, and people typically wear a watch in a consistent location on a particular wrist. Similarly, smartwatch screens do not orient themselves as they rotate (much in the way that a smartphone would). While seemingly insignificant, this guarantees that the smartwatch has consistent orientation every time it is put on (or the screen would be upside down). With these two properties, smartwatches have solved the orientation and off-body problems without unreasonable constraints on

[1]Fordham University, USA

the user. This suggests that if secure smartwatch-based gait biometric system can be created, it has potential to be the first effective implementation.

This chapter will focus on the considerations a researcher or engineer should undertake when designing a smartwatch-based gait biometric system. The steps are presented in the order in which they should be considered—from hardware to how the ultimate system's fitness should be evaluated. Effort will be made to analyze and explain the effects of varying design decisions using references to papers that reflect the current state of the art. In addition, further reading will be suggested at the end of the chapter; whereas this may be of less value to those focused on constructing real-world implementations (i.e., engineers), it will be of great use to those wishing to conduct further research in the field.

N.B. Given the infancy of smartwatches, there exists little research that uses smartwatches as the data collection device. Therefore, in order to address the topic adequately, many of the papers referenced will focus on smartphones; however, they will be used in context to discuss the construction and evaluation of smartwatch systems.

9.2 Smartwatch hardware

Most modern smartwatches contain MEMS (i.e., MicroElectroMechanical System) sensors, which are a single chip that offers both triaxial gyroscope and accelerometer capabilities. Both gyroscopes, offering instantaneous rotational velocities and accelerometers, offering instantaneous accelerations, can be used on their own as the basis for a biometric system. Despite the prevalence of both sensors on devices, accelerometers are overwhelmingly favored. This is most likely because research that tests both accelerometers and gyroscopes as sources demonstrate that the accelerometer performs significantly better than gyroscopes at all biometric tasks [7].

Smartwatches provide a great deal of resources to developers despite small form factors and a relatively recent introduction to the consumer marketplace. Accordingly, these devices offer limited computational and battery power as compared to smartphones. Although the hardware specifications are suboptimal, they do not pose a great problem to the implementation of biometric systems, as current smartwatches exist in an "ecosystem" with a paired smartphone, allowing the watch to offset some of the computational and communicational load onto the more powerful phone. Therefore, many biometric systems (such as [7]) will use the smartwatch solely to collect data and transmit the data to the paired smartphone. The smartphone is then responsible for either processing the data itself if the model is located on the smart device or transmitting the data to a centralized system if the model is stored remotely.

9.3 Biometric tasks: identification and authentication

Neither smartwatches nor biometrics itself is limited to authentication; identification (in both closed- and open-set form) is a task which smartwatch-based systems are readily capable of performing. Understanding the differences between the types of systems (and what type of system an implementation will model) is a crucial factor in the decision making process in later steps of the implementation. Here, these types of systems are described in the context of constructing smartwatch-based systems for each task.

9.3.1 Identification

Identification is the task of recognizing a user from a greater population of users. True to the mathematical definitions of "closed" and "open" sets, "closed-set" identification involves recognizing a user from a set of known users. In contrast, "open-set" identification (also known as the "watchlist" task) involves identifying a user from a pool of users from whom it is not possible to obtain samples from everyone else.

The suggested use of identification-style models is detecting which owner of the device is the current user. For smartwatches, this may have less use than for traditional mobile device systems, as watches may not be as sharable as phones or other devices. Consequently, identification might be limited to determine if the current user is a legitimate one or if the device has been stolen. Given that potential future applications might find greater use for identification-based tasks, both are discussed.

9.3.1.1 Closed-set identification

Closed-set identification is the easier of the identification tasks. In this task, samples from every user within the population are received. Thus, the identification system is a classification system with the number of classes equal to the number of people within the population (i.e., the "users") of the system. In these systems, it is logical that the ability of the system to classify a user correctly is proportional to the size of the population. At the time of writing, the largest closed-set identification system that uses gait utilized data from 59 users [7] collected on a single day. Figure 9.1 shows the effect on accuracy of adding users (5 at a time) to the population. Note that accuracy does not decrease significantly; this suggests that closed-set identification may be possible for a population size of a few hundred users, while maintaining an acceptable level of accuracy. Note that for a sufficiently advanced feature set, accuracy does not rapidly diminish to unusable levels as large populations (i.e., those over 50 users) are reached. Although research has yet to be done on larger populations, this suggests that larger scale identification is possible.

Given that all users within the population provide data, there are no limitations on the types of models required. Specifically, models ranging from simple [e.g., Naïve Bayes and k-Nearest Neighbor (k-NN)] to the very complex (e.g., Random Forest, Multilayer Perceptron, etc.) are all suitable for the task.

9.3.1.2 Open-set identification

Open-set identification is unarguably more challenging because by assumption only a selected few users (i.e., those on the "watchlist") actually provide data. Therefore, the

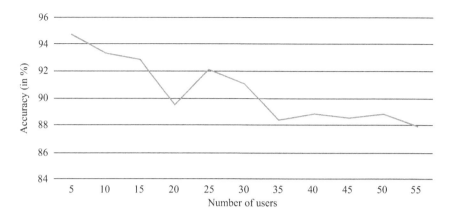

Figure 9.1   Identification accuracy versus population size

task of recognizing these individuals alone becomes much more challenging. In this scenario, there are two approaches: constructing an artificial imposter or similarity-based approaches. The artificial imposter approach relies on a classifier with a number of classes equal to one more than the number of people on the watchlist, with the additional class representing the "imposter." The imposter is constructed with data from a variety of users that are not on the waitlist; this merged imposter can be constructed either from data collected from users not on the waitlist or taken from publically available gait databases. In essence, this is a generalized form of the authentication model described in [7], where a metaimposter (i.e., an imposter created by taking multiple instances from other random users) was created and the classifier had to discern between a genuine user and the imposter. Alternatively, a system could use a k-NN style approach using a particularly distance metric. Such a system would be trained only with the data from those on the watchlist. Classification would involve examining the distance between the testing instance and the closest neighbor. If this distance exceeded some predefined measure, the testing data would be classified as not matching anyone on the watchlist. The accuracy of such a system is harder to assess, as it depends greatly on the choice of distance metric, number of users on the watchlist, and the particular threshold chosen.

Gait-based open-set identification, much like authentication, suffers from a potential problem of scaling. The semiperiodic nature of gait means that an open-set system could be overfit to recognize the general motion of walking as opposed to individual-specific traits. Further, since smartwatches are personal devices, the uses of open-set identification may be limited. If a smartwatch-based open-set system is to be constructed, there must be a careful consideration of features to ensure that the data can be collected from each individual of interest and that the data can be processed in such a way as to remove as much of the periodic nature from the signal as possible.

9.3.2 Authentication

Authentication is the most popular task in gait biometrics, with the primary goal being to verify the claimed identity of a given user. It is therefore a more constrained version of closed-set identification, as the goal of the system is to determine the similarity between a test sample and a single user; with identification, the system must determine the similarity between the test sample and all users within the population. Authentication is the most easily applicable use of smartwatch-based gait biometrics, as the security of the device is a significant concern and the small form of the device does not lend to more traditional authentication methods such as passwords.

Many systems implement authentication in a style that fails to exploit this additional constraint, which should make the task easier (and thus increase the baseline accuracy). These studies create a single "template" from the entirety of a user's training data. In the simplest form, these systems then compare the similarity of the test data, formed into a similar "template," to the learned template. If it is within a specific threshold, then the user is accepted. Although this strategy may be optimal in certain domains (e.g., fingerprints, irises, etc.) where the biological medium lends itself to a system where a single template is logical, it is not an effective strategy gait data. Gait is a time-domain signal, and thus lends itself to the creation of multiple instances over a singular template. Some studies [4] even go as far as to create a feature vector and then construct a template from averaging each feature. From an information theoretic standpoint, it is most likely detrimental to quantize a signal needlessly (which is the essence of constructing a template from raw data), as each level of quantization involves a level of information loss. From a statistical perspective, averaging a set of features is a poor choice unless the data has low variance for similar reasons (i.e., the template may not actually reflect the data from which it was constructed). Therefore, strategies that are more effective instead rely on training a system using a multitude of instances.

Similarly, certain studies [5] create systems that lose effectiveness as the number of users participating in the authentication system increases. Although this is a grim necessity for closed-set identification, a well-constructed authentication system should not suffer from the same problem. An effective approach is the construction of individual authentication models for each user. As discussed previously, each model can be tasked with determining if test data originate from the given user (the "genuine" user) or from a "metaimposter" created from data randomly selected from a few other users. This type of system does not require an excessive amount of other users to construct an adequate metaimposter; one study achieved very strong results (i.e., an EER of 1.4%)[1] using only four other users to construct the metaimposter [7]. Nevertheless, the style of system is not perfect; it suffers from the "cold-boot" problem in that a new system may not have enough users to draw four random users (or at the very least not have enough users to select four users in a manner that is sufficiently random). This problem can only be counteracted with the usage of publically available smartwatch-based gait databases, or by only allowing the system into "production"

---

[1]For a discussion of the evaluation metrics, see "Selecting an Evaluation Metric" later in the chapter.

(i.e., let end users start using the system) once a sufficient number of users have been collected. Ultimately, the choice in authentication strategies will depend on an organization's security requirements and deadlines.

9.4 Data preprocessing

Data preprocessing is an often overlooked yet fundamental part of the smartwatch gait biometric systems. Although the field of signal process offers numerous techniques for extracting valuable information from a potentially noisy signal, many cannot be used (or must be used with extreme caution). By definition, gait-based authentication and identification work on the principle that there exist unique features within an individual's gait, despite gait being a semiperiodic motion. Many signal processing techniques focus on smoothing a signal in order to highlight the most dominant frequencies; if this is done for a gait signal, all of the unique characteristics will be lost and the system performance will suffer.

This is not to suggest that all signal processing techniques should be neglected. Some noise, present as high-frequency data, can be removed. It is important to note that Android Wear, the operating system present on many smartwatches, does not guarantee consistent sampling rates [6]. Consequently, many developers choose a very high rate (or configure Android to sample the sensor at the maximum rate allowed). Although this is an acceptable choice, it will likely introduce significant noise. The Nyquist–Shannon Theorem tells us that a signal can be perfect reproduced at B Hz if it contains no frequencies higher than half of B [15]. Thus, sampling at high frequencies is only necessary if the system designer believes that there is significant high-frequency motion within an individual's gait. Although this may be the case, strong results can be obtained with a sampling rate of only 20 Hz [7]. If research is the intended usage of the data, it may be worthwhile to collect data at a high frequency and apply different Butterworth filters in order to test which frequency is optimal for a given system. For practical applications, a low sampling rate offers increased battery life on the smartwatch and most likely contains all relevant information.

Interpolation is a signal processing strategy that may be used to counteract the inconsistencies incurred through the use of consumer hardware. It is very possible that the smartwatch fails to provide data for every point, or provides data for a time point that is close to the expected one (as determined by sampling frequency and start time). Interpolation in various forms can be used to approximate the value at the correct time. Although interpolation would most likely harm accuracy if used to an extreme, in isolated situations (such as missing data points) it could potentially benefit the system. There is a dearth of research on the effect of the imprecise nature of consumer-grade sensors, but it is valuable to consider, especially for systems (such as cycle-based methods or segmentation-based methods with very small windows) that rely on a specific number of frames in each instance.

Smartwatches face specific challenges with regard to cleaning data that are not present in other mobile device systems. Phone-based systems can be affected if the user handles the device during collection or if the user ceases walking early.

Smartwatch systems are further affected by the user doing anything that would interfere with the natural swing of the arm, such as checking the watch, putting hands in their pocket, or if the user's arms are constrained by thick clothing such as heavy jackets. In a research scenario, these examples can be rejected, but in deployed systems, this level of control over data collection may not be possible. More generally, samples provided by a user could be rejected if they appear to have abnormal levels of movement. An alternative to this approach would be the use of larger segments (discussed below), which would be less affected by minor deviations in the natural movement of the arm.

9.4.1   Segmentation

Segmentation is the most critical aspect of preprocessing as it will ultimately determine the types of features used and model selection. The two primary methods are segmentation-based (e.g., window) approaches and cycle detection. Each takes a different approach as to how walks are best analyzed: segmentation-based methods focus on a period of activity, whereas cycle-based methods attempt to analyze gait through finding pairs of steps (i.e., one step with each leg).

Segmentation-based methods start with determining a window size as a fixed length of time. Although longer segments are more resistant to small problems (e.g., the user pausing very briefly), they necessitate longer training times and require more testing data. If a longer window is preferred, overlapping (where one window uses data used in a previous "window") may be preferred to increase the number of instances derived for training; typically, a 50% overlap is selected. Segmentation offers strong results without significant computation required as opposed to other strategies, but at the cost of being invariant to the number of steps in any given window. Therefore, changes in pace will greatly affect the features generated.

Cycle-detection methods aim to be invariant to pace by standardizing the number of steps as opposed to the amount of time represented in each instance. The challenge with this method is selecting a strategy that detects the beginning and end of each cycle. Strategies range from simplistic (finding minimum points exceeding a particular threshold value [5]) to much more complicated (convolving the signal with specific wavelets [2]). Further, cycle detection is normally used in conjunction with a process to select "good" cycles, either by determining if motion was significant through the use of thresholds or through manual evaluation. Consequently, a significant amount of training data provided by an individual may be eliminated if not enough valid cycles are detected, thus requiring multiple training sessions. Further, some cycle-detection methods fail if the user changes their pace or maintains too quick a pace [8], which negate some of the usefulness of this method. Cycle detection can obtain results as strong as segmentation-based methods (as high as 99.4% [8]) in shorter time periods, but may not be a good choice for decentralized systems, as discussed in the next section.

9.4.2   Segment selection

Especially, if the user must be prompted to begin walking, it is likely that the user will not be walking for the entirety of the data provided. In such situations, segments should be evaluated to determine if the user is in fact walking. This is particularly important for training, as a system trained on poor quality data will likely underperform. A simplistic approach to segment selection is to discard the first (chronologically) and last segments, as users may have started walking after collection began or stopped prematurely. If the design of a system is such that it is possible that users might stop walking during training sessions, segments can be evaluated to determine the total amount of motion present in the segment, discarding those that do not have a significant level of motion (as determined by a threshold).

9.5   Selecting a feature set

A wide variety of features can be utilized to form the gait biometric system. Although features from many domains can offer a system increased accuracy, there are computational and memory concerns that must be understood before determining the feature vector.

Notably, the ultimate location of the model (i.e., on-device or a centralized server) plays a major role in what features are chosen. Mobile devices offer limited battery life and computational power, and a realistic system must allow a user to authenticate frequently (potentially many times a day) and not pose an undue strain on system resources. For researchers, a system that is too much of a strain could result in failing to collect data from a volunteer and add significant delays to the research process. Similarly, certain features are too computationally intensive to be processed on a mobile device in a reasonable amount of time.

9.5.1   Statistical features

Statistical features provide a strong level of accuracy without a great strain on system resources. Features such as averages, variances, and other statistical moments are simplistic to compute and can be used alone to obtain very strong accuracies [7]. Further, these features are easy to calculate on a single axis or can be extended using multidimensional statistics to factor in multiple axes at once (e.g., covariance). Similarly, "zero crossings," or the number of times the signal crosses changes from positive to negative values (or vice versa) are a common feature. Statistical features, while not as encompassing as other methods (in terms of representing the diversity of the data), offer reasonable accuracy without considerable strain. Consequently, they are often used in conjunction with one or more of the more advanced families of features.

9.5.2   Histogram-based features

A computationally inexpensive way of representing the diversity of the data is through the use of histogram-based features. Usually constructed on a per-axis level or across multiple instances (in cases where multiple instances are fused together before training a system), histograms serve to bin the data and can better represent outliers than some statistical features (e.g., averages). The primary cost of histogram-based features is

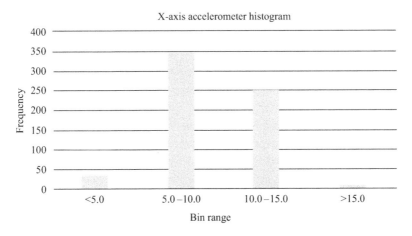

Figure 9.2   A visualized histogram feature for the x-axis of a watch accelerometer

the increased memory usage of storing the feature vector. Further, depending on the size of each "bucket" in the histogram, this feature-generation strategy may result in a sparse feature vector and thus negative impact system accuracy. Many systems that use histogram-based features also incorporate statistical features (such as [7]). A visualized example of a histogram feature for the *x*-axis of a watch accelerometer is shown in Figure 9.2.

Note that this type of feature inherently captures the shape of the distribution of the values.

9.5.3   Cycle-based features

For cycle-detection methods, a common strategy is the usage of raw cycles to train. Either the system is trained with multiple instances of data from the user (i.e., each cycle is provided as a separate instance) or an average template is constructed. Average templates are constructed by taking the per-feature average and result in a single instance per user. Average templates, while having a low memory cost, are usually only effective in systems that utilize similarity scores or distance metrics as other systems typically require a greater amount of training data than a single instance. To illustrate this concept, Figure 9.3 shows four steps or two full cycles, with the horizontal line demarcating the mean and the vertical line separating the two cycles. This graph was generated from the *x*-axis of a watch-based accelerometer.

The mean line demonstrates a simplistic strategy for detecting cycles. The cycle can be identified by finding four consecutive crossings of the mean line. Although this does not find the exact end of the cycle, it serves as an adequate approximation.

9.5.4   Time domain

Alternatively, the data itself, from either a single axis or a fusion of multiple, could be utilized as the features. This approach lends itself to models that assume the

Cycle detection

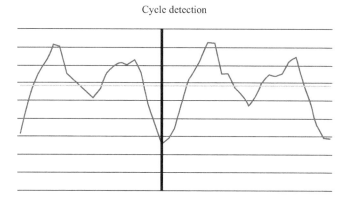

Figure 9.3   Two cycles with mean line

Table 9.1   Selected studies demonstrating accuracy of different feature sets

Citation	Features	Accuracy (in %)
[1]	Correlation of normalized cycles	93.6
[8]	Covariance of frequency-domain features from extracted cycles	99.4
[9]	FFT coefficients from fixed windows	93
[11,14]	Hidden Markov Model with raw data	90
[7]	Statistical and histogram features	100

features are provided in some sort of significant order, such as Hidden Markov Models. Although this strategy requires almost no computation, such systems could be prone to overfitting both to users and a specific pace; thus, careful preprocessing would likely be necessary to make such a system effective [11,14].

Given the low computational cost of this form of feature, it is suitable for systems that intend to do all of the processing and prediction on the watch itself or intend to do continuous authentication.

9.5.5   Summary

Given the diversity of feature methods and combinations, it is hard to assert that a single set of features is suitable for all system types and applications. Therefore, Table 9.1 describes different studies and the types of features they used, as well as the best performance of their authentication system. For clarity, the papers are referred to by their citation number rather than authors and title.

9.6 System evaluation and testing

9.6.1 Selecting an evaluation metric

The choice of evaluation metric will be the result of the model and the specific security concerns the system is designed to address. Equal error rate (EER) is a common choice among the biometric community, but it may not be the best for certain systems. Specifically, EER requires that a parameter of the system (e.g., a threshold) that can be varied in order to generate an EER curve. Many studies in gait biometrics do not utilize threshold-based methods, so EER is not as well defined (occasionally, particularly with systems with very low error, EER can be determined because the error rates are very close to equal). Consequently, TER (total error rate) or HTER (half total error rate, the mean of the False Acceptance Rate and False Rejection Rate) might be choices that are more suitable. Similarly, system accuracy (the mean of true acceptance and true rejection rates) works for all chosen methods. Accuracy, HTER, and TER all provide good intuition as to the general performance of the system at the cost of not giving insight into the overall balance of the system.

9.6.2 Single-instance evaluation and voting schemes

Depending on the amount of data a user is expected to provide to use the system (i.e., be authenticated or identified), it is possible that a system will have more than a single instance of data to authenticate a particular user. In these cases, a variety of strategies can be employed in order to boost the accuracy of the system. This scenario is relatively unique to gait-based systems depending on how data is processed; multiple examples can be derived from a small amount of data. Further, such strategies could work to combine results from different sensors or multiple devices if data is collected on both the watch and phone.

Single-instance evaluation is the standard mode of evaluation for a system. It requires a single instance of data to identify or authenticate, and thus is the most convenient on the part of the user. The cost of this convenience is that if the instance is corrupted in any way, the user may be false accepted or reject. Single-instance evaluation assumes that every instance provided as a test adequately reflects the user who provided it, thus it is the least resistant to error.

Quorum voting is usually applied in systems that are resistant to impersonation but struggle to accept valid users. In these situations, quorum voting requires only a small percentage (anywhere from 20% to 40% [13,11,14]) of instances supplied in a test to classify as valid to authenticate a user. Although this strategy may improve the accuracy of systems that are particularly imbalanced, in practice it will make the system far less resistant to imposters, as the threshold to qualify as valid is significantly lowered. Quorum voting is likely only acceptable in situations where the acceptance of an imposter comes at a low cost (e.g., when it is only a single factor of a multifactor system).

Majority voting is a more conservative approach to boost accuracy. Since majority voting requires most of the instances to concur on a single class, it is less likely to harm the system's ability to detect imposters. The primary problem with majority voting is that it becomes ill defined in identification systems, as the probability of the system being in an "undecided" state is higher (e.g., if all instance classify as different users). Thus, majority voting is more suited to authentication systems.

Weighted majority voting is the most complex strategy, but can be used to boost the power of a system significantly. Weighted majority voting sums the probabilities that the test instances belong to each class (sometimes called a "confidence") and selects the class with the largest sum.

Weighted majority voting is less likely to produce a "tie" than majority voting (as two sums would have to be precisely equal) and allows the system to perform more ably in situations where the system is "confused" (when the system is split between two or more classes). This performance boost results from the ultimately chosen class needing to be only slightly more likely than other classes. The challenge associated with the use of weighted majority voting is in systems where the probability of a class or confidence cannot be easily calculated (e.g., when more complicated neural networks are used). Thus, any system utilizing this method must choose its underlying methods appropriately.

9.7 Template aging: an implementation challenge

There are a number of outstanding challenges in gait biometrics that may hinder the development of a practical system. Although smartwatches resolve many of the outstanding issues (i.e., the position and orientation problems) in gait biometrics, the issue of template aging (sometimes called the "memory problem") is still an open question. Template aging is the problem where models constructed from data collected at some point in the past is not reflective of the user at the current time. Optimally, a system would only need to be trained once, or would require retraining so infrequently as to not present a hindrance to users.

Many gait biometric systems presented in research are trained and tested on data collected on the same day (or within the same session). This can lead to optimistic evaluations of accuracy. Some systems have been trained on a particular day's data and tested with data from a later day. Although some of these systems offer reasonable accuracy [9], many suffer from accuracy that has degraded past the point of suitability for a realistic implementation. Further, those who achieve high accuracy do so through placing strong constraints on their participants, such as wearing the same shoes both days [3].

The problem with template aging is likely most affected by users maintaining different paces in subsequent data collections. Many features (especially those based on the accelerometer) will naturally vary greatly as an individual changes pace. Thus, an implemented system must rely on features that are pace-independent. Alternatively, the system must implement a form of preprocessing that will normalize pace. It is worthy of note that systems based on cycle extraction methods suffer significant losses in accuracy when used against different paces [12]. Thus, segmentation-based methods with larger windows may prove to be more effective in combating problems with pace.

It is not currently known how much shoe choice affects template aging. It is likely that a dramatic change in footwear (e.g., stilettoes versus snow boots) will result in very different templates, but no research has been done (at the time of writing) that attempts to quantify the differences. Further, there is a possibility that a system need not be resistant to such dramatic changes in footwear. If the system is designed to serve a specific purpose (e.g., controlling the lock on a door in an office building), individuals might wear the same type of footwear consistently, and thus negate the problem.

Similarly, there is no research that explores how the template ages over extreme lengths of time. Studies using multiday data to evaluate the system usually collect data for subsequent days over very short periods (i.e., a week or less). Therefore, it is unknown that if gait changes over long spans of time; for an implementation to be successful, it would have to be capable of ignoring or adjusting to such changes.

Although all of these problems have yet to be formally solved, it is possible that model choice can negate, if not mitigate, the effect of these problems. Certain classifiers (such as tree-based classifiers and some Bayesian models) are "updatable," in that the exact parameters of the model can be adjusted even after it is constructed. Ostensibly, these models could be utilized to retrain systems every time a user authenticates (or identifies) successfully. Thus, gradual changes in an individual's gait over time should be accommodated. This flexibility does come at the consequence of potentially making a system less resistant to attacks, as persistent imposters might be more successful over time (assuming that they are falsely authenticated a small percentage of the time). In such cases, it would make sense to pair the system with another more standard form of authentication or identification.

9.8 Conclusion

Although a reliable implementation of a smartwatch-based gait biometric system would usher in a new era of transparent authentication, the creation of such a system would not be without its own ethical problems. A primary problem of passwords is that users tend to reuse a small set of passwords (or, more realistically a single password [10]); a primary responsibility (at least in the eyes of the commercial cybersecurity world) of biometrics is to replace these systems for that very reason (among others). The problem with an effective gait biometric system—much like many other forms of biometrics—is that once a user's training data is provided to the system that system can effectively impersonate that user; in essence, "reuse" is unavoidable. Thus, it becomes the responsibility of those designing gait biometric systems to create systems that are resistant to replay-style attacks. Similarly, work has to be done to ensure that the sensitive data collected from the smartwatch's sensors is not left in memory (where it could be recovered forensically or by a malicious application); much as traditional systems work to secure passwords. Gait biometrics might be the future of transparent authentication, but that does not allow conventional cybersecurity wisdom to be forgotten.

References

[1]   Ailisto, H. J., Lindholm, M. and Mantyjarvi, J., 2005. *Identifying People from Gait Pattern with Accelerometers.* Orlando, SPIE, pp. 7–14.

[2]   Crouse, M. B., Chen, K. and Kung, H., 2014. Gait recognition using encodings with flexible similarity measures. *Proceedings of the 11th International Conference on Autonomic Computing.* USENIX.

[3]   Derawi, M., Nickel, C., Bours, P. and Busch, C., 2010. *Unobtrusive user-authentication on mobile phones using biometric gait recognition.* s.l., IEEE.

[4]   Gafurov, D., Helkala, K. and Søndrol, T., 2006. Biometric gait authentication using accelerometer sensor. *Journal of Computers,* 1(7), pp. 51–59.

[5]   Gafurov, D., Snekkenes, E. and Bours, P., 2007. *Gait authentication and identification using wearable accelerometer sensor.* s.l., IEEE.

[6]   Google Android, n.d. *Sensors Overview,* Mountain View: Android.

[7]   Johnston, A. H. and Weiss, G. M., 2015. *Smartwatch-based Biometric Gait Analysis.* Washington DC, IEEE.

[8]   Juefei-Xu, F. and Bhagavatula, C., 2012. *Gait-ID on the Move: Pace Independent Human Identification Using Cell Phone.* Washington D.C., IEEE.

[9]   Mäntyjärvi, J., Lindholm, M. and Vildjiounaite, E., 2005. *Identifying Users of Portable Devices from Gait Patterns with Accelerometers.* Philadelphia, IEEE.

[10] Munroe, R., 2010. *XKCD #792.* [Online] Available at: https://xkcd.com/792/ [Accessed 15 12 2015].

[11] Nickel, C., Brandt, H. and Busch, C., 2011. *Benchmarking the performance of SVMs and HMMs for accelerometer-based biometric gait recognition.* s.l., s.n.

[12] Nickel, C. and Busch, C., 2011. *Classifying accelerometer data via Hidden Markov Models to authenticate people by the way they walk..* s.l., s.n.

[13] Nickel, C. and Busch, C., 2013. Classifying accelerometer data via hidden Markov models to authenticate people by the way they walk. *IEEE Aerospace and Electronic Systems Magazine,* October, pp. 29–35.

[14] Nickel, C., Busch, C., Rangarajan, S. and Möbius, M., 2011. *Using hidden Markov models for accelerometer-based biometric gait recognition.* s.l., s.n.

[15] Shannon, C. E., 1949. Communication in the presence of noise. *Proceedings of the IRE,* 37(1), pp. 10–21.

[16] Yano Research Institute, 2013. *Global HMD (Head Mounted Display) and Smartwatch Market: Key Research Findings 2013,* Tokyo: Yano Research Institute.

[17] Zhong, Y., Deng, Y. and Meltzner, G., 2015. *Pace-Independent Mobile Gait Biometrics.* Washington DC, IEEE.

Chapter 10
Toward practical mobile gait biometrics

Yu Zhong[1] and Yunbin Deng[2]

Abstract

Gait is the unique human locomotion due to individual specific biophysical and behavior habits. With ubiquitous mobile devices in people's daily life nowadays, accelerometers and gyroscopes provided in these devices directly capture the dynamic motion characteristics and thus have great potential for nonobtrusive gait biometrics. In fact, inertial sensors have been exploited to perform highly accurate gait analysis under controlled experimental settings. However, their performance in realistic scenarios is unsatisfactory due to variations in data measurements affected by physiological, environmental, and sensor-placement-related factors. Practical mobile gait biometric algorithms need to be robust to these variations to achieve high authentication performance in the field. It is the focus of this chapter to address some of these issues for in-the-wild mobile gait biometrics applications. First, we propose a novel gait representation called gait dynamics image (GDI) for accelerometer and gyroscope data sequences. GDIs are constructed to be both sensor-orientation-invariant and highly discriminative to enable high-performing gait biometrics for real-world applications. Second, we show how to further compute walking pace-compensated GDIs that are insensitive to variability in walking speed. Third, we adopt the i-vector paradigm, a state-of-the-art machine learning technique widely used for speaker recognition, to extract gait identities using the proposed invariant gait representation. Fourth, we demonstrate successful fusion of accelerometer and gyroscope modalities for improved authentication performance. Performance studies using both the naturalistic McGill University gait dataset and the large Osaka University gait dataset containing 744 subjects have shown dominant superiority of this novel gait biometrics approach compared to state-of-the-art. Additional performance evaluations on a realistic pace-varying mobile gait dataset containing 51 subjects confirm the merit of the proposed algorithm toward practical mobile gait authentication.

[1]Systems & Technology Research, USA
[2]AIT, BAE Systems, USA

10.1   Introduction

Recent advancements in mobile-computing technology have led to the ubiquitous proliferation of mobile devices, such as smartphones, tablets, smart watches, fitness bracelets, etc. They have rapidly become an indispensable part of everyday life for storing and accessing personal emails, communications on social media, online browsing history, payment information, etc. It is of primary importance to integrate security measures into mobile devices to protect the private, sensitive, and even critical information.

The traditional way to secure mobile phones is the use of passwords or PINs. However, these tokens are at the risk of hacking and malicious attacks. A user needs to memorize the password and change them often because of their insecurity. As a result, alternative authentication approaches are needed to provide the security with ease of use for mobile device users.

Biometric authentication, which confirms a person's identity based on his or her biological, physiological, or behavioral traits, provides a more natural access control procedure than the use of passwords or tokens. Biometric modalities, including fingerprints, iris, face, etc., have been popularly used to identify criminals and to guard customers' access to ATMs and bank payments.

Modern mobile devices are embedded with a suite of advanced sensors that can collect a rich stream of data from the user and their surroundings, including voice, keystrokes, touch dynamics, fingerprints, GPS locations, etc. These data can be analyzed to extract a user's physical, physiological, and behavioral traits, thus enabling mobile biometric solutions of various complementary modalities [1–4]. Mobile biometrics can either be used by themselves, or in combination with traditional security procedures, such as passwords or tokens, to authenticate users in many different applications, including access control, person identification, and online banking.

There has been emerging research on the use of mobile biometrics including face, voice, keystroke dynamics, touch dynamics, etc., for continuous active authentication [1]. Among the rich set of mobile biometric modalities, gait biometrics [5,6] using inertial sensors offers a superb compromise between *collectability* and *performance* [7,8]. Gait is the characteristic pattern of human locomotion and is fairly unique to an individual due to one's specific musculoskeletal biomechanics. Humans can often effortlessly recognize acquaintances from a distance by the way they walk. Automatic gait biometrics, which studies gait using sensory data, has been an active research area receiving increasing attention over the years [9–17].

Luckily, for today's mobile device users, it is possible to continuously acquire high-quality gait data using embedded inertial sensors and to perform gait-biometric-based security procedures. Accelerometers and gyroscopes were originally introduced in smartphones for improved user experience, by automatically adjusting the screen display layout to landscape or portrait mode based on the way the device is held using the direction of gravity calculated from inertial sensors. These sensors have since been used for many other applications, including motion analysis, while being small in size, lightweight, and power efficient. They continuously record fine motion measurements including acceleration and angular rotation rate at a high frame rate

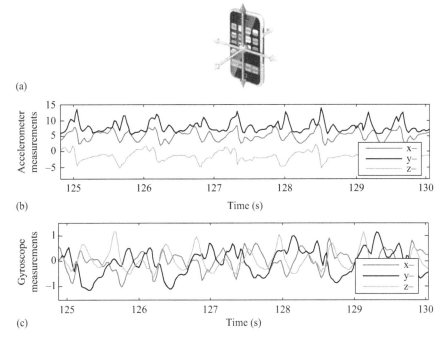

*Figure 10.1   Modern mobile sensors are often embedded with inertial sensors
measuring 3D acceleration and rotation along three orthogonal axes.
(a) Typical reference 3D axes for a smart phone; (b) 3D acceleration
time sequence recorded by the accelerometer in a mobile phone
carried by a walking user; (c) 3D rotation rate time sequence recorded
by the gyroscope in the same phone. The repetitive and unique
locomotion patterns of a mobile device user captured by these inertial
sensors can be exploited to compute highly accurate gait biometrics*

while the user is on the go. Locomotion data can be constantly collected for gait
analysis as a mere consequence of a user carrying a device around, allowing for
nonintrusive and continuous gait-based authentication and identification. These measurements often consist of projections of acceleration forces and angular rotation rates
of the sensor on three orthogonal axes and readily capture the locomotion characteristics of a user, as shown in Figure 10.1. Accelerometer-based body motion analysis
[18] has been intensely researched for gait biometrics [19,20] and activity analysis
[21,22]. Lately, gyroscopes have also been explored for body motion analysis [23,24].
While earlier studies have employed dedicated inertial sensors, more recent work has
used accelerometers and gyroscopes embedded in commercial mobile devices such
as smart phones or watches [25]. Gait identification results of a 99% verification
rate using mobile phones have been published for accelerometer-based gait analysis in controlled experimental settings indicating great potential for authentication
applications [26].

Similar to more traditional biometrics such as fingerprints and face recognition [7], mobile gait biometrics in-the-wild still faces immense challenges in dealing with variations typical in practical applications despite promising results in controlled experimental settings. As a behavioral biometric, gait biometrics exhibits far more variability than physiological biometrics, such as fingerprint or iris biometrics. A person's gait is influenced by his/her physical or psychological status, such as emotion, fatigue, well-being, etc. For example, one may walk quickly with excitement and plod slowly with fatigue. As a matter of fact, fluctuations in walking pace are a constant presence in a person's gait [27]. In addition, external factors such as clothing, footwear, carried loads, and ground conditions may also affect gait [28–30].

For inertial sensor-based gait analysis, a huge amount of variations in gait data arises from the data collection process. An inertial sensor only measures motion local to where it is worn, which varies at different body parts due the articulating nature of a human body [31]. This problem is further complicated by the fact that measurements made by inertial sensors depend not only on where they are placed but also on how they are placed, because such a sensor measures accelerations or rotations projected on its sensor axes, yielding different measurements depending on the sensor orientation even when body motion stays the same. For practical mobile gait biometrics using consumer mobile devices, the phone may be held in the hand, or be placed in a purse. These common usages further complicate the gait analysis process.

To build a robust mobile gait biometrics system to use in real world, it is necessary to tackle all the variations in the gait data in order to obtain high accuracies in authentication and identification tasks. Many existing studies have conducted experiments under ideal experimental settings to suppress variability in gait data. Some research works have bypassed the problem of data dependency to sensor placement by enforcing restrictions on the location and orientation of the sensors, such that they are worn in a fixed way during data collections; this way the data variations due to sensor placement are minimized, making it possible to match gait signatures by direct comparison of data measurements. However, for practical gait biometrics using mobile phones, the device is usually casually carried in any of the pockets, with fairly arbitrary headings. With the growing demand for more practical and robust mobile gait biometrics, there has been an increasing trend in recent years to relax the data collection requirements that better resemble realistic usage, to explore gait-affecting factors in order to understand their influences [30,32] and to develop advanced algorithms utilizing invariant feature extraction or matching that are robust to some of the variations [6,26,33–35].

The focus of this chapter is to investigate accurate mobile gait biometrics for real-world applications. In particular, we directly address two major factors of variation in practical inertial sensor-based gait analysis, namely

1. Data dependency on sensor orientation and certain configurations of sensor placement and
2. Fluctuations in gait cycle.

To compute gait biometrics that is robust to variations in sensor orientation, we have proposed a novel gait representation, called GDI [35], that is invariant to sensor rotation, while preserving highly discriminative temporal and spatial gait dynamics and context information. This GDI representation is also invariant to symmetry transforms. As human locomotion is usually left–right symmetric up to a phase shift, this invariance in symmetry transforms achieves insensitivity to mirrored placement of a sensor on the left and the right side of a human body, further relaxing the placement constraints on sensors. We further overcome the challenge of varying walking speed by computing instantaneous gait cycles from GDIs using an efficient energy minimization scheme. These estimated gait cycles are then used to normalize GDIs for the varying walking pace [36]. Consequently, we obtain a highly discriminative mobile gait representation that is both invariant to sensor orientation and insensitive to varying walking speed, applicable to both accelerometer and gyroscope sensory data measurements.

While we explicitly design a gait representation that is invariant to both sensor orientation and walking speed, we mitigate the influence of the remaining noise in the data using advanced machine-learning methods. In particular, we employ the i-vector ("i" stands for "identity") approach [36,37], a state-of-the-art voice biometrics method, to extract gait identity features from the GDI invariant gait representations. The i-vector approach is built using the Gaussian Mixture Models (GMMs) in the GDI feature space in order to be highly versatile. It encapsulates a set of feature vectors of arbitrary size using a single "supervector" of a fixed length, capturing the "influence" of the set on the GMM by concatenating the accumulated first-order statistics of the feature vector set to each cluster of the GMM. This high dimensional supervector, which encodes rich data statistics of the feature vector set, is reduced to a lower dimensional i-vector using factor analysis to be robust to variations in data for accurate gait classification.

In the past, accelerometers have been predominantly used in research works on gait analysis [38–41]. Gyroscopes, which measure rotation rate around sensor axes, offer complementary motion cues to accelerometers. In this chapter, we also look into fusing gait data from both inertial sensor types and demonstrate for the first time that gyroscopes can be used to boost the gait biometrics accuracy using accelerometers.

The remainder of the chapter is organized as follows: Section 10.2 provides a survey of the literature. Section 10.3 introduces our novel GDI gait representation that is invariant to sensor rotation and walking pace. Section 10.4 details our approach to pace independent mobile gait biometrics. Section 10.5 presents our experiments and performance studies. Section 10.6 draws conclusions and suggests areas for future research.

10.2   Related work

Accelerometer-based gait analysis has been a popular research area since the pioneering work by Ailisto, Mantyjarvi *et al.* in 2006 [5,20]. In the early work, multiple motion sensors were attached to various human body parts to provide localized movements and bio-kinematics analysis. Data from a single sensor at a fixed position, including the feet, hips, or waist, etc., were also exploited [42–44]. More recent work has used accelerometers in commercial off-the-shelf (COTS) mobile devices for gait classification [45–51]. Sprager and Juric gave a comprehensive review of the literature of gait biometric using accelerometers and gyroscopes [52].

Three-axis accelerometers commonly embedded in mobile phones capture accelerations along three orthogonal axes of the sensor. Given a multivariate time series of the acceleration data, feature vectors are usually extracted for signal windows corresponding to each detected gait cycle [26,44] or for windows of a prespecified size [47]. These windowed signals are compared and matched based on template matching [42], using either the correlation method or dynamic time warping (DTW). Alternatively, statistical features including mean, standard deviations, time span between peaks in windows [47], histograms [20,47], entropy, higher order moments [42], and cumulants in spatial domain are also used [53]. FFT and wavelet coefficients in frequency domains are also studied to compare longer sequences [20,26]. Classifiers, including nearest neighbor classifier, SVM [26], and Kohonen self-organizing maps [54], have been investigated. In some cases, preprocessing, such as weighted moving average, is applied to suppress noise in the raw sensory data [44].

However, the majority of existing researches were conducted in controlled lab settings. For example, there are often strict constraints on where and how the sensors are placed. These constraints help to reduce variation and noise in the data. In some cases, the sensors are placed in a specific way, such that intuitive meanings can be assigned to the data components and be exploited for analysis [20].

Unfortunately, it is unrealistic to assume fixed placement of the sensor or the mobile devices. The mobile devices are often carried casually in pockets or hands without constraints in orientation. As such, depending on the sensor orientation, the same locomotion can result in different measurements. For the technology to be useful, it is critical to derive gait biometrics that is invariant to sensor placement and rotation. Magnitude sequences of 3-axis accelerometer measurements have been popular due to their invariance to sensor orientation changes [26,31,45]. However, the use of univariate resultant series of the raw 3D multivariate series results in information loss and ambiguity artifacts undesirable for highly accurate gait biometrics. Mantyjarvi *et al.* used both principal component analysis (PCA) and independent component analysis (ICA) to discover "interesting directions" for computing gait features for activity analysis [20]. Bours *et al.* also used PCA to compute "Eigen steps" for gait biometrics [55]. Unfortunately, the underlying assumption of identical data distributions for both training and testing data is unlikely to hold for realistic applications. Iso *et al.* approached this challenge by augmenting the training set with simulated data at multiple sensor orientations via artificially rotating available training data [56]. In their approach, the significant artificial sampling needed to tessellate 3D rotational space creates an unbearable computational and storage burden with the additional risk of degraded classifier performance. In [57], orientation invariant features were extracted using the power spectrum of the time series. However, it suffered shortcomings common to frequency domain methods: loss of temporal locality and precision, as well as vulnerability to drifting in gait tempo. Sprager *et al.* used a built-in gyroscope sensor to calibrate accelerometer data to the upright posture to reduce the influence of noise in sensor orientation [53]. However, their approach requires calibration prior to every data collection, expects the sensor not to rotate during data collection, relieves noise only in the vertical direction, and makes unrealistic assumptions that all poses are upright. Ngo *et al.* proposed to explicitly estimate the rotation between two 3D acceleration time series [33]. Their approach does not use constraint on sensor orientation or depend on ad-hoc methods. However, their method requires iterative optimization to both align in time and register in space for each pair of 3D sequences to be compared. As a result, it is too computationally intensive to be applicable to large gait databases. In a similar way, Subramanian *et al.* used Kabsch alignment to perform pose registration of the 3D time sequence before matching the gait patterns for sensor orientation invariance [34].

Another major obstacle for realistic mobile gait biometrics is variations in walking speed [6,26,27,44,54,58–60]. Pace variation persists in real life, as gait is influenced by a person's physical and psychological status, and the duration of a gait cycle can change from step to step. Consequently, accurate gait cycle detection is important for high-performing gait biometrics. The problem of characterizing gait with varying walking speed has been a focus of computer vision-based gait biometrics research [61–64]. It has been found that accurate gait cycle detection improves gait authentication accuracy even for data collected on subjects walking at a normal pace [44]. A majority of existing accelerometer-based gait analyses proceed by extracting windows corresponding to gait cycles [60,65] and then matching these detected cycles [26]. In controlled laboratory studies, it is possible to place the sensor such that gait cycle detection is simplified. In [51], the phone is attached to the trouser pocket with the y-axis pointing upward and the z-axis pointing outward, which allows individual gait cycle extraction by detecting dominant peaks or valleys [2,26,51] in the data. Other work [44,66,67] extracts a small window of data as a template and performs a sliding window match in the entire search window to compute a distance trajectory between the template and the overlapping subwindow. The cycle length is subsequently estimated as either the displacement between a pair of consecutive minima in the distance function or the average of these displacements. Although this method is less sensitive to sensor placement, the quality of the detections depends heavily on the chosen subwindow. Furthermore, this approach is vulnerable to gait drifting over time. Seitz and Dyer investigated real cyclic motion with irregularities and proposed a motion representation called "period trace" which consists of the estimated instantaneous periods over time [68]. This "period trace" is computed as a "smooth" snake that minimizes distance between motion measurements as the signal repeats itself at each period. The computed trace is then used to detect irregularities in motion. Makihara *et al.* studied phase estimation of quasi-periodic signals by optimizing a combination of data fit, smoothness in local phase evolution, and a monotonic increasing constraint on the phase function [69]. They first computed a period using normalized correlation and then estimated the phase correspondence using DTW [58]. In a subsequent work on vision-based gait analysis, Aqmar *et al.* not only utilized the phase registration work to suppress influence of gait fluctuations but also exploited consistently presented interperiod gait fluctuation patterns to improve gait recognition performance [27].

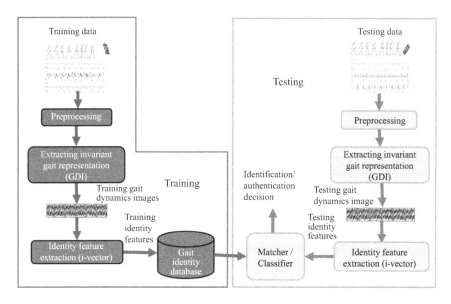

Figure 10.2   Overview of the proposed mobile gait biometric system

With the ubiquity of consumer smart devices causing our life to become increasingly mobile and the urgent demand in secured access with these devices, there is a need for accurate mobile biometrics for real-world applications. This is reflected in the emerging trend in recent research work in mobile gait biometrics where a number of variation factors in gait data have been explored. In this chapter, we investigate some of such factors, including sensor orientation/placement, and walking pace. Different from existing approaches that seek to explicitly register and align gait patterns before they are matched, we directly apply novel invariant feature extraction principles to extract gait representations that are independent of sensor rotation and walking speed. We mitigate the remaining variations in the gait data using the factor analysis based i-vector feature extraction, an advanced machine-learning method popularly used in voice biometrics, to effectively extract identity signatures from the invariant gait representations. Figure 10.2 illustrates an overview of our mobile gait biometrics approach.

10.3   GDI gait representation

Accelerometers and gyroscopes are capable of recording fine motion, in terms of acceleration and rotation rate, at high frame rate. Mobile gait data is highly sensor orientation dependent. This is illustrated in Figure 10.3, where the accelerometer measurements in x, y, z time sequences before and after a mobile phone rotation are shown in Figure 10.3(a), and the corresponding gyroscope measurements are shown in Figure 10.3(c). As we can see, these motion measurements of the same subject differ drastically before and after the change. Note that these changes in motion measurements are not due to differences in walking style, but rather, caused

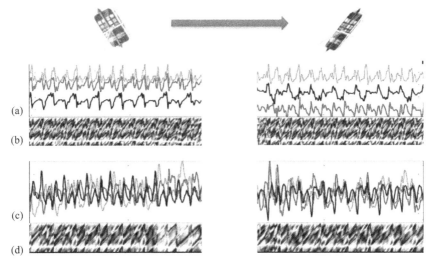

Figure 10.3   *While raw acceleration and rotation rate measurements of locomotion depend heavily on the sensor placement (orientation), their corresponding gait dynamics image (GDI) feature representations are invariant to sensor rotation. We show here the raw data time sequences and their GDIs captured by a phone from a walking subject, before (left) and after (right) a phone rotation. (a) x-, y-, and z-acceleration components from an accelerometer embedded in a mobile phone carried by a walking subject; (b) the corresponding GDI feature representation for the raw acceleration sequence in (a); (c) raw x-, y-, and z-rotation rate measurements from the embedded gyroscope; (d) the corresponding GDI for the raw gyro sequence in (c). The gravity component is removed from the acceleration time series before the computation of GDIs. Both sensors capture distinguishing locomotion patterns characteristic of a person's gait. However, the captured gait patterns change drastically after rotation of the phone due to data dependency on sensor orientation; The GDIs are invariant to sensor orientation and thus show much better consistency before and after the sensor repositioning. The horizontal paths of high intensity in both the GDIs indicate peak self-similarity as the time lag reaches the length of a local gait cycle*

by the nuisance artifacts of sensor placement. Practical gait analysis has to take into account the influence of sensor rotation on the data measurements. Recently, [33,34] addressed this issue by explicitly estimating the rotation of the multivariate gait sequence in space and aligning two sequences in space and time before they are matched. Alternatively, we propose a novel gait representation, called GDI, that is independent of sensor variations while being highly informative and discriminative.

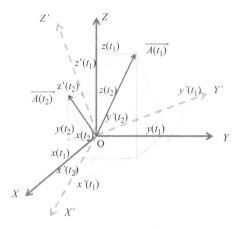

Figure 10.4   While two acceleration vectors $\overrightarrow{A(t_1)}$ at times t_1 and $\overrightarrow{A(t_2)}$ at time t_2, each projecting differently on the axes of a reference frame OXYZ and a rotated reference frame OX'Y'Z', their relationships, e.g., the angle between the two vectors, remain constant regardless of the choice of reference frame

We further take advantage of the new gait representation to estimate instantaneous gait cycles and compute a pace-compensated gait dynamics image representation that is robust to variations in both sensor rotation and walking speed.

10.3.1   Gait dynamics images

We approach the challenge caused by variation in sensor orientation by exploring gait features that characterize the distinguishing locomotion signature while staying invariant to sensor orientation. Although the individual acceleration data does depend on sensor placement, we proposed to extract relationships between pairs of observations that are sensor orientation invariant. In a nutshell, we compute features using these sensor orientation invariant quantities inside each gait cycle to capture the gait dynamics and derive new discriminative and robust representations for gait biometrics.

10.3.1.1   GDI for accelerometer data

Let's represent the acceleration measurement vectors at times t_1 and t_2, as $\overrightarrow{A(t_1)} = [x(t_1) \quad y(t_1) \quad z(t_1)]'$ and $\overrightarrow{A(t_2)} = [x(t_2) \quad y(t_2) \quad z(t_2)]'$, based on an accelerometer with reference frame OXYZ. After the orientation change of the accelerometer, these forces are captured by a new accelerometer measurement, but based on a new reference frame OX'Y'Z', as shown in Figure 10.4. These two new vectors are represented as $\overrightarrow{A'(t_1)} = [x'(t_1) \quad y'(t_1) \quad z'(t_1)]'$ and $\overrightarrow{A'(t_2)} = [x'(t_2) \quad y'(t_2) \quad z'(t_2)]'$. Let's assume the rotation between the two readings is R. This can be written as $\overrightarrow{A'(t_1)} = R\overrightarrow{A(t_1)}$ and $\overrightarrow{A'(t_2)} = R\overrightarrow{A(t_2)}$. Although the original readings depend on the sensor orientation, the relationships between the two vectors, such as the angle between them, remain constant. A new orientation invariant feature can be defined using a pair of motion vectors at times t_1 and t_2 as follows:

$$
\begin{aligned}
<\overrightarrow{A'(t_1)}, \overrightarrow{A'(t_2)}> &= \;<\overrightarrow{RA(t_1)}, \overrightarrow{RA(t_2)}> \\
&= \overrightarrow{A(t_2)}^T R^T \overrightarrow{RA(t_1)} \\
&= \;<\overrightarrow{A(t_1)}, \overrightarrow{A(t_2)}>
\end{aligned}
\tag{10.1}
$$

Thus, the inner product is invariant to sensor rotation:

$$
I_{\text{inp}}\left(\overrightarrow{A(t_1)}, \overrightarrow{A(t_2)}\right) = \;<\overrightarrow{A(t_1)}, \overrightarrow{A(t_2)}>
\tag{10.2}
$$

This invariant quantity can be interpreted as the projection of an acceleration data vector on the other. This project stays the same regardless different reference frame. As a special case, i.e., when $t_1 = t_2$, we have

$$
<\overrightarrow{A(t)}, \overrightarrow{A(t)}> = \|x(t)^2 + y(t)^2 + z(t)^2\|,
$$

which turns out to be the commonly used magnitude series. In more general case, where t_1 *and* t_2 are different, we computed correlation among accelerometer readings at different time instance. Thus, this new feature is able to create feature information much richer than the commonly used magnitude feature.

Furthermore, from the basic inner product invariants, additional invariant features can be derived with normalizing effects. One useful new feature is the normalized cosine similarity measure among accelerometer readings defined as follows:

$$
I_{nc}\left(\overrightarrow{A(t_1)}, \overrightarrow{A(t_2)}\right) = \frac{<\overrightarrow{A(t_1)}, \overrightarrow{A(t_2)}>}{\|\overrightarrow{A(t_1)}\| \|\overrightarrow{A(t_2)}\|}
\tag{10.3}
$$

This invariant is the cosine of the angle between two 3D acceleration readings. It only depends on the relative rotation between the two 3D vectors regardless of the choice of the reference frame that is used to measure the acceleration data.

10.3.1.2 Orientation invariants for gyroscope data

Different from accelerometer, the gyroscopes measure the rotation, or rotation rate, of the sensor's axes. Similarly, orientation invariants from a pair of raw gyro sensor data can be extracted using appropriate rotation representations. We represent a 3D rotation at time t using a 3D rotation axis vector $\overrightarrow{u(t)}$ and a scalar rotation angle $\theta(t)$ around the axis. With this rotation representation, the angle $\theta(t)$ is invariant to the choice of reference coordinate system. The coordinates of rotation axis $\overrightarrow{u(t)}$, on the other hand, do depend on the choice of reference frame. The reference invariant features can be extracted using a pair of 3D rotation readings $\overrightarrow{u(t_1)}$ and $\overrightarrow{u(t_2)}$ as follows:

$$
\begin{aligned}
<\overrightarrow{u'(t_1)}, \overrightarrow{u'(t_2)}> &= \;<\overrightarrow{Ru(t_1)}, \overrightarrow{Ru(t_2)}> \\
&= \overrightarrow{u(t_2)}^T R^T \overrightarrow{Ru(t_1)} \\
&= \;<\overrightarrow{u(t_1)}, \overrightarrow{u(t_2)}>
\end{aligned}
\tag{10.4}
$$

Although the rotation axes depend on its sensor orientation, the inner product between a pair of the 3D rotation axis vectors is orientation invariant. This new quantity is only related to the angle between the two measurements, which stays the same regardless of the reference frame.

Let $\overrightarrow{\omega(t)} = [\omega_x(t) \quad \omega_y(t) \quad \omega_z(t)]^T$ be the output of a 3-axis rate gyroscopes measurement. These are the rotation rate about the x-, y-, z-axes, usually expressed in rad/s. On the basis of the theory of Direction Cosine Matrix IMU [70], given a small time dt, the combination of three small rotations about the x-, y-, z-axes can be characterized by angular rotation vectors $\overrightarrow{\omega_x(t)}, \quad \overrightarrow{\omega_y(t)}, \quad \overrightarrow{\omega_z(t)}$. The $\overrightarrow{\omega_x(t)} = [\omega_x(t) \quad 0 \quad 0]^T$, $\overrightarrow{\omega_y(t)} = [0 \quad \omega_y(t) \quad 0]^T$, and $\overrightarrow{\omega_z(t)} = [0 \quad 0 \quad \omega_z(t)]^T$ are approximately equivalent to one small simultaneous rotation characterized by angular rotation vector $\overrightarrow{\omega(t)} = [\omega_x(t) \quad \omega_y(t) \quad \omega_z(t)]^T$, which is a rotation around the axis defined by $u(t) = [\omega_x(t) \quad \omega_y(t) \quad \omega_z(t)]^T$, with an angle equal to the magnitude $\|[\omega_x(t) \quad \omega_y(t) \quad \omega_z(t)]^T\|$.

On the basis of the previous discussion, the magnitude of $\overrightarrow{\omega(t)}$, which is the rotation angle around the 3D rotation axis, is invariant to the sensor axis. In addition, the inner product of a pair of $\overrightarrow{\omega(t)}$ is also invariant to the sensor reference frame. Thus, a new invariant features can be extracted for a pair of angular rate readings $\overrightarrow{\omega(t_1)} = [\omega_x(t_1) \quad \omega_y(t_1) \quad \omega_z(t_1)]^T$ and $\overrightarrow{\omega(t_2)} = [\omega_x(t_2) \quad \omega_y(t_2) \quad \omega_z(t_2)]^T$ from gyroscopes as follows:

$$I_{\text{inp}}(t_1, t_2) = \;<\overrightarrow{\omega(t_1)}, \overrightarrow{\omega(t_2)}> \tag{10.5}$$

Similar to the accelerometer GDI, there is a special case when $t_1 = t_2$. This results in the magnitude of the rotation:

$$I_{\text{inp}}(t, t) = \;<\overrightarrow{\omega(t)}, \overrightarrow{\omega(t)}> \; = \|\overrightarrow{\omega(t)}\|.$$

Likewise, normalized cosine similarity invariants can be computed for gyroscope readings:

$$I_{\text{nc}}(t_1, t_2) = \frac{<\overrightarrow{\omega(t_1)}, \overrightarrow{\omega(t_2)}>}{\|\overrightarrow{\omega(t_1)}\| \; \|\overrightarrow{\omega(t_2)}\|} \tag{10.6}$$

In the next few subsections, we illustrate and analyze these new features to gain more insight about its discriminative and robustness to sensor orientation variation.

10.3.1.3   Gait dynamics images

We exploit these invariant motion interactions to extract features that characterize the locomotion dynamics and which are robust to variations in sensor orientation. Given a 3D acceleration or rotation rate time series of size n sampled at regular time intervals $\left\{\overrightarrow{A(1)}, \overrightarrow{A(2)}, \overrightarrow{A(3)}, \ldots, \overrightarrow{A(n-1)}, \overrightarrow{A(n)}\right\}$, we define a two-dimensional matrix, which we call GDI, to capture invariant motion dynamics over time and interactions within each gait cycle. Let the invariant feature computed using a pair of data vectors $\overrightarrow{A(t_1)}$ and $\overrightarrow{A(t_2)}$ be $I(t_1, t_2)$, using either (10.2) or (10.3) for acceleration measurements, and (10.5) or (10.6) for rotation measurements. The gait dynamics image of a gait measurement sequence is defined as follows:

$$\mathbf{GDI}(i,j) = I(j, i + j - 1), \tag{10.7}$$

$i = 1, \ldots, l$ and $j = 1, \ldots, n - l + 1$, where l is the range of the time delay for concerning pairwise motion interactions, for which we choose to encode context within a typical gait cycle.

Gait dynamics images encode rich dynamics and context information characterizing the unique gait of an individual. The ith row of the GDI contains all pairwise interactions of time delay $i - 1$ over time, while the jth column consists of interactions between the motion at time j and all its successors up to time lag $l - 1$ to capture local context (see Figure 10.5). In particular, the first row of the inner product GDI image, which are the inner products of observation pairs with time lag 0, corresponds to the magnitude sequence. The magnitude sequence has been shown to be advantageous to the raw component acceleration features in cell phone-based gait ID studies [26] and has been often used in existing research to handle the variations in sensor placement. The remaining rows contain the interactions at varying time lags that contribute to additional discriminating information of gait dynamics, which makes GDIs extremely powerful representations for gait biometrics.

The cosine similarity GDIs can be considered a normalized form of the inner product GDIs by taking out the effects of the magnitudes. These GDIs only depend on the angles between the observation vectors. This normalization may improve the robustness to noisy magnitudes in the data. In summary, the GDI, although built on the sensor rotation-dependent raw acceleration measurements, achieves a view invariant representation of the governing dynamics in the original multivariate time

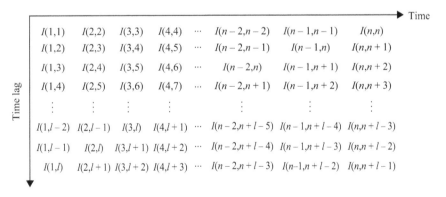

Figure 10.5   *A gait dynamics image consists of pairwise interactions within a gait cycle to encode rich dynamics and context information characterizing the unique gait of an individual. The ith row of the GDI contains all pairwise interactions of time delay i − 1 over time, while the jth column consists of interactions between the motion at time j and all its successors up to time lag l − 1 to capture local context*

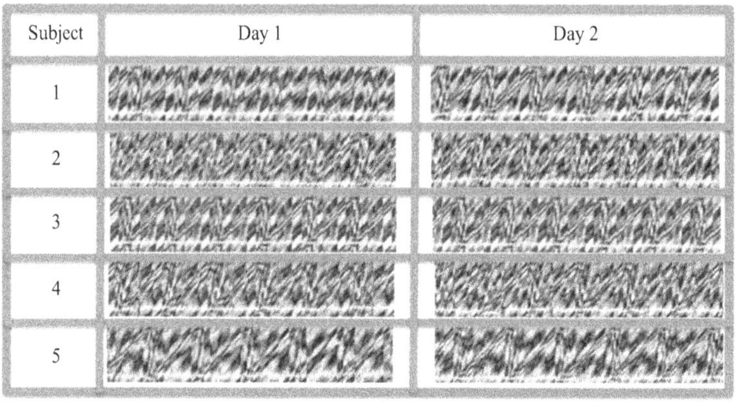

Figure 10.6   Gait dynamics images exhibit good intrasubject consistency and notable intersubject distinctions despite of sensor orientation variations. We show here GDI images for five subjects, taken using mobile phones with casual placement from two separate days

series for robust gait analysis. Furthermore, it preserves the local interactions and contextual information within each gait cycle essential for discriminative motion analysis to enable highly accurate gait biometrics. We show in Figure 10.3(b) and (d) the corresponding cosine similarity GDIs for the raw acceleration measurements in (a) and rotation rate measurements in (c). As expected, the GDIs exhibit much better consistencies between the two collections than the raw time series. We also show in Figure 10.6 the cosine similarity GDIs for five subjects from two sessions. As one can see, GDIs exhibit good intrasubject consistency and notable intersubject distinctions despite of sensor orientation variations.

10.3.1.4   Symmetry invariance properties of gait dynamics images

In addition to invariance to sensor rotation, GDIs are also invariant to symmetry transforms or any concatenation of symmetry and rotation transforms in the motion measurements. This is apparent as both (10.1) and (10.4) hold if we replace the rotation matrix R with any symmetry transformation matrix or concatenations of rotation and symmetry transforms. As the laterally symmetric human anatomy typically creates symmetric locomotion for the left and right sides of the body, it is possible to match GDIs from a phone placed in one pocket to GDIs from phones in the opposite side pocket thus further ease the sensor placement constraint.

10.3.2   Pace-compensated gait dynamics images

GDIs encode both dynamics for the time series and the local interactions. In addition, they provide useful information to estimate instantaneous gait cycles that can be used to compute pace insensitive gait biometrics as described in this subsection.

10.3.2.1   Instantaneous gait cycle estimation using GDIs

Human locomotion for gait is quasi-periodic by nature. This characteristic is readily indicated in the patterns in GDIs. As the motion measurements approximately repeat themselves due to the quasi-periodic nature of human gait, the time lagged self-similarity responses encoded in a GDI peak when the time lag l reaches a gait cycle. This results in a continuous horizontal path of high intensity across the gait dynamics image where the row index of an entry on the path approximates the length of the local gait cycle, as shown in Figures 10.3(b) and (d) and 10.6.

   We take advantage of this property of gait dynamics images to estimate gait cycles by computing the horizontal path across the gait dynamics image that maximizes the sum of similarity responses on the path. This is achieved by dynamic programming search on an energy function defined below. Local gait cycle durations are then extracted from the path and used to normalize the gait dynamics image to a prespecified gait length for pace-compensated gait matching.

10.3.2.2   Energy function

We propose an energy function comprising of three terms, which are as follows:

1. A prior term favoring gait cycles of a prespecified length, usually derived from prior knowledge on gait cycle lengths or training data, to bias toward gait cycles mean;
2. A regularization term enforcing smoothness as gait cycles fluctuate in time, such that abrupt gait cycle changes are penalized; and
3. A likelihood term drawing the path to the maximum response in the GDI image.

Given a GDI image $I(l, t)$ with lag interval $l = 0, \ldots, L - 1$ and time sample $t = 0, \ldots, T - 1$, we denote a path of length T across from the first column to the last column of the GDI as $p(t), t = 0, \ldots, T - 1$ which stores the lag index of the path at time t. Assuming the expected length of a gait cycle to be T_{cycle}, we use the following prior term to penalize gait cycles that deviate from the expected cycle length. It is minimized when the estimates equal the expected cycle length:

$$E^p(p(t)) = \sum_{t=0}^{T-1}(p(t) - T_{\text{cycle}})^2 \tag{10.8}$$

The second regularization term penalizes changes in the gait cycle to ensure its smoothness, which is minimized when gait cycles stay constant:

$$E^r(p(t)) = \sum_{t=0}^{T-2}(p(t) - p(t+1))^2 \tag{10.9}$$

The likelihood term locks the path to points of high responses in the GDI. This term should be maximized, so adding a negative term as the final objective function is to be minimized:

$$E^l(p(t)) = -\sum_{t=0}^{T-1}I(p(t), t) \tag{10.10}$$

We estimate the cycles by minimizing the following energy function on the path:

$$\arg \min_{p(t)} \alpha E^p(p(t)) + \beta E^r(p(t)) + E^l(p(t)) \tag{10.11}$$

where α and β are parameters reflecting the importance of the prior and regularization terms w.r.t. the likelihood term.

10.3.2.3   Efficient instantaneous gait cycle estimation using dynamic programming

The optimal path $p(t)$ which minimizes (10.5) can be efficiently computed using dynamic programing by scanning the GDI image once, left to right and top to bottom. As it scans each node, it computes and stores the score of the optimal path that reaches it starting from the leftmost column, by examining the scores of nodes to its left that connect to it. The preceding node on this optimal path is recorded as well. Once the scan is completed, we find the node with the best score in the rightmost column, and backtrack to recover the optimal path across the GDI. The following is the pseudocode of this algorithm:

% I(L,T) is the input gait dynamics image
% the following code returns the optimal path for (10.5)
% S(l,t) stores the score for the optimal path reaching node (l,t) from the leftmost column
% M(l,t) stores the move from the previous node on the optimal path that reaches node (l,t):
up, down, or level, for back tracking

initialize a 2D array of scores S(L,T) to 0
initialize a 2D array of moves M(L,T-1) to 0

% scan and initialize scores for left most column
t = 0;
for l = 0 to L-1
    S(l, t) = α(l − T_cycle)² − I(l, t)
end
% scan and update subsequent columns based on previous column
for t = 1 to T-1
    for l = 0 to L-1
        M(l,t − 1) = arg min_{k=−1,0,1} S(l + k,t − 1) + βk²
        S(l, t) = α(l − T_cycle)² − I(l, t)
                        + min_{k=−1,0,1} (S(l + k,t − 1) + βk²)
    end
end

% find the node on the rightmost column with the best score
p(T − 1) = arg min_{l=0,1,...,L−1} S(l,T − 1)
% start from the node, backtracking the optimal path
for t = T − 2 : 0
    p(t) = p(t + 1) − M(p(t + 1),t)
end

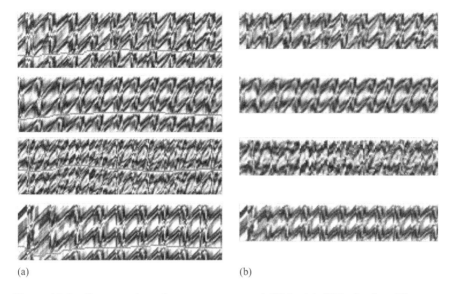

(a)   (b)

Figure 10.7   Computation of pace-compensated GDIs. (a) GDIs for four 3D
acceleration sequences, with gait cycles detected and marked in red;
(b) corresponding pace normalized GDIs. The normalization
dramatically reduces cadence variation within gait sequences while
maintaining uniqueness of each gait sequence

10.3.2.4   Pace-compensated gait dynamics images

Once we have detected the optimal path $p(t)$ across the gait dynamic image which corresponds to instantaneous gait cycle estimates, we then unwarp and rectify the GDI image such that the resultant GDI image is pace independent. This is achieved by first rectifying the original time axis $\vec{t}$ into a standardized time axis $\vec{\tau}$ such that gait cycles are constant w.r.t. the compensated time axis $\vec{\tau}$. Without loss of generality, we set the gait cycle equal to 1. We rectify the time axis as follows:

$$d\tau_i = \tau_{i+1} - \tau_i = (t_{i+1} - t_i)/p(t_i) \qquad (10.12)$$

$$\tau_i = \tau_0 + \sum_{j=0}^{i-1} d\tau_j \quad i = 1, 2, 3, \ldots \qquad (10.13)$$

Now that we have mapped the input time to a new time dimension where the gait cycles are constant, we can uniformly sample the new time and time lag dimensions and compute the GDI responses using bilinear interpolation of the original GDI. This allows us to compute pace-compensated GDI. For the rectified GDI, we use the length of one gait cycle as the maximum lag. Note that this rectification is performed in both the time dimension and the lag dimension of GDIs by rectifying the shared time axis. The pace normalization effect of the algorithm on GDI is illustrated in Figure 10.7. Figure 10.7(a) shows GDI without pace normalization and Figure 10.7(b) shows pace normalized GDI. Gait data from different walking speeds can then be compared using these rectified GDIs for pace-compensated gait biometrics authentication.

Note that similarly to [68,69], we compute instantaneous motion cycles using energy minimization. However, our energy function is more generalized to include prior information on gait cycles. While [68] used iterative snake approach to compute a locally optimal estimate with the risk of being trapped to local minima, our method uses linear dynamic programing to efficiently compute the globally optimal solution using only one scan of a GDI. As pointed out in [69] that there is inherent ambiguity in using DTW for phase estimation and an extra bias correction procedure has to be proposed to remove the ambiguity; our approach bypasses this problem by rectifying the time axes using estimated instantaneous gait cycles to compute simple and effective pace-compensated gait biometrics.

10.4   Gait identity extraction using i-vectors

We use the i-vector approach, a state-of-the-art speaker biometrics method [37], for gait authentication. Although speech and gait authentication represent different application domains, these two problems are similar in nature: both need to extract subject-specific signatures from sensory data confounded with variations from a range of irrelevant sources (such as speech content and microphone channel effect for speaker authentication; footwear and pocket size for mobile gait authentication). The i-vector feature is also closely related to "Fisher vectors" [71], a highly successful feature extraction method for computer vision research: both are based on GMMs built for the data and both use statistics from these empirical distributions to compute features. The i-vector method for gait identity extraction using GDIs has the following advantages:

1.   Use of a large pool of subjects to learn a distribution characterizing the feature vector space, which is typically unavailable in other methods
2.   Unsupervised learning to result in better generalization capability
3.   A single-vector representation of gait-print to scale well to a very large population of subjects
4.   Allowing use of any machine-learning methods on i-vector, such as linear discriminative analysis to further improve the authentication performance.

We first enhance the pace-compensated GDI features to compute delta GDI features, following the common practice in speaker biometrics. We then build a GMM-based universal background model using all training feature vectors (which can be independent of the probe data). A supervector M is computed for each gallery and probe gait sequence by concatenating the Baum–Welch statistics for each component in the GMM using all gait feature vectors in the sequence. The low-rank i-vector w is extracted for a gait sequence using factor analysis in the supervector space with a simplified linear model to encode the identity information:

$$M = m + Tw \qquad (10.14)$$

where m is a subject-independent component, T is a low-rank rectangular matrix called total variability matrix computed during the training session.

The similarity between two gait sequences is computed using the cosine distance [37] between their corresponding i-vectors as the Cosine distance has shown better performance in speaker identification compared to other methods, such as the support vector machine [36]:

$$d(w_1, w_2) = \frac{< w_1, w_2 >}{\|w_1\| \, \|w_2\|}. \tag{10.15}$$

10.5 Performance analysis

To evaluate the effectiveness of the proposed algorithm for practical gait biometrics, we have conducted performance analysis of our gait biometrics algorithms on three datasets:

1. The naturalistic McGill University accelerometer gait dataset collected using mobile phones in a relaxed setting across days;
2. The Osaka University gait dataset, which is the largest publicly available gait dataset containing both accelerometer and gyroscope data, but in a more constrained setting; and
3. Our own accelerometer gait dataset of 51 subjects collected using mobile phones in a relaxed setting, at multiple walking speeds [6]. Details of each experiment are illustrated in the following subsections.

In all experiments, the raw data time sequences are preprocessed to remove the effect of gravity and sensor biases. We apply a moving average estimator to compute the data mean in a time window and subtract the mean from the raw data time series. The size of the window is selected to equal to the length of one or several typical gait cycles. This simple process effectively removes the gravity component in the accelerometer measurements.

10.5.1 McGill University naturalistic gait dataset

The goal of this experiment is to investigate the effectiveness of GDIs for robust gait biometrics. We used the real-world dataset for gait recognition from McGill University [45] because its relaxed data collection protocol resembles more realistic gait biometrics applications than other datasets. Data were collected using HTC Nexus One phones on two different days with little constraint on the placement of the phone except that it was put in a pocket on the same side of a subject during the two data collections.

We extract GDIs corresponding to 50 s of raw signals with a time lapse of up to 1.2 s. A simple correlation-based classification method is used to assess the effectiveness of the GDI representations. The similarity between two gait sequences is computed by aggregating peak correlation coefficients between GDI windows of 2.4 s. Nearest neighbor classifier is used. We compare the recognition accuracy using

Table 10.1 Gait ID accuracy on the McGill University gait dataset using gait dynamics images

Approach	Identification accuracy	
	Same day (%)	Separate days (%)
Magnitude (baseline)	67.5	32.5
Inner product GDI	87.5	61.3
Cosine similarity GDI	85.0	66.3

the GDIs to a baseline using only the magnitude series (which is the first row of the inner product GDI) because of its popularity in existing studies and superior performance to other features [26]. There are two scenarios: (1) training and testing using nonoverlapping data collected on the same day and (2) training and testing using data from separate days are examined. Obviously, the latter is more challenging as the attire, carried loads, footwear, and most importantly the phone placements are all subjected to change in addition to the variations in the same-day scenario.

Table 10.1 shows the accuracies for the gait identification algorithms. Although both the magnitude series and GDIs are robust to orientation variations, GDIs contain much more information on context and interactions in gait cycles, allowing them to offer powerful discrimination and perform significantly better than the magnitude features. This advantage is even more drastic for the challenging separate-day scenario with more variations, where we obtain an accuracy of 66.3%—more than doubling the accuracy using magnitude features. Though all methods performed worse for the separate-day scenario, the methods using GDIs degraded much more gracefully thanks to their rich discriminating gait dynamics.

10.5.2 Osaka University largest gait dataset

The McGill dataset is a realistic dataset. However, it only includes accelerometer data with 20 subjects. Thus, we conducted experiments on the largest publicly available gait dataset—the Osaka University dataset [24] that consists of 744 subjects with both gyroscope and accelerometer data. In this dataset, each subject has two walking sequences—one for training and one for testing.

We apply the i-vector technology to the GDI feature sequence for gait authentication. The UBM was modeled by a GMM of 800 components, and the final i-vector was 60 and 40 dimensions for the accelerometer and the gyroscope modality, respectively. The extracted i-vectors were matched using the cosine similarity to perform authentication. We conducted exhaustive authentication tests using the i-vector modeling tool in [72]. We have performed gait identification experiments using inner product GDIs and cosine similarity GDIs respectively, on gyroscope data and accelerometer data alone, and fused. We have also performed a "mega" fusion combining both data modalities and the two GDI construction methods. The fusion was performed using

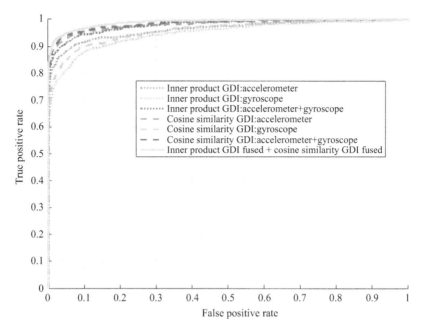

Figure 10.8   I-vector gait authentication results on the 744-subject Osaka University gait dataset, using inner product GDIs and cosine similarity GDIs with accelerometer, gyroscope, and sensor fusion

score averaging. The authentication results are plotted in Figure 10.8. We have three findings from this study:

1. It is evident from the analysis that accelerometer data works better than gyroscope data for gait biometrics, confirming findings reported previously.
2. The normalization effect in cosine similarity GDIs proves to be beneficial as they outperform inner product GDIs for both sensor modalities and also for the fusion case. On the basis of this finding, we only use cosine similarity GDIs in the remaining experiment.
3. Sensor and data fusion helps improving accuracy of the gait biometrics. To the best of our knowledge, this is the first study to demonstrate that gyroscope data can be used to improve the performance of accelerometer-based gait biometrics. It is also interesting to note that combination of cosine similarity GDIs and inner product GDIs outperforms either GDI method.

We also compare the performances of the proposed algorithms to four existing gait authentication algorithms. As shown in Table 10.2, our novel GDI + i-vector approaches have significantly reduced EERs, compared with recently published results on the same data set of 744 subjects, using the same experimental protocol.

Table 10.2   Gait authentication performance (EER in %) on the
Osaka Univ. gait dataset containing 744 subjects. We
compare EERs for the proposed algorithm (bottom two
rows: using inner product GDIs and cosine similarity
GDIs combined with i-vector modeling) with those of
four published algorithms (top four rows, as reported in
[24] Table 2) using the same dataset

Approach	Sensor modality		
	Acce.	**Gyro.**	**Fused**
Gafurov *et al.* [54]	15.8 [24]	NA	NA
Derawi *et al.* [44]	14.3 [24]	NA	NA
Liu *et al.* [73]	14.3 [24]	NA	NA
Trung *et al.* [24]	13.5 [24]	20.2 [24]	NA
Inner product GDI + i-vector	8.9	11.3	7.1
Cosine similarity GDI + i-vector	6.8	10.9	5.6

10.5.3   Mobile dataset with multiple walking speed

Because a mobile gait data set with multiple walking speeds collected using casually placed mobile phones is not publically available, we conducted our own data collection effort. We collected data from a total of 51 subjects (all company employees) of both genders, with ages ranging from 21 to 67, height ranging from 60 to 76 in. and weight ranging from 105 to 260 lbs. Each subject was outfitted with six Google Nexus 5 phones. One phone was placed in each of the two front pants pockets, and in each of the two rear pants pockets. In addition, one was carried in a backpack and another was held in the subject's hand. In all the cases, restrictions on neither the facing nor the orientation of the phones were imposed. The subjects were asked to walk in a prescribed loop path for four loops, with short pauses between the loops. They were instructed to walk normally in the first two loops, slowly in the third loop, and quickly in the fourth loop. The accelerometer signals were collected at 100 Hz using a custom Android app. For the purposes of this analysis, only data from the phone placed in the front right pocket was used.

Using the collected data, we extracted a subset of data roughly from the second half of each session. This set is used as the data for testing conditions under varying walking speed. We extracted another subset of data from the first half of each data session. This normal walking speed subset is further randomly divided into an enrollment set and a testing set. On average, each subject has 70 s of normal speed data for enrollment, 43 s normal walking speed data for testing, and 105 s varying walking speed data for testing.

We first compute the regular GDIs and pace normalized GDIs for all data sequences in the enrollment and test sets. We then divide each subject's GDIs into nonoverlapping 5-s segments for both the enrollment and the testing sets and compute an i-vector for each segment. In other words, we compute multiple i-vector templates for each subject based on each 5-s data. The gait authentication decisions are based on the 5-s gait datasets.

To extract the i-vector from GDI gait representations, we used a correlation of the current sample frame to every 8th frame until the 60th frame, i.e., the extraction GDI feature dimension is 8 at each time step. Similar to speech feature extraction methods used in speech and speaker recognition, we apply delta and delta–delta features to the GDIs to capture feature dynamics based a frame window size of 2. The final input to the i-vector GMM is thus a 24-dimensional feature vector.

For i-vector modeling, we choose an UBM consisting of 400 GMMs. This results in a super-vector of $24 \times 400 = 9,600$ dimensions for each 5 s of gait data. We then perform factor analysis to extract an identity vector of 100 dimensions for every 5-s segment of gait data. During testing, each testing i-vector is evaluated against all i-vectors of a hypothesis model. Nearest Neighbor classifier using Cosine distance is used to classify each 5-s test segment based on its i-vector. The evaluation is exhaustive, i.e., all testing segments are evaluated against all subject models. The test on the normal speed case has a total of 24,888 comparisons and the test on the varying speed case has a total of 59,313 comparisons.

We have conducted tests for two scenarios: using testing data with normal walking speed and with multiple walking speeds respectively. The enrollment is performed on the training set with normal walking speed for both testing scenarios.

We use the collected dataset to evaluate the proposed algorithm as well as the GDI based gait biometrics algorithm described in [35] as a baseline for comparison. Figure 10.9 shows the performance curves for the four tests, where we plot the true positive rate against the false acceptance rate (FAR). We also show in Table 10.3 the equal error rate (EER) where FRR equals FAR and its standard deviation [24] for each test. For the less challenging test with few pace variations, both algorithms work very well and their performances are statistically equivalent, even though the data was collected with casually placed mobile phones. When gait data with different walking speeds from the training data is tested, both approached degrade. However, the proposed algorithm, normalized using detected gait cycles, is able to achieve an ERR of 7.22% with a performance improvement of 37% over the regular GDI approach.

10.6 Conclusions and future work

The common challenge shared by mobile gait biometrics and other biometrics modalities, including face recognition, fingerprinting, iris recognition, and voice identification is to work reliably in an unconstrained environment [7,8]. After years of accelerometer-based gait biometrics research under controlled settings there has been a shift toward a more relaxed environment that allows for realistic variations typical in practical applications, as consumer mobile devices embedded with inertial sensors become an indispensable part of people's daily life. We have joined this effort

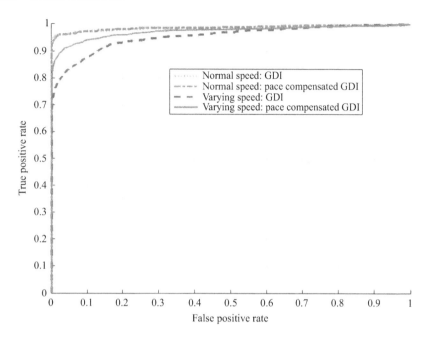

Figure 10.9   *Gait authentication performance using 5-s gait segments on our*
51-subject mobile gait dataset, using regular cosine similarity GDIs
and pace normalized cosine similarity GDIs, on testing data with
normal walking speeds, and varying walking speeds. Training data
were acquired with normal walking speed. Both regular GDIs and
pace-compensated GDIs work very well on testing data of normal
walking speed, with performances that are statistically equivalent.
Varying walking speed presented a major challenge as the
performances degrade for both GDI methods. However,
pace-compensated GDIs prove to be much more robust than regular
GDIs in the presence of variations in gait cycles and significantly
outperform the latter on the test dataset of varying walking speed

by proposing a mobile gait biometrics algorithm meant for use in real world scenarios.
We directly address two major variations in practical mobile gait biometrics, namely
data dependency on sensor rotation, and varying walking speed, to propose a gait
representation that is invariant to sensor orientation and pace insensitive. In addition,
this gait representation is also invariant to symmetry transforms, further relaxing the
sensor location constraint thanks to the left-right symmetry exhibited in the locomotion of most human bodies. We also use the i-vector method, a factor analysis based
voice biometrics approach, to model additional variations in the data and extract subject specific identity signatures for accurate gait biometrics. Performance evaluations
on three mobile gait datasets have validated the effectiveness and robustness of the

Table 10.3   *Gait authentication equal error rate (σ of EER) in % using 5-s*
              gait segments on our 51-subject realistic mobile gait dataset,
              using regular GDIs and pace normalized GDIs, on testing
              data with normal walking speeds, and varying walking speeds

Gait biometrics algorithm	Test scenario	
	Normal walking speed	**Multiple walking speeds**
GDI approach [35]	3.88 (0.54)	11.53 (1.6)
Proposed algorithm	3.89 (0.54)	7.22 (1.0)

proposed algorithms, demonstrating significant performance improvement compared to state-of-the-art gait authentication approaches.

The proposed work moves gait biometrics one step forward from existing laboratory studies towards more widespread and casual use in people's daily lives. This is still an emerging field full of open challenges to develop robust, reliable, and computationally efficient gait biometrics for realistic use. We need to better understand, quantify, and cope with other motion variations, including those influenced by sensor location, footwear, and floor conditions, etc. It is desirable to collect gait data over an extended time period under normal daily activities and assess the performance of gait biometrics over time. There is great potential in integrating gait biometrics from sensors from multiple locations, such as mobile phones and smart watches, or fusing gait with other biometric modalities available to mobile devices such as touch, keystroke dynamics, etc. to form a comprehensive security solution.

Acknowledgments

We would like to thank Thanh Trung Ngo and Jordan Frank for providing the gait datasets used in the performance study, and for answering our questions. We would also like to thank the anonymous reviewers for their constructive suggestions.

References

[1]   V. M. Patel, R. Chellappa, D. Chandra, and B. Barbello. Continuous user authentication on mobile devices: recent progress and remaining challenges. *IEEE Signal Processing Magazine*, 33(4): 49–61, 2016.

[2]   D. Crouse, H. Han, D. Chandra, B. Barbello, and A. Jain. Continuous authentication of mobile user: fusion of face image and inertial measurement unit data. In *IEEE Conf. on Biometrics*, 2015.

[3]   Y. Deng and Y. Zhong. Keystroke dynamics advances for mobile devices. In *Recent Advances in User Authentication Using Keystroke Dynamics*, edited by Y. Zhong and Y. Deng, Science Gate Publishing, 2015.

[4]   A. Primo, V.V. Phoha, R. Kumar, and A. Serwadda. Context-aware active authentication using smartphone accelerometer measurements. In *Computer Vision and Pattern Recognition Workshops (CVPRW), 2014 IEEE Conference on, IEEE*, pp. 98–105, June 2014.

[5]   H.J. Ailisto, M. Lindholm, J. Mantyjarvi, E. Vildjiounaite, and S.M. Makela. Identifying people from gait pattern with accelerometers. In *Defense and Security*, pp. 7–14. International Society for Optics and Photonics, 2005.

[6]   Y. Zhong, Y. Deng, and G. Meltzner. Pace independent mobile gait biometrics. In *IEEE Conf. BTAS*, 2015.

[7]   A.K. Jain, R. Bolle, and S. Pankanti (editors). *Biometrics: Personal Identification in Networked Society*, Kluwer Academic Publishers, 1999.

[8]   A.K. Jain, S. Pankanti, S. Prabhakar, L. Hong, and A. Ross. Biometrics: a grand challenge. *Proceedings of the International Conference on Pattern Recognition*, 2: 935–942, 2004.

[9]   S. Sarkar, P.J. Phillips, Z. Liu, I.R. Vega, P. Grother, and K.W. Bowyer. The human ID gait challenge problem: data sets, performance, and analysis. *IEEE Transactions on Pattern Analysis and Machine Intelligence*, 27(2): 162–177, 2005.

[10] T. Lee, M. Belkhatir, and S. Sanei. A comprehensive review of past and present vision-based techniques for gait recognition. *Multimedia Tools and Applications*, 2013.

[11] M.S. Nixon and J.N. Carter. Advances in automatic gait recognition. *IEEE International Conference on Automatic Face and Gesture Recognition*, 2004.

[12] A. Kale, N. Cuntoor, B. Yegnanarayana, A.N. Rajagopalan, and R. Chellappa. *Gait Analysis for Human Identification, Audio- and Video-Based Biometric Person Authentication*, pp. 706–714. Springer Berlin Heidelberg, 2003.

[13] A. Alvarez-Alvarez, G. Trivino, and O. Cordón. Human gait modeling using a genetic fuzzy finite state machine. *IEEE Transactions on Fuzzy Systems*, 20(2): 205–223, 2012.

[14] L. Lee and W.E.L. Grimson. Gait analysis for recognition and classification. In *Proc. 5th IEEE Int'l Conf. Automatic Face and Gesture Recognition*, pp. 148–155, 2002.

[15] T.B. Moeslund and E. Granum. A survey of computer vision-based human motion capture. *Computer Vision and Image Understanding*, 81(3): 231–268, 2001.

[16] G. Trivino, A. Alvarez-Alvarez, and G. Bailador. Application of the computational theory of perceptions to human gait pattern recognition. *Pattern Recognition*, 43: 2572–2581, 2010.

[17] M. Hu, Y. Wang, and Z. Zhang. Cross-view gait recognition with short probe sequences: from view transform model to view-independent stance independent identity vector. *International Journal of Pattern Recognition and Artificial Intelligence*, 27(06), 2013.

[18] A. Muro-de-la-Herran, B. Garcia-Zapirain, and A. Mendez-Zorrilla. Gait analysis methods: an overview of wearable and non-wearable systems, highlighting clinical applications. *Sensors*, 14(2): 3362–3394, 2014.

[19] M.O. Derawi. Accelerometer-based gait analysis, a survey. *Norwegian Information Security Conference*, 33–44, 2010.

[20] J. Mantyjarvi, M. Lindholm, E. Vildjiounaite, S.M. Makela, and H. Ailisto. Identifying users of portable devices from gait pattern with accelerometers. *IEEE International Conference on Acoustics, Speech, and Signal Processing*, 2, 2005.

[21] L. Bao and S. Intille. Activity recognition from user-annotated acceleration data. In *Pervasive Computing*, pp. 1–17. Springer Berlin Heidelberg, 2004.

[22] A.U. Deshmukh, N.D. Deshmukh, and R.R. Maharana. State of art in real time gait analysis system for healthy ambulation. *International Journal of Computer Science and Applications*, 8(2), 2015.

[23] K. Aminian, B. Najafi, C. Büla, P.-F. Leyvraz, and P. Robert. Spatio-temporal parameters of gait measured by an ambulatory system using miniature gyroscopes. *Journal of Biomechanics*, 35(5): 689–699, 2002.

[24] T.T. Ngo, Y. Makihara, H. Nagahara, Y. Mukaigawa, and Y. Yagi. The largest inertial sensor-based gait database and performance evaluation of gait recognition. *Pattern Recognition*, 47(1): 228–237, January 2014.

[25] A. Johnston and G. Weiss. Smartwatch-based biometric gait recognition. In *BTAS*, 2015.

[26] F. Juefei-Xu, C. Bhagavatula, A. Jaech, U. Prasad, and M. Savvides. Gait-ID on the move: Pace independent human identification using cell phone accelerometer dynamics. In *IEEE Fifth Int'l Conf. on Biometrics: Theory, Applications and Systems (BTAS)*, pp. 8–15, 2012.

[27] M.R. Aqmar, Y. Fujihara, Y. Makihara, and Y. Yagi, Gait recognition by fluctuations. *Computer Vision and Image Understanding*, 126: 38–52, 2014.

[28] M. Bachlin, J. Schumm, D. Roggen, and G. Toster. Quantifying gait similarity: user authentication and real-world challenge. *Advances in Biometrics*, 2009.

[29] I. Bouchrika and M.S. Nixon. Exploratory factor analysis of gait recognition. In *8th IEEE International Conference on Automatic Face & Gesture Recognition*, 2008.

[30] Y. Watanabe. Toward application of immunity-based model to gait recognition using smart phone sensors: a study of various walking states. *Procedia Computer Science*, 60: 1856–1864, 2015.

[31] G. Pan, Y. Zhang, and Z. Wu. Accelerometer-based gait recognition via voting by signature points. *Electronics Letters*, 45(22): 1116–1118, 2009.

[32] Y. Watanabe. Influence of holding smart phone for acceleration-based gait authentication. In *Emerging Security Technologies (EST), 2014 Fifth International Conference on, IEEE*, 2014.

[33] T.T. Ngo, Y. Makihara, H. Nagahara, Y. Mukaigawa, and Y. Yagi. Orientation-compensative signal registration for owner authentication using an accelerometer. *IEICE Transactions on Information and Systems*, 97(3): 541–553.

[34]   R. Subramanian, S. Sarkar, M. Labrador, *et al.* Orientation invariant gait matching algorithm based on the Kabsch alignment. In *Identity, Security and Behavior Analysis (ISBA), 2015 IEEE International Conference on, IEEE*, pp. 1–8, 2015.

[35]   Y. Zhong and Y. Deng. Sensor orientation invariant mobile gait biometrics. In *Int'l Joint Conf. on Biometrics*, September 2014.

[36]   N. Dehak, R. Dehak, P. Kenny, N. Brummer, P. Ouellet, and P. Dumouchel. Support vector machines versus fast scoring in the low-dimensional total variability space for speaker verification. In *INTERSPEECH*, September 2009.

[37]   N. Dehak, P.J. Kenny, R. Dehak, P. Dumouchel, and P. Ouellet. Front-end factor analysis for speaker verification. *IEEE Transactions on Audio, Speech and Language Processing*, 19(4), 2011.

[38]   S. Choi, I.H. Youn, R. LeMay, S. Burns, and J.H. Youn. Biometric gait recognition based on wireless acceleration sensor using k-nearest neighbor classification. In *Computing, Networking and Communications (ICNC), 2014 International Conference on, IEEE*, pp. 1091–1095, February 2014.

[39]   M.L. McGuire. An overview of gait analysis and step detection in mobile computing devices. In *Intelligent Networking and Collaborative Systems (INCoS), 2012 4th International Conference on, IEEE*. For Health and Clinical Applications, pp. 648–651, September 2012.

[40]   S. Sprager and M. Juric. An efficient HOS-based gait authentication of accelerometer data. *IEEE Transactions on Information Forensics and Security*, 10(7), 2015.

[41]   Y. Zhang, G. Pan, K. Jia, M. Lu, Y. Wang, and Z. Wu. Accelerometer-based gait recognition by sparse representation of signature points with clusters. *IEEE Transactions on Cybernetics*, 2014, doi:10.1109/TCYB.2014.2361287.

[42]   D. Gafurov, K. Helkala, and T. Søndrol. Biometric gait authentication using accelerometer sensor. *Journal of Chemical Physics*, 1(7), 2006.

[43]   W. Tao, T. Liu, R. Zheng, and H. Feng. Gait analysis using wearable sensors. *Sensors*, 12(2), 2012.

[44]   M.O. Derawi, P. Bours, and K. Holien. Improved cycle detection for accelerometer based gait authentication. In *6th IEEE Int'l Conf. Intelligent Information Hiding and Multimedia Signal Processing (IIH-MSP)*, 2010.

[45]   J. Frank, S. Mannor, and D. Precup. Data Sets: Mobile Phone Gait Recognition Data. http://www.cs.mcgill.ca/~jfrank8/data/gait-dataset.html, 2010.

[46]   J. Frank, S. Mannor, J. Pineau, and D. Precup. Time series analysis using geometric template matching. *IEEE Transactions on Pattern Analysis and Machine Intelligence*, 35(3), 2013.

[47]   J.R. Kwapisz, G.M. Weiss, and S.A. Moore. Cell phone-based biometric identification. In *Biometrics: 4th IEEE Int'l Conf. Theory Applications and Systems (BTAS)*, 2010.

[48]   M. Muaaz and R. Mayrhofer. An analysis of different approaches to gait recognition using cell phone based accelerometers. In *Proc. Int'l Conf. on Advances in Mobile Computing & Multimedia*, ACM, 2013.

[49] C. Nickel, M.O. Derawi, P. Bours, and C. Busch. Scenario test of accelerometer-based biometric gait recognition. In *3rd Int'l Workshop on Security and Communication Networks*, 2011.

[50] Y. Ren, Y. Chen, M.C. Chuah, and J. Yang. User verification leveraging gait recognition for smartphone enabled mobile healthcare systems. *IEEE Transactions on Mobile Computing*, 14(9), 2015.

[51] H.M. Thang, V.Q. Viet, N.D. Thuc, and D. Choi. Gait identification using accelerometer on mobile phone. In *IEEE Int'l Conf. on Control, Automation and Information Sciences (ICCAIS)*, 2012.

[52] S. Sprager and M.B. Juric. Inertial sensor-based gait recognition: a review. *Sensors*, 15(9): 22089–22127, 2015.

[53] S. Sprager. A cumulant-based method for gait identification using accelerometer data with principal component analysis and support vector machine. In *Sensors, Signals, Visualization, Imaging, Simulation and Materials*, pp. 94–99, 2009.

[54] D. Gafurov, E. Snekkenes, and P. Bours. Improved gait recognition performance using cycle matching. In *IEEE 24th Int'l Conf. Advanced Information Networking and Applications Workshops (WAINA)*, 2010.

[55] P. Bours and R. Shrestha. Eigensteps: a giant leap for gait recognition. In *Proceedings of the 2nd International Workshop on Security and Communication Networks (IWSCN)*, Karlstad, Sweden, pp. 1–6, 26–28 May 2010.

[56] T. Iso and K. Yamazaki. Gait analyzer based on a cell phone with a single three-axis accelerometer. In the *8th Conf. on Human-Computer Interaction with Mobile Devices and Services*, pp. 141–144, 2006.

[57] T. Kobayashi, K. Hasida, and N. Otsu. Rotation invariant feature extraction from 3-D acceleration signals. In *IEEE Int'l Conf. on Acoustics, Speech, and Signal Processing*, 2011.

[58] N. Boulgouris, K. Plataniotis K, and D. Hatzinakos. Gait recognition using dynamic time warping. *Proc. IEEE 6th Workshop on Multimedia Signal Processing*, 2004.

[59] R. Tanawongsuwan and A. Bobick. Performance analysis of time-distance gait parameters under different speeds. In *Proc. Int. Conf. Audio–Video-Based Biometric Person Authentication*, pp. 715–724, 2003.

[60] H. Ying, C. Silex, A. Schnitzer, S. Leonhardt, and M. Schiek. Automatic step detection in the accelerometer signal. In *4th International Workshop on Wearable and Implantable Body Sensor Networks*, 2007.

[61] Y. Makihara, A. Tsuji, and Y. Yagi. Speed-invariant gait recognition. In *Signal and Image Processing for Biometrics*, pp. 209–229, Springer Berlin Heidelberg, 2014.

[62] M.Q. Aqmar, M. Rasyid, K. Shinoda, and S. Furui. Robust gait-based person identification against walking speed variations. *IEICE Transactions on Information and Systems*, 95(2): 668–676, 2012.

[63] G.B. Del Pozo, C. Sánchez-Avila, A. De-Santos-Sierra, and J. Guerra-Casanova. Speed-independent gait identification for mobile devices.

International Journal of Pattern Recognition and Artificial Intelligence, 26(8), 2012.

[64]   R.G. Cutler and L.S. Davis, "Robust real-time periodic motion detection, analysis and applications," *IEEE Transactions on Pattern Analysis and Machine Intelligence*, 22(8), 2000.

[65]   C. Nickel, C. Busch, S. Rangarajan, and M. Mobius. Using hidden Markov models for accelerometer-based biometric gait recognition. In *Proc. of the IEEE 7th International Colloquium on Signal Processing and its Applications (CSPA)*, Penang, Malaysia, pp. 58–63, 4–6 March 2011.

[66]   M.O. Derawi, C. Nickel, P. Bours, and C. Busch. Unobtrusive user-authentication on mobile phones using biometric gait recognition. In *IEEE 6th Int'l Conf. on Intelligent Information Hiding and Multimedia Signal Processing (IIH-MSP)*, 2010.

[67]   K. Holien. *Gait Recognition under Non-Standard Circumstances*, M.S. thesis, Gjøvik University College – Dept. of Computer Science and Media Technology, 2008.

[68]   S.M. Seitz and C.R. Dyer. Detecting irregularities in cyclic motion. In *Proc. IEEE Workshop on Motion of Non-Rigid and Articulated Objects*, 1994.

[69]   Y. Makihara, M.R. Aqmar, N.T. Trung, *et al.* Phase estimation of a single quasi-periodic signal. *IEEE Transactions on Signal Processing*, 62(8): 2066–2079, 2014.

[70]   W. Premerlani and P. Bizard. Direction Cosine Matrix IMU: Theory. http://gentlenav.googlecode.com/files/DCMDraft2.pdf.

[71]   J. Sanchez, F. Perronnin, T. Mensink, and J. Verbeek. Image classification with the Fisher vector: Theory and practice. *International Journal of Computer Vision*, 105(3): 222–245, June 2013.

[72]   D. Povey, A. Ghoshal, G. Boulianne, *et al.* The Kaldi speech recognition toolkit. *Proceedings of IEEE ASRU*, 2011.

[73]   R. Liu, J. Zhou, and X. Hou. A wearable acceleration sensor system for gait recognition. In *2nd IEEE Conf. Industrial Electronics and Applications*, 2007.

Chapter 11

4F™-ID: mobile four-fingers biometrics system

Asem Othman[1], Francis Mather[1] and Hector Hoyos[1]

11.1 Introduction

Smartphones are personal communication tools that have become ubiquitous devices in our daily lives. We are carrying them at all times to facilitate day-to-day activities, including checking emails, shopping, entertainment, expressing ourselves as well as communicating with others via phone calls and social networks. The rapid proliferation of mobile phones[2] has also played a major role in accelerating this trend. Now, users access and store sensitive information, such as photos, email addresses, and phone numbers on their smartphones. Further, smartphones become the quick gateways to access cloud storage and financial accounts (e.g., bank and credit card accounts). With the increasing functionality and services accessible via smartphones, an attack on the mobile device, or the loss of it, can have devastating consequences, such as the intrusion of privacy, the opportunity to impersonate users, and even severe financial loss. Therefore, there is a need for supplantation of passwords and 4-digit personal identification numbers (PINs) that have traditionally been used; because they are cumbersome (i.e., these schemes do not always provide a satisfactory user experience[3]), extremely vulnerable to shoulder surfing attacks [2], and easily cracked—through sharing, reuse, or using weak passwords [3].

One of the principal alternatives is the use of biometric traits for a secure mobile authentication to verify the identity of a person accessing the device. Biometric traits such as fingerprints, iris, face, and voice can be acquired to verify identities [4] which ensures the physical presence of the user while offering a significantly higher security barrier. Therefore, recently commercial phones have provided an integrated fingerprint sensor for authentication. However, it is of paramount importance to investigate the possibilities to use regular smartphones for biometrics authentication by exploiting the capabilities of the built-in sensors instead of additional hardware.

---

[1]Hoyos Labs LLC, USA

[2]It is estimated that over 1 billion mobile phones were sold worldwide in 2014 alone. Source: http://www.gartner.com/newsroom/id/2996817

[3]Recent market research concluded that U.S. customers are increasingly using their mobile devices to conduct financial transactions and they will be more comfortable if alternative methods are utilized to secure their mobile accessing and authentication [1].

Most of smartphones have a rich set of input sensors, including cameras, microphones, touch screens, and GPS. Biometric authentication methods using these sensors could offer cheap alternatives to password schemes because the users are familiar and already use these sensors for a variety of mobile tasks.

In comparison to the extensive development of biometrics research, the research on mobile biometric authentication is still quite poor. Although it may seem obvious to inherit techniques and strategies used in previous deployed biometric systems, the development of mobile biometrics is facing two main challenges:

1. Selecting a biometric trait that balances the trade-off between usability and performance. In other words, selecting a biometric trait that it is easy to quickly capture a good quality sample of it without making the process cumbersome. Therefore, the developed methods have to be designed with the consideration that the interaction with the mobile device tends to be brief [5] and interruption driven [6].

2. The development of algorithms that are suitable for the limited storage and computational resources associated with mobile devices. Although the recent advancement in the hardware design of mobiles, still the mobile devices' resources cannot be considered equivalent to those available to traditional deployed systems.

In this chapter, we discuss a mobile fingerprint systems denoted as $4F^{TM}$-ID system. This system utilizes the built-in rear camera of the smartphone to capture a photo of four fingers, as shown in Figure 11.1. This new touchless fingerprint recognition system requires no extra hardware due to the use of fingers photo to perform authentication.

The $4F^{TM}$-ID system is an ideal candidate for mobile biometric recognition for the following reasons:

1. Fingerprint is one of the most accepted biometric traits [4], due to many reasons, such as fingerprints are pose, expression, occlusion, and aging invariant. Therefore, the $4F^{TM}$-ID is particularly useful in scenarios when other biometric traits fails in mobile environments: (a) traditional face recognition systems fail due to occlusions and illumination variation, (b) iris recognition systems fail due to visible illumination and dark pigmentation of some irises, and (c) voice recognition systems fail due to the ambient noise and variation in microphone channel characteristics across devices.

2. No extra hardware is necessary, therefore, it is a great advantage over traditional touch-based fingerprint recognition system, which can only be used on a few mobile phones containing fingerprint sensors.

3. $4F^{TM}$-ID acquisition process does not present particular constraints because the user can use the rear camera of a smartphone while maintaining a suitable and comfortable pose to capture the region-of-interest (ROI). Based on a usability study, which is done by Trewin *et al.* [6] on the usage of biometrics for mobile phones, capturing images for authentication are faster than writing passcodes, and more convenient and private than speaking passcodes.

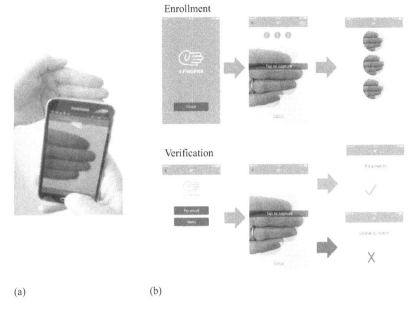

Figure 11.1 *The proposed 4F™-ID mobile biometric system. (a) An illustration of the image acquisition using 4F™-ID mobile system and (b) a diagram of the enrollment and verification steps*

Although, the merits of utilizing 4F™-ID as a mobile biometrics, there are many challenges that are posted.

- There are some smartphone cameras that cannot focus and capture the ridge patterns of finger tips due to the limitation of the camera's depth of field. Therefore, the fingers should be away from the camera, but this reduces the effective usable resolution of the finger images.
- There are many potential variations of the orientation angel, pitch angle, position of the fingers, and the distance between fingers and camera's lenses.
- There are unlimited number of unpredictable lighting conditions and backgrounds associated with capturing fingerphotos using the rear camera of a smartphone.
- Fingerphotos captured by smartphone cameras may exhibit a low contrast between the ridges and valleys of the finger tips which increase the challenge of extracting accurate feature representations of the fingers.

The rest of this chapter is organized as follows: In Section 11.2, we review some previous efforts carried out on mobile finger authentication using fingerphotos. Section 11.3 presents the 4F™-ID system to use four fingers for mobile biometrics authentication. Conducted experiments are presented and discussed in Section 11.4. Finally, Section 11.5 concludes the chapter.

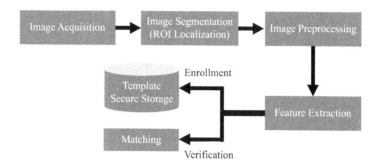

Figure 11.2 A block diagram of the steps involved in the different fingerphoto authentication systems

11.2 Related work

The main components of traditional fingerprint recognition systems [7] include image acquisition, image enhancement, feature extraction, and matching. However, the nature of fingerphoto images captured by cameras of mobile phones is different from images obtained with legacy contact-based devices. New methods of image analysis and enhancement must be implemented to provide additional flexibility for specific applications and customers. Therefore, besides traditional components, segmentation step is required after acquiring a fingerphoto with a mobile camera to define the ROI which is the fingertip, where the uniqueness characteristics of a finger are presented, i.e., ridges orientation and minutiae. The performance of fingerphoto systems depends on segmenting the fingers from varied backgrounds and determining accurately the (ROI). As shown in Figure 11.2, in addition to the traditional components (i.e., acquisition, preprocessing, feature extraction, secure storage, and matching) for any fingerprint system [7], the segmentation step is included.

In this chapter, we discuss the state-of-art systems that utilized mobile fingerphoto images for the purpose of authentication. Table 11.1 summarizes the steps of these systems along with the processing time if the system deployed in a mobile environment, and the list of used mobile devices for fingerphotos collection.

There are systems that have been developed for acquiring contactless fingerprint using digital cameras, but most of these contactless systems use complex setups, such as fixed cameras and finger placements. These constrained acquisition settings are different than the case of unconstrained acquisition conditions that are usually associated with mobile fingerphoto systems. Because dealing with special acquisition setup is out of the scope of our chapter, we encourage the reader to read [8] for a quick overview of 2D and 3D contactless fingerprint recognition systems.

11.2.1 Finger segmentation (ROI localization)

Available state-of-the-art schemes rely on capturing a single finger with a controlled background. Also, most of the published work assume that the acquired finger is

Table 11.1   A high-level comparison of the purposed mobile fingerphoto systems

Research	Segmentation	Preprocessing	Feature extraction and matching	Deployment/time	Devices used for data collection
Stein et al. [9]	Red-channel thresholding	Geometric normalization of the ROI followed by median and local Binarization filters	Minutiae based	Yes/50 s (enrollment) −22 s (authentication)	Nexus S and Galaxy Nexus
Raghavendra et al. [10]	MSS followed by the computation of metrics to distinguish the finger sub-regions	Scaling ROI followed by a Wiener filter to deblur the ROI	Minutiae based	N/A	Apple iPhone 4, Samsung Galaxy S1 and Nokia N8
Sankaran et al. [11]	Adaptive skin color thresholding followed by a connected component scheme to isolate the ROI	Histogram equalization followed by a high-pass filter	ScatNet features	N/A	Apple iPhone5
Tiwari and Gupta[12]	Adaptive skin thresholding followed by morphological operations and a connected component scheme to isolate the ROI	Computing the principle X axis based on PCA to rotate the ROI followed by adaptive histogram equalization	SURF	N/A	Samsung Galaxy Note "GT- N7000"
4F™-ID	Foreground segmentation followed by fingers separation and ROI localization	Resolution normalization followed by band-pass filter and adaptive histogram equalization scheme	Minutiae based	Yes/25 s (enrollment) −10 s (authentication)[a]	Apple iPhone 5C, Apple iPhone 6, Samsung Galaxy S4, Samsung Galaxy S5, etc.

[a]For the recent statistics, please contact us at info@hoyoslabs.com

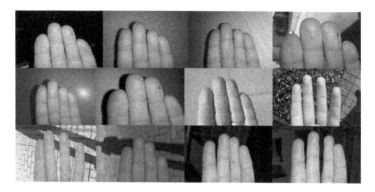

Figure 11.3   Sample images captured by 4F™-ID technology shows the different background and illumination variations

covering and occupying most of the image [9–12]. These assumptions led to a series of approaches that segment the finger based on the skin color information. Stien *et al.* [9] utilized the red-channel of the acquired image to determine the finger by tracking pixels in the red-channel until a predetermine value, i.e., a threshold occurs. This approach has been designed to find the whole finger and depends on the assumption of controlled backgrounds and a specific skin color reflection. In [11,12], skin adaptive thresholding has been applied to find the finger regions followed by a connected component scheme to isolate the ROI of the acquired finger. These approaches still depend on the assumption of a controlled background exhibiting colors away from the skin color.

Raghavendra *et al.* [10] utilized the color information by applying a course segmentation technique, i.e., mean shift segmentation (MSS). A set of metrics has been computed from the segmented regions to find regions that have the high likelihood to be from the acquired finger. Further, they located the fingertips by finding the thick valley that separates the fingertips from the rest of the finger. In order to determine this thick valley, in [10], they examined all the pixel that surround the midline of the detected finger region. Although this approach can accurately segment fingertips in realistic situations and backgrounds, using MSS to segment the original captured image, and then computing different metrics for finger parts detection may consume the mobile resources and increase the computational time.

Finally, it is worth mentioning that in slap fingerprint recognition [7], where the user leaves an impression of four fingers on a contact-based fingerprint sensor, the segmentation of a slap image into four separate single fingerprints is known as *slap segmentation*. Many researchers proposed different approaches solving the slap segmentation problem [7], though the proposed slap segmentation approaches *cannot* be adopted in fingerphoto systems. Fingerphoto images exhibit background and illumination variations that are not present in the case of slap fingerprint systems. Figure 11.3 illustrates the background variations that occur in the case of capturing fingerphoto using mobile cameras.

11.2.2 Image preprocessing and enhancement

In traditional systems (i.e., touch-based systems), image preprocessing can be done to help the feature extractor, which is rarely needed because fingerprint sensors have been designed for biometric use. However the main challenge of any fingerphoto system that smartphone cameras are not designed for fingerprint collection. Therefore, applying image preprocessing and enhancement on the segmented ROI is essential to normalize the effect of illumination and distance variations in addition to improve the contrast between ridge and valley structure in a fingerprint.

In [9,10], the fingerphoto systems are based on identity comparison techniques that require fixed resolution images (e.g., minutiae matching algorithms), hence a scaling step on the estimated ROI is performed. However, in [12], the rotation step is required due to the use of a nonminutiae-based algorithm, i.e., Speeded Up Robust Features (SURF) [12].

As enumerated in Table 11.1, after scale or rotation normalization step if needed, most of the proposed approaches apply a series of different image filters in order to enhance the contrast between the ridges and valleys along with the noise removal step.

11.2.3 Feature extraction and matching

Following the image preprocessing and enhancement step, the enhanced fingerphoto has a ridge-valley texture which can be utilized effectively to extract fingerprint features from it.

In most of the work done on fingerphoto systems, such as in [9,10], the use of minutiae-based features is explored along with the use of the traditional matching algorithms. In [11,12], the authors explored nontraditional feature extraction schemes.

Sankaran *et al.* [11] proposed Scattering Networks (ScatNet) features as the representation of the ridge-valley pattern. ScatNet representations of the fingerphoto are computed by applying a filterbank of wavelets on the enhanced ROI. To match two ScatNET feature sets, Random Decision Forest has been used as a supervised-based classifier to classify match and not-match cases [11].

Meanwhile, in [12], Tiwari and Gupta proposed the use of a keypoint-based features, i.e., SURF followed by nearest neighbor ratio to perform matching on the extracted SURF keypoints from the query and gallery images.

Although these nonminutiae-based approaches are designed for fingerphoto authentication systems, the generated feature sets are not compatible with legacy databases, i.e., databases of minutiae templates from contact-based sensors. These databases are considered as legacy databases because they comply with the standards [13,14]. Standards have been released to specify the format of fingerprint templates to guarantee the interoperability of different fingerprint systems (e.g., the template created by any feature extractor must be acceptable to different vendors' matchers). Therefore, it became customary for fingerprint systems to adopt standard minutiae formats such as ISO/IEC 19794-2 [13] and ANSI/NIST-ITL 1 [14].

Therefore, we believe that these nonminutia feature extractors [11,12] cannot currently be used in an interoperable scenario. For instance, using fingerphoto system for collecting and generating fingerprint template that should be send to a government database (e.g., the U.S. Automated Fingerprint Identification System) for the matching purposes. The worldwide large-scale deployment of mobile authentication systems such as fingerpohto systems demands a new generation of accurate and highly interoperable algorithms; therefore, the development of minutiae-based matching algorithms that compliant with fingerprint template standards should be the main objective of any fingerphoto system.

11.2.4 *System deployment*

Finally, as highlighted in Table 11.1, most of the aforementioned approaches used the smartphone merely for collection purposes, except Stein *et al.* [9] deployed their system in Android platform devices and reported the processing time for enrollment and authentication steps.

11.3   4F^{TM}-ID system

In the literature, see Table 11.1, the proposed mobile fingerphoto systems rely on capturing an image of a single finger for the authentication purpose, and most of these systems have not been deployed in different mobile platforms or tested in a large-scale deployment. Further, the performance of a single fingerprint based on mobile fingerphoto is still not reliable enough for applications requiring higher security such as medium- to high-value bank transactions and the authentication to an enterprise system. Henceforth, we proposed 4F^{TM}-ID system to provide a reliable and ubiquitous mobile fingerprint system.

4F^{TM}-ID uses the presence of multiple, independent pieces of biometrics (i.e., four fingers) which offers the reliability and security advantages over the single mobile fingerprint systems; either fingerphoto systems (see Table 11.1) or systems using embedded sensors as in recent Apple and Samsung smartphones. A combination of uncorrelated modalities (e.g., four fingers of a person) is expected to result in a better performance than a single finger recognition system [15]. This accuracy improvement happens due to two reasons. Firstly, the fusion of biometric evidences from different fingers effectively increases the dimensionality of the discriminatory feature space and reduces the overlap between the feature distributions of different users. In other words, a combination of multiple fingers is more discriminative to an individual than a single finger. Secondly, the noise (caused by factors like dirt) and imprecision during the acquisition of a subset of the fingers can be compensated by the discriminatory information provided by the remaining fingers.

The principal aim of the 4F^{TM}-ID system is to provide a reliable and secure means of finger recognition on mobile devices that convenient to use. 4F^{TM}-ID performs four finger recognition using cameras that are typically present on mobile devices, so that no additional volume, cost, or weight is incurred on the design and usage can be ubiquitous.

To accomplish the reliability and ubiquitous objectives, the 4F^{TM}-ID system uses the rear camera and the full power of the LED flash while a hand guide encourages the user to position their hand in a position that is optimal for capturing the ridge reflection from the flash. Afterward, the 4F™-ID system detects the fingertips, enhance these ROIs for an accurate rendering of the finger ridges. The extracted feature sets are minutiae-based features. Therefore, any off the shelf fingerprint matcher can be utilized. The component of the 4F™-ID system is discussed in the following.

11.3.1   *4F*™*-ID image acquisition*

4F™-ID system acquires the users' biometrics by capturing four fingers of a hand using the rear camera of a mobile device. In order to capture the ridges pattern of the four fingers, there are some elements that have to be considered during the acquisition step: (a) camera resolution, (b) LED illumination, and (c) camera focus.

The resolution of the mobile device camera is the most important criterion. The sharpness level of the captured images relies on the resolution, which must be high enough to detect the fingers' ridges. Therefore, based on our initial tests, the rear camera resolution is preferred to be greater than 5 megapixels in order to capture the uniqueness features of fingers (i.e., minutiae and ridge orientations). During the capture process, the LED is switched on to illuminate the captured fingers which alleviate the lighting conditions by generating a homogeneous illumination on the fingers and get rid of shadows. Further, the fingers will be brighter than the background which assist the segmentation step. Regarding the camera focus, the camera should use the closest possible focus which can be achieved by setting the camera focus to the "macromode," which is available in most of the recent smartphones.

However, after considering the aforementioned elements of the camera, the capturing process is still a challenging process due to the various potential poses of the fingers and distances from the camera lens. Also, most of smartphone cameras cannot focus on the necessary close range to capture the pattern ridges of the fingers. This is due to the limitation of the smartphone cameras' depth of field. Without a proper focus on the fingers, it is impossible to detect the pattern ridges of the fingers. However if the fingers is too far away from the camera, the effectively usable resolution of the fingerphoto is reduced and the risk that fingers cannot be detected increases. Additionally, the low amount of configuration possibilities of the smartphone cameras tightens the conditions for the fingerphoto recognition.

Therefore, we proposed the use of hand guide that is a four fingers outline overlaid on the camera image preview screen (see Figure 11.1). This hand guide has been designed to make sure the captured objects, i.e., fingers are in appropriate distance from the camera for the best focus. The users would know that their fingers are at an appropriate distance from the camera when they fill the outline on the camera preview as possible as they could with their fingers. The hand guide of the 4F™-ID system also guides the users to hold their fingers together rather than having them spaced apart. By using this hand guide, the users will more likely hold their fingers at similar distances in different instances to fill the guide and keep their fingers tight to each other which will minimize the orientation variation of the fingers.

Because the user may not position their finger precisely in the position specified by the guide, the image segmentation step has a significant role in order to identify the background and then detect the ROI of each finger.

11.3.2   4F^{TM}-ID image segmentation

During the image acquisition, an outline of the hand is previewed to prompt the users to place their fingers at an optimal location and orientation with respect to the position of the illuminating light source and camera. However, the ROI localization (ROI is the finger region where the uniqueness features reside) is still a challenge that has to be tackled due to the different hand sizes and background variation. Further, in the case of the 4F^{TM}-ID system, localizing four finger tips and classifying them into four classes is a problem that has not been discussed in the state-of-the-art schemes because most of them only collected a single finger and assume that this finger is covering most of the image area. These challenges motivate us to propose a new scheme for automatic fingers segmentation and localization.

11.3.2.1   Foreground and background segmentation

The first step of our scheme is segmenting the acquired image into homogeneous regions to distinguish between the fingers and nonfingers regions. To accomplish this, we utilized the MSS algorithm [16] as in the proposed scheme of Raghavendra *et al.* [10]. However, in [10], they applied MSS algorithm on the original captured image. Segmenting the captured image can be successful in the case of capturing a single finger along with the assumption that the finger cover most of the captured image and less variation in the background. For our 4F^{TM}-ID system, to aid the MSS algorithm and to increase the probability of segmenting the whole hand as one region, the nonlinear decomposition based on local total variation (LTV) [17] is utilized to decompose the image into two components, viz cartoon and texture components.

A 4F^{TM}-ID image, f, is decomposed based on LTV [17] as a sum of two components: $f = u + v$, where u represents the cartoon component of f and v represents the texture component of f. Based on the characteristic of a cartoon image that its total variation does not decrease by low-pass filtering, Buades *et al.* [17] proposed a fast nonlinear decomposition method. Figure 11.4 shows 4F^{TM}-ID image and its two components. Then, the cartoon component of the 4F^{TM}-ID image is segmented by employing the MSS algorithm. As illustrated in Figure 11.5, applying the MSS on the cartoon component results in less number of segments to be examined than applying it on the original image as in [10]. Next, based on the size of the segmented regions, the largest area, which is located closer to the center of the image, is classified as the four fingers region in the original 4F^{TM}-ID image and the other segments are masked (see Figure 11.6).

11.3.2.2   Fingers separation

After segmenting and masking the nonfingers area, each finger region has to be identified. The objective is splitting the segmented image of four fingers into four

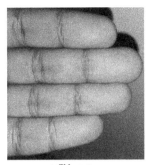

(a)   A 4F^{TM}-ID image

(b)   The cartoon image

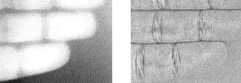

(c)   The texture image

Figure 11.4   Cartoon-texture decomposition using a total variation method [17]

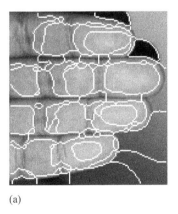

(a)

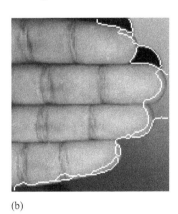

(b)

Figure 11.5   Illustration of the results of applying MSS on (a) A 4F^{TM}*-ID image*
and (b) cartoon component of 4F^{TM}*-ID image*

individual finger images by carving seams [18] that define the darker lines of
shadowed skin between adjacent fingers. Figure 11.7 illustrates the overall procedure
of identifying the four fingers regions based on complex filtering [19] that locate
reference points, where there are perturbations in the ridges orientations, followed

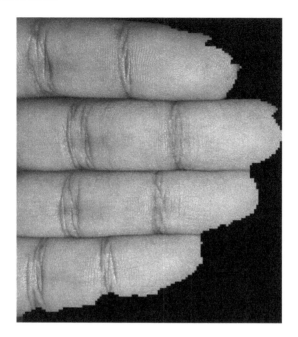

Figure 11.6   An example of the masked 4F^{TM}-ID image

by k-medoids clustering [20] to determine four centroids of the determined reference points. Then, seam carving scheme [18] is utilized to separate the 4F^{TM}-ID image into four images. Fingers separation can be structured using the following steps.

1. Determining reference points:
    One of the unique features of the fingerprint images is their ridge patterns that have unique global features such as singular points. Singular points are the most important global characteristics that can be deduced from of the orientations of fingerprint ridges. Singular points are points around which the ridges and valleys are wrapped. These singular points have been used as intrinsic points of reference in the literature to reduce the computational complexity of matchers. Complex filtering [19] has been known to deliver both the position and direction of singular points in a single separable filtering step. This technique relies on detecting the parabolic and triangular symmetry associated with core and delta points, then pick local maximum points as the singular points. In [19], the complex filtering is done on the complex images associated with the orientation image. In this chapter, the orientation image O of the segmented 4F^{TM}-ID image can be determined via the least mean-square method [21]. Then, to apply the complex filter, the orientation in the complex domain, Z, has to be obtained:

$$Z = \cos(2O) + i \sin(2O). \tag{11.1}$$

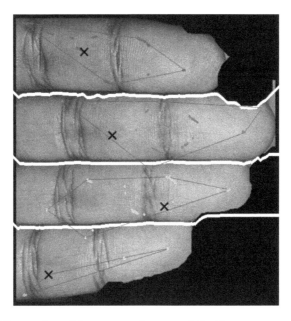

Figure 11.7   *Illustration of the proposed approach for fingers separation. The colored points are the determined reference points and the black marked points are the centroids that have been determined by the clustering algorithm. The white seams are the computed seams to separate fingers*

Then, the certainty map of reference points is defined as:

$$C_{ref} = Z * \overline{T}_{ref}, \tag{11.2}$$

where "$*$" is the convolution operator and $\overline{T}_{ref}$ is the conjugate of

$$T_{ref} = (x + iy) \cdot \frac{1}{(2\pi\sigma^2)} e^{-\frac{x^2+y^2}{2\sigma^2}}, \tag{11.3}$$

which is the kernel of the reference points detection where $\sigma = 1.5$ as in [19]. Then, an improved certainty map [22] is calculated as:

$$C'_{ref} = \begin{cases} C_{ref} \cdot \sin(Arg(C_{ref})) & \text{if } Arg(C_{ref}) > 0 \\ 0 & \text{otherwise} \end{cases}, \tag{11.4}$$

where $Arg(z)$ return the principal value of the argument of z (defined from $-\pi$ to π).

Finally, the reference points are the points, where the amplitude of C'_{ref} is local maximums. Figure 11.7 shows an example of the located reference points.

2. Clustering references points:

A k-medoids algorithm is utilized to cluster the determined references points into four clusters representing the four fingers. The k-medoids algorithm is a clustering algorithm related to the k-means algorithm and the medoid shift algorithm.

Both the k-means and k-medoids algorithms partition a dataset up into k clusters known a priori and both attempt to minimize the distance between points labeled to be in a cluster and a point designated as the center of that cluster. In contrast to the k-means algorithm, k-medoids is more robust to noise and outliers because it chooses datapoints as centers (medoids) and works with an arbitrary metrics of distances between datapoints instead of a sum of squared Euclidean distances [20].

In this work, we deigned a special distance function, D, to compute the distance matrix that will be used in the k-medoids clustering algorithm:

$$D(p_1, p_2) = c.e^{-|x_1 - x_2|} + e^{|y_1 - y_2|},$$ (11.5)

where $p_1 = \{x_1, y_1\}$ and $p_2 = \{x_2, y_2\}$ are two reference points and c is constant number that can be determined empirically. This special function results in smaller distance measures to points that locate on the same finger even if they are horizontally far from each other with respect to the Euclidean distance. The result of the k-medoids clustering algorithm is four cluster along with their medoids that represent the four fingers. Each medoid can be defined as the object of a cluster whose average distance to all the reference points in the cluster is minimal, i.e., it is a most centrally located reference point in the finger. Figure 11.7 shows an example of clustering predetermined reference points from the previous step, where the medoids are highlighted as black marks. These four medoids are utilized to define the area between the adjacent fingers.

3. Seam carving:

   The segmented 4F^{TM}-ID image is divided into four individual finger images by carving seams [18] between each of the adjacent fingers via the areas between fingers that have been defined in the aforementioned step. These seams are horizontally connected pixels that mark the darker lines of shadowed skin between fingers based on their importance. The importance of a pixel is defined by an importance map that evaluates the importance of each pixel based on its contrast with its neighbors. Extracting seams based on the concept of importance map has been proposed in [18] for content-aware resizing of images.

   Therefore, the last step in fingers separation is locating the least significant pixels in the area between fingers to mark them as boundaries to separate between adjacent fingers.

   This is done by assigning a value to every pixel, where higher values mean higher importance. In this work, every pixel in the area between fingers, I, will have a corresponding value in the importance map IM, which will be the absolute sum of both gradient components at that pixel:

$$IM = \left| \frac{\partial I}{\partial x} \right| + \left| \frac{\partial I}{\partial y} \right|.$$ (11.6)

Therefore, once the importance map is calculated, the next step is to find the optimal horizontal seam. A horizontal seam s_h is an eight-connected path of pixels that runs from the left of the image to the right with one pixel in each column. The eight-connected property means that for each pixel, during the backtracking process, only its eight adjacent neighbors are considered. Based on this definition of a seam and given that I is a defined image of the area between adjacent fingers of size $n \times m$, where n is the number of rows and m is the number of columns, a horizontal seam is defined as [18]:

$$s_h = \{(S_h(j),j)\}_{j=1}^{m}, s.t. \forall j, |S_h(j) - S_h(j-1)| \leq 1, \tag{11.7}$$

where S_h is a mapping function of s_h and $S_h(j) = 1, \ldots, n$.

The importance of a seam s is defined as the sum of the associated importance values of pixels lying on that seam in the importance map. Hence, given the importance map IM, an optimal horizontal seam s_h^* is the seam with a mapping function S_h that minimize its importance:

$$s_h^* = \min_{S_h} \sum_{j=1}^{m} IM(S_h(j),j). \tag{11.8}$$

To find horizontal optimal seams in order to mark them for separation, the dynamic programming concept is utilized, and maximum cumulative importance maps are created [18].

Hence, to locate the optimal horizontal seam, the horizontal cumulative importance map CM_h is computed. The first column in the importance map IM is copied to the first column in CM_h and then:

$$CM_h(i,j) = IM(i,j) + \min\{IM(i-1,j-1), IM(i,j-1), IM(i+1,j-1)\}. \tag{11.9}$$

The optimal horizontal seam is determined by simply backtracking on CM_h the minimum entry. Finally, the marked horizontal seams between adjacent seams are used to separate the original 4F™-ID image into four images as shown in Figure 11.7.

11.3.2.3   ROI localization

The final step in image segmentation is locating the ROI in each of separated finger images. For each finger image, the ROI is the region from the tip of the finger to the thick valley between the intermediate and distal phalanges. In this work, we adopt Raghavendra *et al.*'s work [10]. The first step in [10] is obtaining a binary finger image by setting a threshold based on the pixel values of the finger image. However, applying a threshold based on the skin color may not be reliable due to the background variations. Therefore, in this work, Hong *et al.*'s approach [21] is utilized to find the region mask based on the ridges orientation, which is used as the corresponding binary finger image. By utilizing this binary mask, the outer boundary and the midline of the finger can be determined as in [10] which can be used to locate the fingertip point. Finally, Raghavendra *et al.*'s approach, that is based on finding the local minima followed by a k-means clustering, can be used to find the thick valleys of the finger and to select one of them as the valley between the intermediate and distal phalanges.

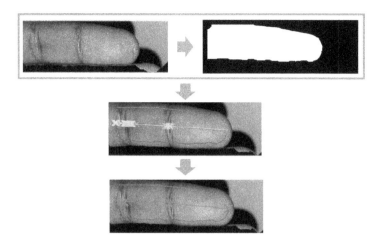

Figure 11.8   A diagram of the adopted scheme [10] to locate the ROI

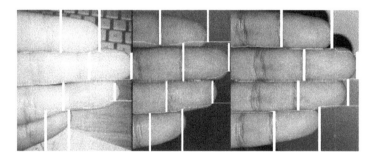

Figure 11.9   Examples of 4F^{TM}-ID images after ROI localization

Figure 11.8 shows the detected upper boundary, lower boundary, and midpoint (or axis) of the whole finger along with the tip and the two determined valleys. These results can be used for ROI localization in the 4F^{TM}-ID images and Figure 11.9 shows examples of located ROIs of different 4F^{TM}-ID images. As shown in Figure 11.9, the proposed scheme for 4F^{TM}-ID images segmentation is robust enough to accurately find the four fingers and extract the fingerprint regions.

11.3.3   *4F^{TM}-ID image preprocessing*

Once the relevant regions of the fingers are identified (i.e., ROIs), we proceed further to perform the image preprocessing step which can be divided into two task: resolution normalization and enhancement.

In this work, the identity comparison is based on minutiae matching algorithms, and this requires fixed resolution images. Although, a hand guide has been used during the 4F^{TM}-ID image capture process, distance variations may occur and each sample may be acquired at a different distance. Hence, a resolution normalization step

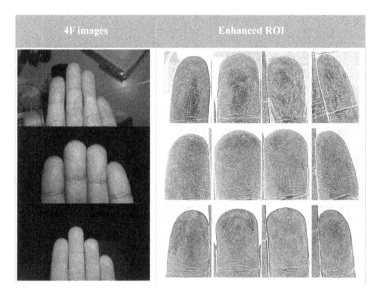

Figure 11.10   Examples of 4F^{TM}-ID images and their enhanced ROIs

is performed by scaling the localized ROIs of the 4F^{TM}-ID images to a predetermined image resolution.

To improve the extraction of discriminatory features from fingertips, it can be advantageous to use enhancement filters to boost the contrast between ridges and valleys. An adaptive histogram equalization [23] algorithm is used to increase the local image contrast by evenly distributing intensities over the possible range of values. However, applying only a contrast enhancement algorithm has the drawback of being indiscriminate and so enhancing background noise as well as the signal of interest. As such it can be beneficial to isolate only those signals of interest by filtering prior to contrast enhancement. Hence, a band-pass filter has been applied to remove signals with frequencies that do not correspond to the expected frequency of fingerprint ridges. A histogram equalization can then be applied to the result of the band-pass filter to attain an optimal image for feature extraction. The enhanced image is more suitable for the feature extraction and matching. Figure 11.10 shows examples of the enhanced ROI images.

11.3.4   Feature extraction and matching

In 4F^{TM}-ID system, the objective is developing a system that extract features compliant with fingerprint template standards and utilizing a minutiae-based algorithm to generate the similarity score between fingerprints. Adopting these standard minutiae formats guarantees the interoperability of 4F^{TM}-ID system with legacy databases, i.e., databases of contact-based fingerprints such as the U.S. Automated Fingerprint Identification System (AFIS).

To this extent, in this chapter,[4] the NIST biometric image software [24] which is an open source software has been utilized to generate feature sets according to the standardized format ISO/IEC 19794-2 [13] and to match these templates by a minutiae-based algorithm.

11.4   Experimental results

The described 4F^{TM}-ID system are deployed in Android and IOS operating systems. We used the deployed 4F^{TM}-ID apps to collect a database of 275 4F^{TM}-ID images from 33 subjects with an average of 8 samples per subject. Our in-house database comprises subjects from different countries (USA, UK, and Romania) and these 4F^{TM}-ID users used a variety of modern smartphones such as Apple iPhone 5C, Apple iPhone 6, Samsung S4, Samsung S5, etc. The only condition is the used smartphone must have a built-in rear camera with a resolution equal or greater than 5 MP. We asked the subjects to capture their 4F^{TM}-ID image in different real-life scenarios and just make sure that they position their hand naturally in the hand guide before the flash photo is taken. Hence, our in-house database contains images with various background, skin pigmentation, indoor and outdoor lighting conditions.

As shown in Figure 11.1, for each subject, three 4F^{TM}-ID images were added to the gallery and the remaining images are considered as probes. Then, our experiment consisted of matching the probes against the gallery entries. Because we have four fingers for each subject from each gallery sample, we proposed to perform multi-finger along with multi-sample score fusion. Then, the equal error rate (EER) was used to observe the matching performance. It is observed that at threshold of 270 results in an EER of 1.67%.[5]

Moreover, the duration of the capture process and the processing time measured. The average duration for all devices of the enrollment process is approximately 25 seconds and for the authentication process is 10 s. The reason for the longer duration of the enrollment is that three samples are acquired during the enrollment.

11.5   Summary

In this chapter, we introduced and discussed the new 4F^{TM}-ID system which is a complete authentication system for different mobile operating systems based on a fingerphoto of four fingers.

We proposed 4F^{TM}-ID system as a novel multimodal mobile biometric system which is more robust than unimodal ones [9–12]. 4F^{TM}-ID gains the advantages of multiple biometric traits to improve the performance in many aspects including accuracy, noise resistance, universality, spoof attacks, and reduce performance degradation in large-scale database applications.

In this chapter, we discussed the experimental results based on Hoyos Labs dataset consisting of 33 subjects from different countries using a variety of modern

[4]Our commercial version of the 4F^{TM}-ID system is using different matching schemes.
[5]For the recent statistics, please contact us at info@hoyoslabs.com

smartphones with various background and illumination reflecting real-life scenarios. A key factor in 4F^{TM}-ID image capture is that we use the full power of the LED flash while the hand guide encourages the user to position their hand in a position that is optimal for capturing ridge reflections from the flash. This enables the highest quality ridge feature extraction later in the process.

Moreover, we proposed a completely automatic approach that includes, robust fingers segmentation and separation schemes for accurate fingertips localization followed by preprocessing algorithms to extract reliable feature sets of the fingerprint regions. These algorithms allow fingerprint matching to be completed using any off the shelf fingerprint matcher. In this chapter, we employed the NIST minutiae extractor and matcher [24] to extract and match the minutiae-based feature sets. Finally, our experiment results showed that 4F^{TM}-ID system is highly effective in addressing the challenges involved in mobile biometric systems.

Our current work is focused on reducing the computational time and improve performance. Furthermore, the 4F^{TM}-ID system may also be used to capture print information from other regions of the hand, including palm prints and hand prints to further increase the reliability of the system. Moreover, the proposed innovation can be combined with an existing mobile face recognition system.

References

[1] "Independent consumer survey finds biometrics more secure than usernames and passwords," http://info.hoyoslabs.com/hacked-for-the-holidays/.
[2] M. Brader, "Shoulder-surfing automated," *Risks Digest*, vol. 19, 1998.
[3] A. Brown, E. Bracken, S. Zoccoli, and K. Douglas, "Generating and remembering passwords," *Applied Cognitive Psychology*, vol. 18, no. 6, pp. 641–651, 2004.
[4] A.K. Jain, A. Ross, and S. Prabhakar, "An introduction to biometric recognition," *IEEE Transactions on Circuits and Systems for Video Technology*, vol. 14, no. 1, pp. 4–20, 2004.
[5] A. Oulasvirta, S. Tamminen, V. Roto, and J. Kuorelahti, "Interaction in 4-second bursts: the fragmented nature of attentional resources in mobile hci," in *Proceedings of the SIGCHI Conference on Human Factors in Computing Systems*. New York, NY, USA: ACM, 2005, pp. 919–928.
[6] S. Trewin, C. Swart, L. Koved, J. Martino, K. Singh, and S. Ben-David, "Biometric authentication on a mobile device: a study of user effort, error and task disruption," in *Proceedings of the 28th Annual Computer Security Applications Conference*. Orlando, FL, USA: ACM, 2012, pp. 159–168.
[7] D. Maltoni, D. Maio, A.K. Jain, and S. Prabhakar, *Handbook of Fingerprint Recognition*, New York, NY, USA: Springer-Verlag New York Inc, 2009.
[8] R. Labati, A. Genovese, V. Piuri, and F. Scotti, "Touchless fingerprint biometrics: a survey on 2d and 3d technologies," *Journal of Internet Technology*, vol. 15, no. 3, pp. 325–332, 2014.

[9]   C. Stein, C. Nickel, and C. Busch, "Fingerphoto recognition with smartphone cameras," in *Proceedings of the International Conference of the Biometrics Special Interest Group (BIOSIG)*, Darmstadt, Germany, 2012, pp. 1–12.

[10] R. Raghavendra, C. Busch, and B. Yang, "Scaling-robust fingerprint verification with smartphone camera in real-life scenarios," in *IEEE Sixth International Conference on Biometrics: Theory, Applications and Systems (BTAS)*, 2013, pp. 1–8.

[11] A. Sankaran, A. Malhotra, A. Mittal, M. Vatsa, and R. Singh, "On smartphone camera based fingerphoto authentication," in *IEEE 7th International Conference on Biometrics Theory, Applications and Systems (BTAS)*. Arlington, VA, USA, 2015, pp. 1–7, IEEE.

[12] K. Tiwari and P. Gupta, "A touch-less fingerphoto recognition system for mobile hand-held devices," in *2015 International Conference on Biometrics (ICB)*, May 2015, Phuket, Thailand, pp. 151–156.

[13] ISO/IEC, "ISO/IEC 19794-2: biometric data interchange formats – part 2: finger minutiae data," Tech. Rep., ISO/IEC Standard, 2005.

[14] ANSI/NIST-ITL 1, "Data format for the interchange of fingerprint facial, & other biometric information (revision of ANSI/NIST-ITL 1-2000)," pp. 500–271, 2007.

[15] A.A. Ross, K. Nandakumar, and A.K. Jain, *Handbook of Multibiometrics*, New York, NY, USA: Springer-Verlag New York Inc, 2006.

[16] P. Kakumanu, S. Makrogiannis, and N. Bourbakis, "A survey of skin-color modeling and detection methods," *Pattern Recognition*, vol. 40, no. 3, pp. 1106–1122, 2007.

[17] A. Buades, T. Le, J. Morel, and L. Vese, "Fast cartoon+ texture image filters," *IEEE Transactions on Image Processing*, vol. 19, no. 8, pp. 1978–1986, 2010.

[18] S. Avidan and A. Shamir, "Seam carving for content-aware image resizing," *ACM Transactions on Graphics*, vol. 26, July 2007.

[19] K. Nilsson and J. Bigun, "Localization of corresponding points in fingerprints by complex filtering," *Pattern Recognition Letters*, vol. 24, no. 13, pp. 2135–2144, 2003.

[20] H.S. Park and C.H. Jun, "A simple and fast algorithm for k-medoids clustering," *Expert Systems with Applications*, vol. 36, no. 2, pp. 3336–3341, 2009.

[21] L. Hong, Y. Wan, and A. Jain, "Fingerprint image enhancement: algorithm and performance evaluation," *IEEE Transactions on Pattern Analysis and Machine Intelligence*, vol. 20, no. 8, pp. 777 –789, 1998.

[22] S. Chikkerur and N. Ratha, "Impact of singular point detection on fingerprint matching performance," in *Fourth IEEE Workshop on Automatic Identification Advanced Technologies*. Buffalo, NY, USA: IEEE, 2005, pp. 207–212.

[23] S. Pizer, E. Amburn, J. Austin *et al.*, "Adaptive histogram equalization and its variations," *Computer Vision, Graphics, and Image Processing*, vol. 39, no. 3, pp. 355–368, 1987.

[24] C. Watson, M. Garris, E. Tabassi, *et al.* "User's guide to NIST biometric image software (NBIS)," Tech. Rep., NIST, 2007.

Chapter 12
Palmprint recognition on mobile devices
Lu Leng[1,2,3]

12.1 Background

"Mobile device" refers to a small computing device. Mobile device, also known as handheld computer, is typically small enough to be handheld. In a broad sense, mobile devices include mobile phones, tablet computers, notebooks, payment devices etc. A mobile device commonly has a display screen with touch input and a miniature keyboard.

With the rapid development of telecommunication and network technologies, the application fields of mobile device are remarkably expanded. On the other hand, due to the advancement of integrated circuit, mobile devices have more powerful capability of data processing at the moment. Thus mobile devices have become integrated information processing platforms with plenty of functions, rather than just simple communication tools as before. Take mobile phone as an example, at present, humans use smart phones not only for wireless communications, like calling, text messaging, but also for E-mail, online payment, telephone banking, Internet surfing, entertainment (like music, movies, mobile games) and so on. It is possible that mobile devices can provide a natural access to any service anywhere anytime.

It follows that mobile devices play a significant role for almost everyone, and hence, one can see that it is vital to protect privacy on mobile devices, including personal data and information. Unfortunately, the users are often racked by the missing or theft of mobile devices. According to Federal Communications Commission, about one out of three robberies involved the theft of a mobile phone. Police data in San Francisco show that one half of all robberies stole mobile phones in 2012. Thus, how to accurately identify/verify users and authorization control are crucial functions of a mobile device.

[1] School of Software, Nanchang Hangkong University, P. R. China
[2] Lane Department of Computer Science and Electrical Engineering, West Virginia University, USA
[3] Shanghai Electronic Certificate Authority Center Co., Ltd, P. R. China

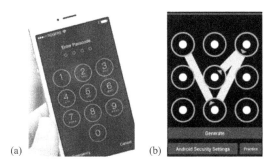

Figure 12.1   Knowledge-authentication on mobile devices: (a) numeric and
(b) grid graph password

12.2   Current authentication technologies on mobile devices

Authentication technologies prevent impostors from abusing genuine users' mobile devices without legal authorization. Unfortunately, current authentication techniques are not secure enough as they are claimed, which can be briefly categorized into two classes, knowledge-authentication and biometric-authentication.

12.2.1   Knowledge-authentication

Knowledge here refers to one secret that is kept confidentially by genuine users. Numeric password and grid graph password are the popular authenticators in iOS and Android systems, as shown in Figure 12.1.

It is not difficult to peep the knowledge authenticators when they are inputted. Once an impostor steals a mobile device and obtains the knowledge, he/she can impersonate the genuine user and use the mobile device illegally. As a result, the genuine user's privacy, economic interests and other rights are all severely violated.

Most users prefer low-entropy, short, simple knowledge authenticators, which are easy to memorize, but are also easy to observed, compromised and stolen. Thus, high-entropy, long, complex passwords are more favorable from a security standpoint. However, these authenticators are difficult to memorize, so a lot of users tend to forget them, especially the rarely used authenticators.

12.2.2   Biometric-authentication

In order to address the aforementioned drawbacks of knowledge-authentication, biometric technologies are paid more and more attention and have become increasingly popular. Biometric refers to metrics related to human physiologic and behavioral characteristics. Biometric is used in computer science as a form of identification or authentication. Biometric should meet the requirements of universality, uniqueness, permanence, measurability (collectability), acceptability and anticircumvention.

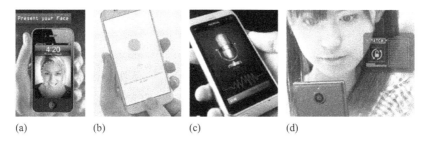

(a)   (b)   (c)   (d)

Figure 12.2   Mobile biometric modalities: (a) face, (b) fingerprint, (c) voice and (d) iris

Table 12.1   Disadvantages of current mobile biometrics

Biometric modality	Disadvantages
Face	Limited accuracy, easy leakage, lack of flash lamp for front camera
Fingerprint	Limited universality, multiple interference factors, easy leakage, relatively low acceptance, additional hardware cost
Voice	Limited accuracy, easy leakage
Iris	Camera requirement, low acceptance

Users do not need to memorize their biometric data. In addition, biometric data have high-entropy, so they are not easy to crack.

Mobile biometric is to develop novel mobile services secured by biometric authentication means [1]. Face [2], fingerprint [3], voice [4] and iris [5] are popular biometric modalities for authentication on mobile devices, as shown in Figure 12.2.

Unfortunately, the current mobile biometrics has several disadvantages, which are summarized in Table 12.1 and described, respectively, as follows.

12.2.2.1   Drawbacks of mobile face

* Limited accuracy
  The accuracies of face recognition algorithms depend on multifactors dramatically, such as expressions, jewelries, face-painting, aging, health condition.
* Easy leakage
  Humans' faces can be easily acquired under unawareness. It is possible to impersonate a genuine user with a mask that is produced according to his/her face images.
* Lack of flash lamp
  Users cannot see the display screen when using rear cameras to capture face images, so they capture their face images by front cameras. The resolution of front camera is relatively low. Moreover, there is no flash lamp for front cameras; consequently, it is impractical to implement mobile face recognition in dark environments.

12.2.2.2   Drawbacks of mobile fingerprint

- Limited universality
  A few users do not have fingerprint inherently or clear fingerprint due to professional reasons, like brick moving dough kneading, etc.
- Multiple interference factors
  Contact sensors are commonly used in fingerprint recognition systems. Users contact their fingers with sensor surface, so pressure force, dryness and cleanliness of finger surface all noticeably affect accuracy.
- Easy leakage
  Fingerprints contact the surface of numerous devices, like keyboard, screen of phone, so they are easily detected and copied. Furthermore, the leaked fingerprint data can be abused to fabricate fingerprint film for impersonating.
- Relatively low acceptance
  As users' hands have to contact the surface of sensors for fingerprint collection, the risks of infectious disease definitely increase. In some traditional cultures, the users of different genders are not allowed to contact identical surface, so fingerprint systems are not popular in these nations or countries.
- Additional hardware cost
  Most smart phones do not have built-in fingerprint sensors. The embedding of fingerprint sensor definitely leads to additional cost.

12.2.2.3   Drawbacks of mobile voice

- Limited accuracy
  The accuracies of voice recognition are severely disrupted by several factors, like aging, health condition, emotions, noise, just to name a few.
- Easy leakage
  Voice can be easily recorded under unawareness, so it is probable to impersonate genuine user with his voice records.

12.2.2.4   Drawbacks of mobile iris

- Camera requirement
  Most iris systems employ specific cameras to capture iris images. To some extent, the accuracies for iris recognition with built-in cameras of smart phone are not as high as those with specific cameras.
- Low acceptance
  It is not comfortable for users to scan or capture their irises.

12.3   Mobile palmprint recognition framework

12.3.1   Introduction on palmprint

Palmprint refers to the features on the palm region of a hand. Palm itself consists of principal lines, wrinkles (secondary lines) and epidermal ridges [6], as shown in

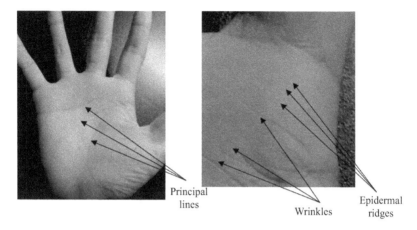

Principal
lines

Wrinkles

Epidermal
ridges

Figure 12.3   Palmprint features

Figure 12.3. Palmprint contains plenty of features, such as texture, indents and marks, which can be used for recognition [7].

12.3.2   Strengths of mobile palmprint

Mobile palmprint has several superiorities to other mobile biometric modalities.

- Plenty of discriminant features
  The area of palm is large, so plenty of discriminant features can be extracted from palmprint, and hence, mobile palmprint recognition can achieve high accuracy.
- Few restrictive conditions
  A number of palmprint features can be extracted accurately even from low-resolution images and abrasive palm surfaces.
- Low cost
  Palmprint images can be acquired directly by built-in cameras in mobile devices. There is no need to spend additional cost for supplementary sensors. Moreover, the accuracies still meet practical requirements even in cheap low-resolution systems.
- Difficult leakage
  It is not easy to acquire palmprints under unawareness. Palmprints are not preserved on the device surface in mobile systems, as palmprints are captured in contactless mode. Therefore, the leakage risk of palmprint is low.
- High acceptance
  The hygiene problems and resistance against contact systems in traditional cultures are relieved by contactless palmprint acquirement systems.

According to the above comparison, mobile palmprint is a favorable authentication technique on mobile devices.

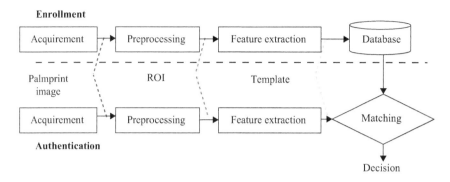

Figure 12.4   General framework of palmprint recognition

12.3.3 Palmprint recognition framework

The general framework of palmprint recognition is shown in Figure 12.4. Four steps, including acquirement, preprocessing, feature extraction, storage/matching, are implemented in sequence.

1. Acquirement: A palmprint image is acquired by capture device.
2. Preprocessing: The aim of preprocessing is to locate and crop the region of interest (ROI) from original palmprint image.
3. Feature extraction: Palmprint features are extracted from ROI and constitute feature template.
4. Storage/Matching: In enrollment stage, the gallery template is stored in database. In authentication stage, the query template and the gallery template stored in database are compared. Finally, according to the similarity/dissimilarity between the two templates, the decision judges whether the two templates are of an identical user, i.e., whether the query template passes authentication successfully.

In mobile palmprint recognition, palmprint images are acquired by built-in cameras of mobile devices, such as smart phone, iPad. The preprocessing, feature extraction and storage/matching are implemented on the platform of mobile device.

12.4 Palmprint acquirement modes

According to acquirement modes, palmprint systems can be briefly categorized into offline and online modes [8].

12.4.1 Offline mode

In offline mode, ink or newsprint is spread on the palm, and then, the palm is put on one piece of paper. The pieces of paper can be scanned so that the palmprints are acquired on digital images [9]. It is not clean to daub ink or newsprint on palm. In addition, some parts of offline palmprint are often missing or blurry, so high-resolution features cannot be extracted from scanned palmprint images. Even worse, offline mode cannot be applied in real-time systems.

12.4.2   Online mode

Digital palmprint images can be captured directly by imaging devices in online mode. Online mode can be further categorized into contact and contactless modes.

12.4.2.1   Contact mode

In contact mode, users' hands and equipment surface are contacted. In some contact systems, even some pegs are used to fix hand position. Both the background and illumination are stable in contact acquisition, so it is easy to segment hand region and locate ROI.

Although contact palmprint recognition systems can achieve high accuracy performance, their practical applications incur several problems as follows.

- Personal hygiene
  Due to the health and personal safety, it is unhygienic to make the users' fingers or palms contact identical sensors or devices. Contact acquisition increases the risk of infectious disease.
- Lack of acquisition flexibility
  The user acceptance is depressed by the fixing devices that degrade acquisition flexibility and convenience.
- Surface contamination
  The surface-of-contact sensor is contaminated easily especially in harsh, dirty and outdoor environments. The surface contamination of contact sensor will degrade the quality and accuracy performance of the subsequently acquired palmprint images.
- Resistance of traditional cultures
  The users in some conservative nations resist placing their hands on the device surface that is touched by the people of the opposite gender.

Thus, the research on palmprint recognition system has been toward contactless mode gradually.

12.4.2.2   Contactless mode

Users' hands do not need to contact any equipment surface in contactless palmprint systems, so the acceptance is improved. Unfortunately, it is impractical to transplant the contact preprocessing methods onto contactless palmprint systems directly due to the severe challenges as follows.

- Uncontrolled hand pose and position
  The appropriate positions of hand placement are different in various contactless palmprint systems. In addition, users can place their hands freely. If the hand is

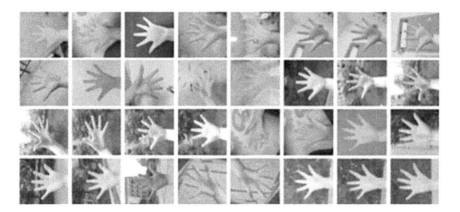

Figure 12.5   Palmprint images of mobile device

too far from the camera, palmprint details are not acquired clearly. Although if the hand is too close to the camera, it is probable that some parts of palm are beyond the imaging area. It is also possible that the cameras cannot focus accurately in a too short distance. Furthermore, the hand can be translated, rotated or revolved without any restriction. Therefore, it is tough work to locate ROI and match templates in contactless palmprint systems.

* Interference of complex background
  There are many interference regions, like wooden furniture with skin-like color, in the complex background, which interrupt preprocessing.
* Illumination variance
  The illumination may be strong but nonuniform in open environments, or too weak in dark environments, so unstable light sources and nonuniform illuminations also hinder preprocessing.

Contactless systems can be placed on a fixed site, such as the location in front of the door of a building; therefore, to some extent, the background and illumination can be partially controlled.

12.4.2.3   Mobile mode

As palmprint images can be captured directly by the built-in cameras of mobile device, palms do not need to contact the equipment surface, and accordingly mobile mode can be considered as one special case of contactless mode.

Figure 12.5 shows several palmprint images of mobile device. Palmprint recognition on mobile devices has tremendous economic and market potential; however, far more technical challenges impede its development and promotion.

* Uncontrolled hand pose and position
  Similar to those in contactless mode, neither hand pose nor position can be rigidly controlled in mobile mode.

Table 12.2   Comparison of palmprint acquirement mode

Mode	Submode	Description	Strengths	Drawbacks/Challenges
Offline	–	Daub palm with ink, put daubed palm on paper, scan paper	–	Dirty, noncomplete, blurry, not real-time
Online	Contact	Contact hands with equipment surface, fix hand position sometimes	Easy hand segmentation and ROI location, high accuracy	Personal hygiene problem, low acquisition flexibility, surface contamination, resistance
	Contactless	Noncontact acquirement	Good acceptance	Uncontrolled hand pose and position, complex background, illumination variance
	Mobile	Contactless acquirement by built-in cameras in mobile devices	Good acceptance, wide application fields	Uncontrolled hand pose and position, complex background, illumination variance, limited hardware resource

- Extremely complex background
  The background in mobile palmprint images is extreme complex, which are captured in indoor or outdoor environments. Many regions have skin-like colors or similar geometries to fingers or hands, which obstruct hand segmentation and ROI location seriously.
- Remarkable illumination variance
  The illumination in mobile environments varies greatly, unlike the illumination strictly controlled in laboratory environments or contact systems.
- Limited hardware resource
  The computation power and storage capacity of mobile device are not comparable to those of desktop computers. In other words, both computation complexity and storage cost of mobile palmprint recognition schemes must be low.

It follows that palmprint recognition is considerably difficult in mobile mode. The palmprint acquirement modes are summarized in Table 12.2. Figure 12.6 shows the palmprint images acquired by the systems in different modes, which are developed by some universities or institutes, including Hongkong Polytechnic University (PolyU), Harbin Institute of Technology (HIT), Peking University (PKU), Multimedia University (MMU), Nanchang Hangkong University (NCHU), Yonsei University.

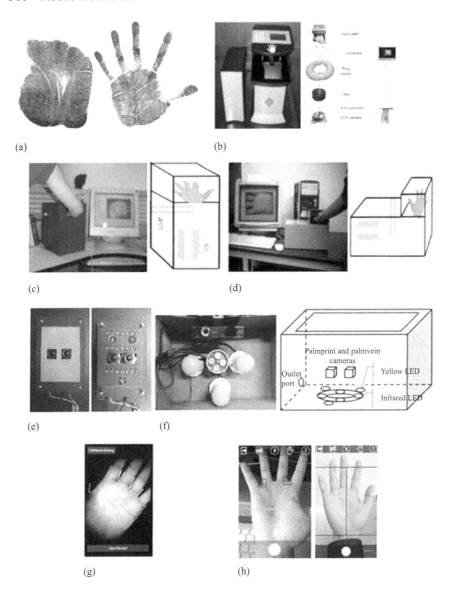

Figure 12.6   Palmprint images in different modes: (a) offline, (b) contact (PolyU, Hong Kong), (c) contact (HIT, China), (d) contact (PKU, China), (e) contactless (MMU, Malaysia), (f) contactless (NCHU, China), (g) mobile (Yonsei University, South Korea) and (h) mobile (NCHU, China)

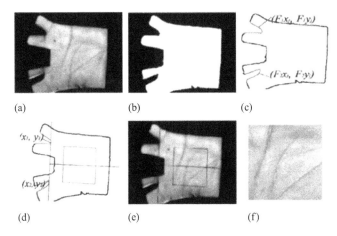

Figure 12.7   Contour tracking for ROI cropping: (a) original, (b) binary,
(c) contour, (d) key-points, (e) coordinate, (f) ROI

12.5   Palmprint acquirement and preprocessing

Preprocessing is critical to palmprint recognition. The aim of palmprint preprocessing is to accurately locate and crop ROI. Contact palmprint preprocessing could not be used for contactless mode directly, but it is the important base of the preprocessing in contactless mode and mobile mode. More restrictions and complexities have to be fully considered in contactless and mobile modes. Owing to the severe challenges in mobile mode, auxiliary acquirement techniques are sometimes necessary to preprocessing. The palmprint acquirement and preprocessing approaches in three modes are elaborated as follows:

12.5.1   Preprocessing in contact mode

It is relatively easy to segment hand region from background in contact mode as the background and illumination can be carefully controlled. In Figure 12.7(a), in enclosed circumstance, the background of original image is almost dark. The brightness of light source can be set to be uniform. Thus, palm and background regions can be divided by threshold segmentation, as shown in Figure 12.7(b). Next task following palm segmentation is how to accurately locate and crop ROI.

12.5.1.1   Contour tracking

In Zhang *et al.*'s method [10], palm contour in Figure 12.7(c) was extracted by edge detection. Two key-points, i.e., the two valley points between index finger and middle finger, ring finger and little finger, were detected by contour tracking, as shown in Figure 12.7(d). The coordinate system in Figure 12.7(e) was established in virtue of the two key-points, which helped locate and crop ROI in Figure 12.7(f). The precision

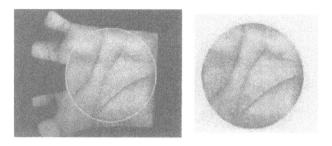

Figure 12.8   Maximal inscribed circle for ROI cropping

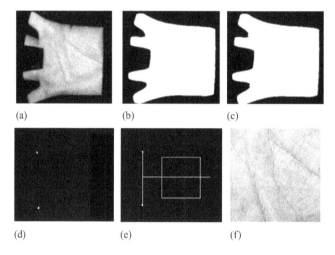

*Figure 12.9   Morphological operator for ROI cropping: (a) original, (b) binary,
(c) closed, (d) two-valley areas, (e) coordinate and (f) ROI*

of key-point detection mainly depends on accurate palm segmentation and contour detection.

12.5.1.2   Maximal inscribed circle

In Liambas and Tsouros's scheme [11], maximal inscribed circle was detected in palm region, as shown in Figure 12.8, which delimited ROI. However, this method does not solve the problems of rotation and size normalization.

12.5.1.3   Morphological operator

Hennings-Yeomans *et al.* [12] employed morphological operators to detect key-points, as shown in Figure 12.9. Two small-valley areas were generated from the difference between the segmentation binary image and its result of close operation (successive implementation of dilation and erosion operations). The following steps are similar to those in contour tracking.

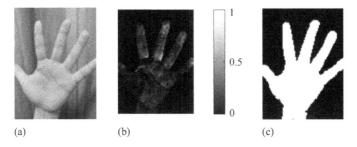

Figure 12.10 Skin-color likelihood threshold segmentation: (a) original image, (b) skin-color likelihood and (c) binary image

Some other palmprint preprocessing methods were also developed for contact mode. Poon *et al.* [13] divided hand region into sectors of elliptical half-rings, which are less affected by misalignment due to rotational error. Li *et al.* [14] established the coordinate system according to the centroid of palm region and three principal lines.

12.5.2   Preprocessing in contactless mode

It is difficult to segment hand from the complex background in contactless mode. Due to imperfect segmentation, robust ROI location is a knotty problem.

12.5.2.1   Skin-color segmentation

Gray level is not enough for hand region segmentation in contactless mode. A color image contains more color information than a gray image, so skin-color model can be used to segment hand region. RGB (red, green, blue) color space is not suitable for skin-color segmentation. In order to overcome illumination disturbance, RGB color space is commonly converted to YCbCr (luminance, blue-color and red-color components) color space so that color and brightness are separated.

Human skin colors are obviously clustered in Cb–Cr color space. The skin color was modeled as a two-dimensional (2D) Gaussian distribution of Cb–Cr [15]. Figure 12.10 shows the likelihood threshold segmentation of hand region. The points on the hand contour were judged whether they were key-points of finger valleys according to some conditions and rules.

12.5.2.2   Conjunction of triple-perpendicular-directional translation residual

The segmented hand regions in contactless mode are not always complete or accurate. Leng *et al.* developed logical conjunction of triple-perpendicular-directional translation residual (TPDTR) to improve the accuracy of key-point detection [16], as shown in Figure 12.11. Triple-perpendicular-directions (TPDs) included the principal direction of hand, i.e., middle-finger direction, and its two perpendicular directions. The TPDs in Figure 12.11 were upward, leftward and rightward directions approximately.

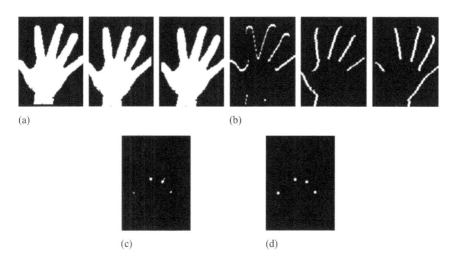

(a)   (b)

(c)   (d)

*Figure 12.11   Logical conjunction of TPDTR for key-point detection: (a) images
translated at TPDs, (b) TPDTR, (c) logical conjunction of TPDTR
and (d) valley points*

The search of four finger valley points was within the four borders of finger valley gaps detected by TPDTR; therefore, the computation complexity is decreased
effectively. Moreover, the antiinterference capacity of region is stronger than that of
point and line, so TPDTR improves the robustness of key-point detection. The performance of TPDTR can be further improved by rotating the principal direction to
vertical direction.

12.5.2.3   Minimum radial distance

In Ito *et al.*'s scheme [17], the rotation angle of a hand was firstly estimated using
Principal Component Analysis (PCA), and then the binarized image was rotated so
as to fit the principal direction to the horizontal axis, as shown in Figure 12.12(a)
and (b). The centroid was on the split line that divided hand region into two parts in
Figure 12.12(c). The minimum radial distances between the contour points of fingers
and reference point (wrist point) were calculated, as shown in Figure 12.12(d) and (e).
The contour points with minimum radial distances were considered as key-point
candidates. Finally, the key-point candidates were optimized to accurately determine
the key-points. This method is effective for both outstretched and adherent fingers.

12.5.2.4   Active appearance model + nonlinear regression

In Figure 12.13, active appearance model was used for hand region segmentation, and
then nonlinear regression, i.e., least squares support vector regression, was used to
fit palm model to locate and crop ROI [18,19]. This model-based approach provides
reliable ROI location that is robust to small segmentation errors.

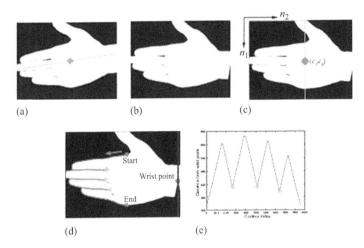

Figure 12.12   Minimum radial distance for key-point detection: (a) rotation angle,
(b) rotated binary image, (c) centroid and split line, (d) finger
contour and (e) minimum radial distance

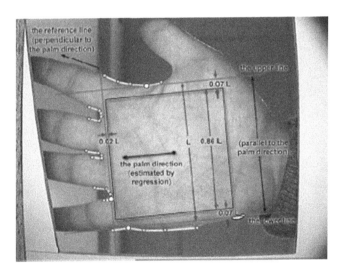

Figure 12.13   Active appearance model + nonlinear regression for ROI location

12.5.3   Acquirement and preprocessing in mobile mode

As mobile palmprint recognition is far more difficult than contact and contactless
palmprint recognition, auxiliary acquirement techniques are sometimes necessary
to preprocessing. The auxiliary acquirement and preprocessing techniques are
elaborated as follows, which are summarized in Table 12.3.

314   *Mobile biometrics*

*Table 12.3   Auxiliary acquirement and preprocessing techniques for mobile
              palmprint*

Ref.	Year	Device	Auxiliary acquirement	Preprocessing
[20]	2007		Root regions of three fingers are aligned with frame top border	
[21]	2009	D810 VGA/ 2MP, SGH-i900 5MP	Black background	Gray thresholding, local area, key-point detection
[22]	2009	Megapixels		Color segmentation, k-means, morphology
[23]	2010		Video	Registration and combination of multiple frames
[24]	2011		Center area (13×13) in hand region	Rough orientation normalization, row line scan
[26,27]	2011	HTC, Sony Ericsson w380i	Blue-color background, synthetic hand images, different conditions (hand opening, distance, rotation)	Gaussian multiscale aggregation
[28]	2012		Black background	Skin-color detection, significant points
[29]	2012	Canon IXUS 950 IS, Motorola ME525, Nokia 5800	Black background	Otsu, radial distance, key-point detection, scale normalization
[30]	2013		Red guide rectangle	Skin-color detection, region growing, radial distance
[31]	2013		Vertical peak + Horizontal peak, Canny edge + 5 peaks and 4 valleys	
[32]	2014		2 points on two side boundaries	3 principal lines
[33]	2014	HTC Desire		Image adjustment and filtering, Otsu, contour detection, key-point detection
[34]	2015	iPhone 3GS		Edge extraction, pairing, key-point detection
[35]	2015			Skin-color detection, Parameter estimation in central area, central point detection
[36]	2015		Hand-shaped guide curves, skin-color test, profile-line test	Key-point detection and verification
[37]	2016			Local normalization

In Han *et al.*'s scheme [20], root regions of index, middle and ring fingers were required to be aligned with the frame top border. Moreover, the left and right borders of the frame were required to be aligned with the corresponding borders of the central region of a palm. It is not easy for the users to align the top, left and right borders synchronously.

Li *et al.* set black background when acquiring mobile palmprint images. The black background is unrealistic in practical applications, so they just studied mobile palmprint matching but without considering the actual background and illumination. Gray thresholding was used to segment hand region. Key-points were detected in local area surrounding the finger valley areas [21].

Sierra *et al.* combined color segmentation and k-means to segment hand region. Opening, as a morphological operation, was performed to make hand region be softer and more suitable. The authors claimed that their scheme was effective for complex background even more challenging background [22].

Methani and Namboodiri developed video-based palmprint recognition on mobile devices. Registration and combination of multiple frames were employed to generate additional texture information [23].

Franzgrote *et al.* supposed that the center area of image with the size of 13×13 was in hand region [24]. The median and variance of the center area in RGB color space were computed, which helped to segment hand region from background. The authors detected numerous valley candidates by the method in [25] to estimate an initial hand orientation. The palmprint images were roughly rotated so that the fingers directed to the top afterwards. As a result of rotation adjustment, the valley points could be found more easily and reliably.

Sierra *et al.* acquired palmprint in blue-color background. A part of natural image was replaced by the hand region segmented from blue background to generate a synthetic hand image. Different conditions, including hand opening and closing, distance, rotation, were evaluated in their experiments [26,27].

Similar to [21], Choraś and Kozik also set black background when acquiring mobile palmprint images, which is unrealistic in practical applications. Skin-color model was employed to segment hand region. The polygon constituted by significant points was used to extract shape feature, whereas the texture features were extracted from ROI [28].

Jia *et al.* studied palmprint recognition across different mobile devices. They also set black background and used Otsu thresholding to segment hand region. Radial distances of the hand contour points were measured to detect key-points. Their main contribution is a robust method to calculate the palm width, which can be effectively used for scale normalization [29].

Aoyama *et al.* used a red rectangle guide window to help users place their hands. The region in guide window must be within palm area. Skin-color thresholding was applied only to the small center region of hand. Afterwards, region growing helped segment the whole hand, whose initiate region was the extracted small region [30]. Radial distances of contour points were measured to detect key-points.

In [31], Ibrahima and Ramlia designed two methods for ROI extraction. In their first method, horizontal highest peak (Hpeak) and vertical highest peak (Vpeak) were detected to locate reference point (Vpeak-507, Hpeak) for ROI cropping. This method requires saturated illumination, smooth hand boundary and appropriate hand placement. In their second method, ROI was cropped according to five peaks and four valleys that represent the tips and roots of the fingers. This method fails if some peaks or valleys are missing or not detected accurately.

Lee *et al.* designed two reference points that are aligned with two points on inside and outside boundaries of palm [32]; however, the two points cannot control hand position satisfactorily as the two points can move on the boundary.

Moço *et al.* used image adjustment and filtering to preprocess palmprint image. Otsu thresholding was performed on gray images to segment hand region; however, in practice, gray information is not enough for hand segmentation as they claimed. Convex hull and convexity defects were computed. A convex hull, as a sequence of points, produces the minimum polygon that encapsulates the target contour. Convexity defects or valleys correspond to the maximum distance between two consecutive hull points. By calculating the convexity defects of a hand, finger valleys and tips were computed for key-point detection [33].

The lines on the boundaries within three finger gaps were paired by Fang and Neera to check whether palmprint was captured validly [34]. Each key-point was detected according to each pair of lines. However, the authors neither fully considered the complex background in segmentation stage.

Javidnia *et al.* employed skin-color detection to segment hand region. They assumed that the center of the hand was always situated around the center of acquired image, so the parameters of skin-color model were estimated in the central area of image [35]. Unfortunately, this scheme cannot accurately segment hand region in challenging scenes.

Kim *et al.* designed hand-shaped guide curves for auxiliary acquirement. Skin-color test and profile line test were performed to check whether hand placement was valid. Key-points were detected in two restricted areas. The detected key-points were further checked to see whether they were incorrectly located in the background [36]. Flexibility and acceptance are degraded as the five-guide curves are not easy to align with synchronously.

Javidnia *et al.* enhanced palmprint images by local normalization so that the mean and variance around a local neighborhood were uniform [37]. However, they performed their experiments on ROI database directly, while did not consider how to accurately crop ROI.

Some auxiliary acquirement techniques are shown in Figure 12.14. Some preprocessing techniques used for ROI location are shown in Figure 12.15.

12.6   Palmprint feature extraction and matching

After ROI is cropped, palmprint features are extracted for matching. The palmprint feature representation techniques can be mainly divided into five categories, including line-based approaches, appearance-based approaches, statistic-based approaches, texture-based approaches and fusion-based approaches [38].

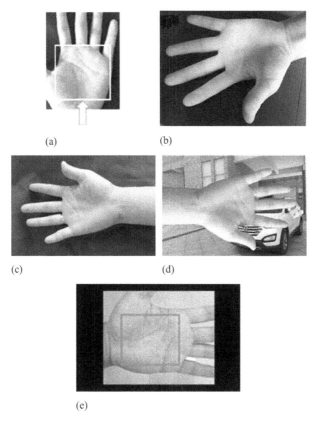

(a)   (b)

(c)   (d)

(e)

Figure 12.14 Auxiliary palmprint acquirement on mobile devices: (a) Han et al.'s
scheme [20], (b) Black background [21,28,29], (c) blue background
[27], (d) synthetic hand image [27] and (e) red rectangle guide
window [30]

The matching is to calculate the similarity/dissimilarity degree of two given palmprint feature templates, and return a dichotomy decision (yes/no). Different matching rules are suitable for various features.

Geometry-based matching finds the geometrical alignment between the enrolled (gallery) and verified (query) templates and returns the maximum number of feature pairs or smallest/largest dissimilarity/similarity degree. Compared with point features, line features are more popular and widely used in palmprint recognition as they are relatively easier to extract, even from low-resolution palmprint images.

Euclidean distance can measure the dissimilarity degree between appearance templates and statistic templates [39,40]. Hamming distance can measure the dissimilarity degree between texture coding templates [41–47]. A multiple-translated matching process is required for the matching between two texture coding templates

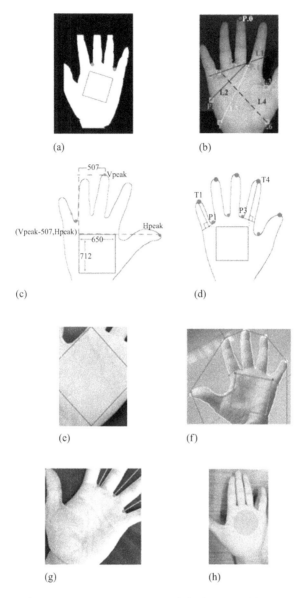

Figure 12.15   Palmprint preprocessing on mobile devices: (a) rotation adjustment [24], (b) significant points [28], (c) vertical peak + horizontal peak [31], (d) five peaks + four valleys [31], (e) two reference points on boundaries [32], (f) convex hull and convexity defects [33], (g) line pairing [34] and (h) central area for parameter estimation [35]

Table 12.4   Feature extraction and matching for mobile palmprint recognition

Ref.	Year	Feature	Matching
[20]	2007	Sum-difference ordinal code	Hamming distance
[22]	2009	Hand silhouette	DNA-based algorithms
[23]	2010	PalmCode	Hamming distance
[48]	2010	2D SAX (Symbolic Aggregate approXimation) conversion	MINDIST
[24]	2011	Competitive Code	Partial (2 rows in the middle) matching
[28]	2012	Three-valued base functions	
[29]	2012	Learning method, correlation method, orientation coding	Peak, peak-to-correlation energy, peak-to-sidelobe ratio
[30]	2013	Competitive Code, Ordinal Code	Affine transformation (geometric correction), band-limited phase-only correlation, peak
[32]	2014	3 principal lines	
[33]	2014	Orthogonal line ordinal code	Hamming distance
[34]	2015	Average filtering, line detection, thresholding, line thinning, line edge map	Hausdorff distance
[35]	2015	LBP, SIFT	Lowe's rule, geometric transformation
[36]	2015	local orientation histogram of line	Chi-square distance
[37]	2016	SIFT	Lowe's rule

to remedy the vertical and horizontal dislocation. Texture coding schemes need neither training nor high-resolution cameras. In addition, both matching computation complexity and storage cost of binary templates are low. Thus texture coding schemes are favorable in palmprint systems. For histogram templates, the dissimilarity degree is commonly measured by chi-square distance.

The techniques of feature extraction and matching for mobile palmprint recognition are summarized in Table 12.4.

Sum-difference ordinal filter, which contains only integer coefficients for the reduction of computational complexity, was performed to generate sum-difference ordinal codes, whose dissimilarity degree was measure by Hamming distance [20].

Sierra *et al.* extracted two feature templates from silhouette and contour. As the sequences of template did not have uniform lengths, the matching between the templates was performed by three principles of DNA-based algorithms: mutation, insertion and deletion [22].

In [23], Methani and Namboodiri used Hamming distance to measure the dissimilarity degree between the binarized templates generated from the real and imaginary Gabor responses [41].

Chen *et al.* converted ROI to 2D SAX (Symbolic Aggregate approXimation) form for palmprint representation and "MINDIST" matching. The author claimed that their approach had very low computational complexity so that it could be efficiently implemented on mobile embedded platforms [48].

Franzgrote *et al.* used Competitive Code [43] for palmprint recognition [24]. They extended the horizontal and vertical translation ranges from $[-2,2]$ to $[-4,4]$. In addition, the matching was restricted to only a small part of the template, i.e., 2 rows in the middle of Competitive Code. Actually, this scheme can accelerate matching speed, but some discriminant information in the template is neglected.

Choraś and Kozik developed 2D masks with three-valued $(-1, 0, 1)$ functions [28]. The three-valued functions refer to Haar-like functions proposed by Viola and Jones [49]. Three-valued masks can be computed fast, but it is not easy to choose the appropriate set of masks for the description of significant palmprint features. Thus three strategies were investigated, namely random masks, manually selected masks and Eigen-palms extraction.

In [29], Jia *et al.* compared different approaches for various device sensors, namely learning method, correlation method, orientation coding. They employed peak, peak-to-correlation energy and peak-to-sidelobe ratio for matching.

Aoyama *et al.* used Competitive Code [43] and Ordinal Code [44] for recognition. Affine transformation was employed for geometric correction in matching. Band-limited phase-only correlation and peak measured the similarity degree between the templates [30].

As neither hand pose nor position could be controlled rigidly in Lee *et al.*'s scheme, they extracted three principal lines as palmprint feature [32]; however, the accuracy of principal line recognition is not high.

Moço *et al.* used Orthogonal Code [44] for palmprint recognition, whose dissimilarity degree was measured by Hamming distance [33].

Fang and Neera extracted lines as palmprint feature. The steps were performed successively, including average filtering, line detection, thresholding, line-thinning and line-edge map generation [34]. Hausdorff distance was calculated as the dissimilarity degree in matching.

In [35,37], Local Binary Pattern (LBP) and Scale-Invariant Feature Transform (SIFT) were extracted as palmprint features. Lowe's rule [50] was used to search correct matchings, and then geometric transformation was applied to find the best pairs of SIFT features.

In [36], chi-square distance was calculated as the dissimilarity degree between local orientation histograms of line.

12.7 Conclusions and development trends

Although some papers on mobile palmprint recognition have been published, the current schemes cannot solve all the severe problems in practical applications satisfactorily. More interference factors and limited resources have to be fully considered for the promotion of palmprint recognition on mobile devices. Some development trends in the future are summarized as follows.

(1) Multimodal biometric recognition on mobile devices

Biometric fusion can achieve more advantages, including higher accuracy performance, improved availability, widened application range, higher degree of freedom and less susceptibility to spoof attacks. Therefore, mobile biometric is developing toward multimodal fusion, which combines several biometric modalities to provide more reliable and flexible biometric recognition.

(2) Remote mobile palmprint authentication

Biometric authentication can be divided into two types: local and remote authentications. In the former type, the biometric data are matched on the user client side. However, local authentication is not enough when a user wants to receive the service in remote environments. Thus remote mobile palmprint authentication has been a hot issue [51,52].

(3) Palmprint security and privacy

The secure and privacy problems plague many biometrics, including palmprint. Palmprint features are immutable, which implies that palmprint templates cannot be revoked or reissued even if compromised. With the widespread usage, palmprint templates are stored diversely in different databases. If one palmprint template in the database with low security level is compromised, the same templates stored in other databases are no longer safe. Users' privacy information, like gene deficiency, health condition and so on, is possible to be leaked from original palmprint features. Thus palmprint template protection is essential to avoid direct disclosure of original palmprint features especially in mobile environments [53,54].

(4) Promotion of acceptance and flexibility

It is remarkably difficult to locate ROI of palmprint images acquired on mobile devices, so some auxiliary acquirement techniques are employed to help preprocess and improve accuracy performance. However, it is possible that auxiliary acquirement techniques lead to uncomfortable experience. How to balance comfort degree and verification accuracy is also an open issue currently.

With the improvement and modification of the related technologies, palmprint recognition on mobile device will be promoted gradually, and accordingly plays an increasingly significant role in advanced authentication fields.

Acknowledgments

This work was supported by National Natural Science Foundation of China (61305010, 61681240391), Science and Technology Project of Education Department of Jiangxi Province (GJJ150715), Voyage Project of Jiangxi Province (201450), Open Foundation of Key Laboratory of Jiangxi Province for Image Processing and

Pattern Recognition (TX201604002), Doctoral Initiating Foundation of Nanchang Hangkong University (EA201620045) and Development Foundation of Shanghai Electronic Certificate Authority Center Co., Ltd (HFK201628002).

References

[1]   H A Shabeer, and P Suganthi. Mobile phones security using biometrics. *International Conference on Computational Intelligence and Multimedia Applications*, 2007:270–274.

[2]   Y C Wang, and K T Cheng. Energy-optimized mapping of application to smartphone platform – A case study of mobile face recognition. *IEEE Conference on Computer Vision and Pattern Recognition*, 2011:84–89.

[3]   C H Lee, S H Lee, J H Kim, and S J Kim. Preprocessing of a fingerprint image captured with a mobile camera. *International Conference on Biometrics*, 2006:348–355.

[4]   J A Markowitz. Voice biometrics. *Communications of the ACM*, 2000,43(9): 66–73.

[5]   D H Cho, K R Park, D W Rhee, Y G Kim, and J H Yang. Pupil and iris localization for iris recognition in mobile phones. *Seventh ACIS International Conference on Software Engineering, Artificial Intelligence, Networking, and Parallel/Distributed Computing*, 2006:197–201.

[6]   D Zhang, W M Zuo, and F Yue. A comparative study of palmprint recognition algorithms. *ACM Computing Surveys*, 2012,44(1):1–37.

[7]   A Kong, D Zhang, and M Kamel. A survey of palmprint recognition. *Pattern Recognition*, 2009,42(7):1408–1418.

[8]   F Yue, W M Zuo, and D Zhang. Survey of palmprint recognition algorithms. *Acta Automatica Sinica*, 2010,36(3):353–365.

[9]   J You, W X Li, and D Zhang, Hierarchical palmprint identification via multiple feature extraction. *Pattern Recognition*, 2002,35(4):847–859.

[10] D Zhang, A W K Kong, J You, and M Wong. Online palmprint identification. *IEEE Transactions on Pattern Analysis and Machine Intelligence*, 2003,25(9): 1041–1050.

[11] C Liambas, and C Tsouros. An algorithm for detecting hand orientation and palmprint location from a highly noisy image. *IEEE International Symposium on Intelligent Signal Processing*, 2007:1–6.

[12] P H Hennings-Yeomans, B V K V Kumar, and M Savvide. Palmprint classification using multiple advanced correlation filters and palm-specific segmentation. *IEEE Transactions on Information Forensics and Security*, 2007,2(3): 613–622.

[13] C Poon, D C M Wong, and H C Shen. A new method in locating and segmenting palmprint into region-of-interest. *17th International Conference on Pattern Recognition*, 2004:533–536.

[14] M Li, C H Yan, and G H Liu. Personal identification system using palm prints. *Journal of Image and Graphics*, 2000,5(2):134–137.

[15] G K O Michael, T Connie, and A B J Teoh. A contactless biometric system using multiple hand features. *Journal of Visual Communication and Image Representation*, 2012,23(7):1068–1084.

[16] L Leng, G Liu, M Li, M K Khan, and A M Al-Khouri. Logical conjunction of triple-perpendicular-directional translation residual for contactless palmprint preprocessing. *11th International Conference on Information Technology: New Generations*, 2014:523–528.

[17] K Ito, T Sato, S Aoyama, S Sakai, S Yusa, and T Aoki. Palm region extraction for contactless palmprint recognition. *International Conference on Biometrics*, 2015:334–340.

[18] M Aykut, and M Ekinci. Developing a contactless palmprint authentication system by introducing a novel ROI extraction method. *Image and Vision Computing*, 2015,40:65–74.

[19] M Aykut, and M Ekinci. AAM-based palm segmentation in unrestricted back-grounds and various postures for palmprint recognition. *Pattern Recognition Letters*, 2013,34(9):955–962.

[20] Y F Han, T N Tan, Z N Sun, and Y Hao. Embedded palmprint recog-nition system on mobile devices. *International Conference on Biometrics*, 2007:1184–1193.

[21] F Li, M K H Leung, and C S Chian. Make palm print matching mobile. *Second Symposium International Computer Science and Computational Technology*, 2009:128–133.

[22] A S Sierra, J G Casanova, C S Ávila, and V J Vera. Silhouette-based hand recog-nition on mobile devices. *43rd Annual International Carnahan Conference on Security Technology*, 2009:160–166.

[23] C Methani, and A M Namboodiri. Video based palmprint recognition. *20th International Conference on Pattern Recognition*, 2010:1352–1355.

[24] M Franzgrote, C Borg, B J T Ries *et al.* Palmprint verification on mobile phones using accelerated competitive code. *International Conference on Hand-Based Biometrics*, 2011: 1–6.

[25] G K O Michael, T Connie, and A B J Teoh. Touch-less palm print biometrics: Novel design and implementation. *Image and Vision Computing*, 2008,26(12): 1551–1560.

[26] A S Sierra, C S Ávila, J G Casanova, and A M Ormaza. Towards hand biometrics in mobile devices. *En:BIOSIG*, 2011:203–210.

[27] A S Sierra, C S Ávila, J G Casanova, and A M Ormaza. Hand biometrics in mobile devices. *Advanced Biometric Technologies, InTech*, 2011:367–382.

[28] M Choraś, and R Kozik. Contactless palmprint and knuckle biometrics for mobile devices. *Pattern Analysis and Applications*, 2012,15(1):73–85.

[29] W Jia, R X Hu, J Gui, Y Zhao, and X M Ren. Palmprint recognition across different devices. *Sensors*, 2012,12:7938–7964.

[30]   S Aoyama, K Ito, T Aoki, and H Ota. A contactless palmprint recognition algorithm for mobile phones. *International Workshop on Advanced Image Technology*, 2013:409–413.

[31]   S Ibrahima, and D A Ramlia. Evaluation on palm-print ROI selection techniques for smart phone based touch-less biometric system. *American Academic & Scholarly Research Journal*, 2013,5(5):205–211.

[32]   S H Lee, S B Kang, D H Nyang, and K H Lee. Effective palm print authentication guideline image with smart phone. *The Journal of Korean Institute of Communications and Information Sciences*, 2014,39C(11):994–999.

[33]   N F Moço, P L Correia, and L D Soares. Smartphone-based palmprint recognition system. *21st International Conference on Telecommunications*, 2014:457–461.

[34]   L Fang, and Neera. Mobile based palmprint recognition system. *International Conference on Control, Automation and Robotics*, 2015:233–237.

[35]   H Javidnia, A Ungureanu, and P Corcoran. Palm-print recognition for authentication on smartphones. *IEEE International Symposium on Technology in Society*, 2015:1–5.

[36]   J S Kim, G Li, B J Son, and J H Kim. An empirical study of palmprint recognition for mobile phones. *IEEE Transactions on Consumer Electronics*, 2015,61(3):311–319.

[37]   H Javidnia, A Ungureanu, C Costache, and P Corcoran. Palmprint as a smartphone biometric. *IEEE International Conference on Consumer Electronics*, 2016:463–466.

[38]   A B J Teoh, and L Leng. *Palmprint Matching in "Encyclopedia of Biometrics"*. New York, NY: Springer, 2014:1–8.

[39]   L Leng, J S Zhang, M K Khan, X Chen, and K Alghathbar. Dynamic weighted discrimination power analysis: a novel approach for face and palmprint recognition in DCT domain. *International Journal of the Physical Sciences*, 2010,5(17):2543–2554.

[40]   L Leng, J S Zhang, J Xu, M K Khan, and K Alghathbar. Dynamic weighted discrimination power analysis in DCT domain for face and palmprint recognition. *International Conference on Information and Communication Technology Convergence*, 2010:467–471.

[41]   D Zhang, A Kong, J You, and M Wong. Online palmprint identification. *IEEE Transactions on Pattern Analysis and Machine Intelligence*, 2003,25(9): 1041–1050.

[42]   A Kong, and D Zhang. Feature-level fusion for effective palmprint authentication. *First International Conference on Biometric Authentication*, 2004: 761–767.

[43]   A Kong, and D Zhang. Competitive coding scheme for palmprint verification. *17th International Conference on Pattern Recognition*, 2004:520–523.

[44]   Z N Sun, T N Tan, Y H Wang, and S Z Li. Ordinal palmprint representation for personal identification. *IEEE International Conference on Computer Vision and Pattern Recognition*, 2005:279–284.

[45] W Jia, D S Huang, and D Zhang. Palmprint verification based on robust line orientation code. *Pattern Recognition*, 2008,41(5):1504–1513.

[46] Z H Guo, D Zhang, L Zhang, and W M Zuo. Palmprint verification using binary orientation co-occurrence vector. *Pattern Recognition Letters*, 2009,30(13):1219–1227.

[47] L Zhang, H Y Li, and J Y Niu. Fragile bits in palmprint recognition. *IEEE Signal Processing Letters*, 2012,19(10):663–666.

[48] J S Chen, Y S Moon, M F Wong, and G D Su. Palmprint authentication using a symbolic representation of images. *Image and Vision Computing*, 2010,28(3):343–351.

[49] P Viola, and M Jones. Rapid object detection using a boosted cascade of simple features. *IEEE Conference on Computer Vision and Pattern Recognition*, 2001:511–518.

[50] D G Lowe. Distinctive image features from scale-invariant keypoints. *International Journal of Computer Vision*, 2004,60(2):91–110.

[51] H Ota, R Watanabe, K Ito, T Tanaka, and T Aoki. Implementation of remote system using touchless palmprint recognition algorithm. *Eighth International Conference on Advances in Mobile Computing and Multimedia*, 2010:33–41.

[52] H Ota, S Aoyama, R Watanabe, K Ito, Y Miyake, and T Aoki. Implementation and evaluation of a remote authentication system using touchless palmprint recognition. *Multimedia Systems*, 2013,19(2):117–129.

[53] L Leng, and A B J Teoh. Alignment-free row-co-occurrence cancelable palmprint fuzzy vault. *Pattern Recognition*, 2015,48(7):2290–2303.

[54] L Leng, and J S Zhang. PalmHash Code vs. PalmPhasor code. *Neurocomputing*, 2013,108:1–12.

Chapter 13
Addressing the presentation attacks using periocular region for smartphone biometrics

Kiran B. Raja[1], R. Raghavendra[1], and Christoph Busch[1]

13.1 Introduction

Acceptance of biometrics in secure access applications is acknowledged by many governmental agencies and border control processes. The choice of a suitable biometric characteristics employed for operational systems varies on the basis of security requirements and the ease of use. Most systems employ face-based biometrics for a wide range of applications [12], while a significant number of applications employ iris characteristics motivated by its robust performance [4] and other access control applications use fingerprint recognition [12]. The success of biometrics has further pushed large-scale projects such as UIDAI in India to employ both face, iris and fingerprint as key biometric characteristics [8]. Success of regular biometric systems has spun a new avenue of secure access control applications on smartphones serving the purposes from low-security level applications to high security level applications [6,18,19,29,32,38]. The security requirements vary from unlocking applications on smartphones to secure authentication for banking applications via smartphones. As an added advantage, the emergence of smartphone biometrics is being supported by high quality of the embedded cameras on the smartphones, which provides biometric image samples of acceptable quality.

Various works have used face [6,19,23,38] and periocular characteristics on smartphones [16,18,19,28,33]. The traditional problems of pose symmetry, expressions and illumination are resulting in degraded performance of any biometric systems employing face characteristics [25]. To address the problem of degraded performance for face-based biometric systems, an earlier work has indicated the preference towards employing the periocular region as a mono-modal biometric characteristic or as a complementary biometric characteristic [25]. Motivated by this, smartphone-based biometric systems have explored the periocular region in many recent works [16,18,19,28,33]. Going a step further, iris has been investigated as a biometric characteristic using smartphone images that are captured in visible spectrum [6,18–20,39,40].

[1]Norwegian Biometrics Laboratory, NTNU Gjøvik, Norway

Further, a biometric system is considered robust not just by its ability to accept a high number of genuine attempts and reject at the same time a high number of zero-effort impostor attempts, but also through its ability to distinguish the normal or Bona Fide presentations from attack presentations. A Bona Fide presentation is defined as the interaction of the biometric capture subject and the biometric data capture subsystem in the fashion intended by the policy of the biometric system – commonly understood as a 'normal/live' presentation. The threat of presentation attacks (a.k.a. spoofing) should not compromise the security of access control procedures using biometrics on smartphones. The vulnerability of biometric systems on smartphones is well illustrated in earlier works [14,15,36]. The vulnerability to attacks is not just relevant to smartphones biometrics but in principle to all regular biometric systems, which was analysed in [3,5,7,9,10,21,41]. Additionally, many works have proposed robust methods to detect targeted presentation attacks in iris and ocular biometric systems [3,5,7,9,10,14,15,21,36,41]. Thus, we can conclude that a smartphone-based biometric system serves the purpose of secure access control in a robust manner only when coupled with strong presentation attack detection (PAD) methods.

Possible presentation attacks span across many different modes, which range from low-cost printed artefacts (as presentation attack instrument) to near-real quality replay attacks through high-resolution screens. Ocular biometric systems, operating either in the visible spectrum (regular cameras on smartphones) or operating in near-infrared spectrum, have multiple times being challenged using simple low-cost printed artefacts [3,5,7,9,10,14,15,21,26,27,30,36,41]. MobILive 2014 (Mobile Iris Liveness Detection Competition) [36] was specifically designed to address presentation attacks using printed images. Further, the ocular biometric systems operating in the visible spectrum via smartphones were challenged with a more advanced attack mechanism of a replay video using high-quality screens [14,15,17].

Figure 13.1 illustrates the architecture of an ocular/iris biometrics system with integrated PAD functionality. A typical conventional ocular/iris biometric system involves all components enclosed in the thick-solid lines without PAD in which the ocular region is localized and subsequently features are extracted from the iris/periocular region and compared against the reference samples present in the enrolment database to verify or reject the claimed identity. Such a system can be attacked by presenting artefact samples, which can compromise the security level of the system. Thus, it is essential to introduce a PAD module that verifies the liveness of a presented ocular/iris characteristics. The intended operational biometric system with PAD ability is enclosed in thick dashed line which operates with PAD.

In this chapter, we will cover recent research on different presentation attacks and PAD methods for periocular/ocular biometrics on smartphones from two perspectives. It has to be noted that the scope of the current chapter is limited to work on ocular biometrics using smartphones. In the first place, we list the state-of-the-art ocular biometric systems on smartphones and discuss their vulnerabilities. This is essential to reveal whether an artefact database is relevant to be used to develop and evaluate PAD methods. We further, systematically provide a summary of various PAD methods based on textural features along with the obtained results.

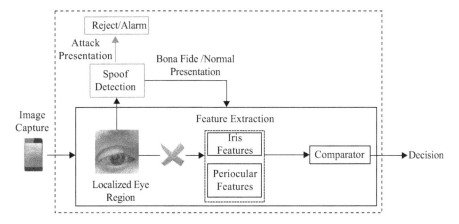

Figure 13.1   *General architecture of an ocular/iris biometric system with incorporated PAD module. Note: The block enclosed in the thick solid lines indicates a conventional biometric system without PAD, and the whole system enclosed in thick dashed line is the intended biometric system with PAD*

In the rest of the chapter, we discuss the available databases in Section 13.2 for ocular biometrics on smartphones. In Section 13.3, we analyse the vulnerability towards possible attacks of known types. A list of all the attack detection techniques along with the obtained results is given in Section 13.4. In Section 13.5, we discuss the major summary of the available works on PAD for ocular biometrics on smartphones along with the remarks.

13.2   Database

This section provides a summary of available ocular databases, which were captured using smartphones, and which were specifically acquired with the intention to investigate presentation attacks. Two publicly available ocular databases are presented, which provide two different attack artefacts species – image-based attacks (either electronic or printed attacks) and video-based attacks to surpass the biometric system. Whereas MobILive 2014 database concentrates on the print attack, the presentation attack video iris database (PAVID) focuses on video-presentation attacks in video-based authentication scenarios. The key advantage of the video-based system is the high number of frames, which can be used to make an aggregate decision.

13.2.1   MobILive 2014 Database

MobILive 2014 dataset was constructed in conjunction with the first Mobile Iris Liveness Detection Competition (MobILive) organized in the context of IJCB2014 [35,36]. The MobILive 2014 dataset was derived from the MobBIO multimodal database [34],

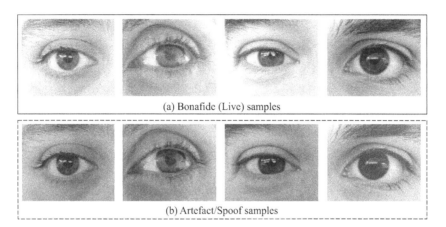

Figure 13.2   Sample images from MobILive 2014 dataset: (a) sample images from
Bona Fide (a.k.a normal/live) presentation and (b) attack presentation
images corresponding to normal images in (a)

which comprises samples of face images of 105 volunteers collected using an Asus EeePad Transformer tablet. The face images were used to crop the ocular region corresponding to the normal Bona Fide image set of MobILive 2013 database.[1] The cropped ocular images correspond to a size of 250 × 200 pixels captured indoors with natural light and room light, with variable eye orientations and occlusion levels [34]. Four ocular samples were used for each eye corresponding to each subject resulting in 800 live ocular images in the dataset.

The artefact dataset of MobILive 2014 database, also referred as MobBIOfake database [36] consists of artefact images collected from 100 subjects. Each of the live ocular image is printed using a colour printer and re-captured using the Asus EeePad Transformer tablet. Thus, the dataset comprises 1,600 samples of which 800 are Bona Fide samples and the rest of 800 are attack samples. The attack samples were obtained from printed images of the original ones captured with the same handheld device and in similar conditions. Figure 13.2 presents the example images from the MobILive 2014 database. The panel (a) presents the images corresponding to Bona Fide images, while the panel (b) in Figure 13.2 presents the attack samples corresponding to the images in the panel (a).

13.2.2   PAVID Database

Another available public database for research on smartphone ocular images is the PAVID. This database is constructed to explore video-based presentation attacks against ocular biometric systems on smartphones operating in the visible spectrum. The PAVID database consists of two parts which correspond to Bona Fide and attack

---

[1]Only 100 subjects from MobBIO dataset were used to create MobILive 2013 dataset.

Table 13.1   Bona Fide iris/ocular videos in the PAVID database

	Smartphone	
	Nokia Lumia 1020	**iPhone 5S**
Number of subjects	76	76
Unique eye instances	152	152
Enrolment videos	152	152
Probe videos	152	152

samples of iris/ocular videos. The Bona Fide videos are the recordings of the live ocular characteristics and the artefact videos are the recordings of video replays. Unlike most other databases, the PAVID database is specifically tailored to address video-based attacks. Video-based ocular biometric systems can be considered superior to image-based system because there are number of frames available. With a higher number of frames available, the decision module can employ decisions on individual frames or aggregated frames using independent classifiers or ensemble methods. Thus, in this chapter, we focus on frame-based independent decision along with the majority voting for a number of 25 frames in the video.

13.2.2.1   PAVID – Bona Fide iris video database

PAVID is a relatively large-scale iris/ocular video database acquired using two smartphones operating in visible spectrum.

The PAVID database [14,15] consists of videos captured from 152 unique eye instances from 76 subjects using two widely deployed smartphones – Nokia Lumia 1020 and iPhone 5S. The total distribution of iris/ocular videos is presented in Table 13.1. Each unique eye instance is captured in two different sessions that correspond to enrolment and verification attempt. In each of the sessions, a video of 1–3 s duration is acquired for each subject. The captured eye videos are processed such that at least 25 frames are obtained between two eye blinks. The videos for each subject are used as the enrolment videos. In a similar manner, the ocular/iris video is obtained for a verification session. Figure 13.3(a) presents the sample frames of the Bona Fide access videos captured using the iPhone 5S and Figure 13.3(b) presents the samples acquired from Nokia Lumia 1020 and the corresponding artefacts for each phone. The reader is referred to [14] for a detailed summary of the database.

13.2.2.2   PAVID – attack iris video database

Further, the PAVID database also consists of an artefact subset, which has videos of iris/ocular samples captured using two smartphones. However, the videos are obtained from the electronic screen of devices, which are capable of displaying high- and near-original-quality videos. To simulate a realistic attack in a verification scenario, the attack database is constructed under the assumption that the video from the enrolment

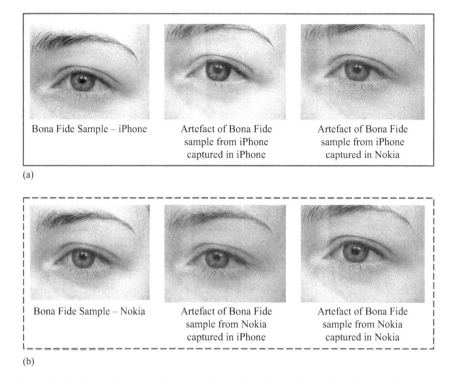

*Figure 13.3   Sample frames from ocular videos from the PAVID dataset. The panel
(a) indicates the live Bona Fide sample (left) captured from an iPhone
along with the corresponding attack samples (replayed using iPad)
captured from iPhone and Nokia. The panel (b) indicates the Bona
Fide sample captured from Nokia along with the attack samples
(replayed using iPad) captured from iPhone and Nokia
correspondingly. Samples with extreme variation are illustrated in the
figure for the sake of exemplification*

database is available to the impostor. In this scenario, the impostor can use the enrolment video to generate the printed artefacts or replay the source video. However,
when the biometric system employs video-based authentication, one has to design
the attacks using video-based approaches. Going by such an argument, the artefact
database is created by replaying the iris/ocular video on the high-quality display
enabled iPad and presenting it to smartphones (i.e., the biometric sensor in this case).
The replay attack database consists of four different attack subsets such that the enrolment videos obtained from iPhone are replayed to iPhone and also to Nokia. Similarly,
the enrolment videos obtained from Nokia are replayed to Nokia and iPhone as well.
Under each replay attack subset, a total of 152 iris/ocular videos are present, which
make a total of 608 artefact iris/ocular videos in total in the PAVID database. Table
13.2 provides an overview of the different subsets in the PAVID database.

Table 13.2   PAVID attack database composition

Source obtained from	Smartphone attacked	Number of videos
iPhone	iPhone	152
	Nokia	152
Nokia	iPhone	152
	Nokia	152

13.3   Vulnerabilities towards presentation attacks

In order to make the ocular/iris biometric systems robust against the presentation attacks, the first step is to assess the vulnerability of a conventional unprotected system towards the attacks. The biometric performance of a biometric system is measured using the false accept rate (FAR) versus the false reject rate (FRR) [11]. Compliant to a general system, the vulnerability towards presentation attacks can be evaluated by presenting artefact samples and measuring the FAR–FRR using a specific baseline system. This analysis indicates the vulnerability of a system in the sense of an impostor attack presentation match rate (IAPMR). We conduct such vulnerability analysis using the PAVID database.

13.3.1   Vulnerability analysis using the PAVID

This section discusses the vulnerability of periocular systems using an analysis of the PAVID database and specifically the artefact samples to simulate an attack verification attempt against a biometric system that is observing the ocular characteristics. To conduct such a simplistic operation, a sample frame from a probe video is compared against a reference frame in the enrolment database. The baseline performance with Bona Fide ocular videos and the baseline performance of a system when ocular videos are used for attacking the biometric sensor are analysed in this section. Such an analysis is based on the idea of measuring initially the genuine and zero-effort impostor scores based on only normal Bona Fide presentations. Furthermore, the analysis comprises measuring the genuine and impostor scores when the artefacts are used to attack the sensor. The genuine score is obtained by comparing the reference frame from enrolment video with all the frames from probe video. These scores are used to obtain the detection error trade-off curves of the system. To simplify the number of comparisons, we consider 1 frame from an ocular video of enrolment as the reference image and 25 frames from the probe video as the probe samples. A similar approach is used to compute the genuine and impostor scores when the replay attack video is used. In this work, the baseline system is evaluated using simple local binary pattern (LBP) [24] feature extractor and sparse reconstruction classifier [15,20,31] on the periocular characteristics.

As indicated in Figure 13.4, the baseline provides an Equal Error Rate (EER) of close to 8% for iPhone 5S which is represented in green curves. Further, the artefact

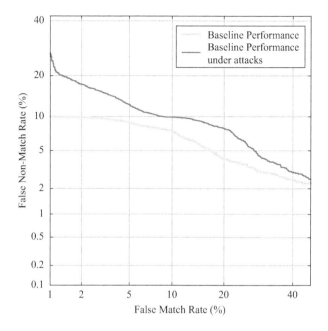

Figure 13.4 Baseline scores and performance of the systems employing iPhone 5S capture smartphone

samples are presented using to the baseline system and the False Match Rate–False Non-Match Rate (FMR–FNMR) is measured which is indicated in the red curve in the Figure 13.4. As it can be noted from Figure 13.4, the artefacts are accepted by the system to a greater extent which results in an EER of around 10% for iPhone 5S. The EER obtained from the attack presentation competes closely with the Bona Fide presentations and thereby indicates the vulnerability of the system. For an operational threshold of FMR = 0.1%, the IAPMR of the unprotected system is 90% indicating high vulnerability.

13.4 PAD techniques

In the backdrop of system vulnerability analysis using the described ocular database, it is essential to devise algorithms that are robust to detect the attacks before they are submitted to the verification pipeline. Thus, this section discusses various state-of-art techniques proposed to address the attacks. Further, in order to decide the suitability of a PAD algorithm, we employ the standardized metrics defined by recent international standardization activities [11], which are described briefly in this section.

As indicated in Figure 13.1, the PAD module can be placed in the verification pipeline between the capture subsystem and the feature extraction subsystem.

Thus, the performance of PAD algorithms are measured independently of the system and are reported explicitly. The brief overview of PAD algorithms is provided in Section 13.4.2.

13.4.1 Metrics for PAD algorithms

Generally, we can consider the PAD module as a stand-alone subsystem, which can provide a binary decision indicating the presentation as a Bona Fide presentation or attack presentation, and thus, quantify the PAD reliability using set of metrics recommended by the international standard developed by ISO/IEC [11]. The detection performance of a PAD algorithm can be benchmarked in accordance to ISO/IEC 30107-3 [11], which defines two metrics (1) attack presentation classification error rate (APCER) that is defined as a proportion of attack presentation incorrectly classified as bona fide (or normal or real) presentation and (2) Bona Fide presentation classification error rate (BPCER) that is defined as proportion of normal presentation incorrectly classified as attack presentation [11]. For practical purpose (but not supported by ISO/IEC standards), we indicate a single performance metric of the PAD algorithm in terms of average classification error rate (ACER) such that:

$$\text{ACER} = \frac{\text{APCER} + \text{BPCER}}{2} \tag{13.1}$$

The lower values of APCER indicate high rejection of attack samples, while lower values of BPCER indicate a very high rate of Bona Fide attempts normally accepted. An ideal system operating in a biometric access control scenario is expected to provide APCER $= 0$ and BPCER $= 0$ resulting in ACER $= 0$. However, practically, a system with lower ACER will be preferred in a technology evaluation in preference over other systems with higher ACER.

13.4.2 Texture features for PAD

In this section, we provide a brief summary of various texture-based features used in PAD algorithms. Further, we also discuss the key advantages and intuition behind each of the feature extraction methods.

13.4.2.1 Image quality features

A key-factor observed with artefact samples is that the samples differ significantly from the normal Bona Fide images. The difference in quality can be attributed to the fact that the attack presentation images are generally printed using the normal low-cost printers which inherently introduces signal artefacts [7]. When these images are re-captured using the biometric sensors, one can observe the exhibited artefacts [7]. Intuitively, the authors have used quality features [7] with both no-reference and full-reference measures with a support vector classifier (Image Quality Metrics–Support Vector Machine (IQM–SVM)) [7]. In this chapter, we evaluate the IQM–SVM to evaluate the performance in detecting attack presentations in ocular biometrics.

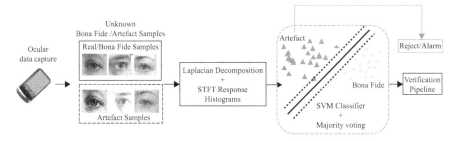

Figure 13.5   (LPFR) features for presentation attack detection

13.4.2.2   Texture features from LBP

Another category of techniques have leveraged on the textural features using hand-crafted features such as LBP [22]. LBPs are well known to encode the discriminative texture features in the image by using the binarization rules in a fixed neighbourhood of pixels. LBP has proven their robustness for face-based presentation attacks [22]. Motivated by that, we explore LBP features in conjunction with a support vector classifier [2] to determine the robustness of the technique in determining the artefact presentation for ocular biometrics on smartphone which is further referred as LBP–SVM [22] in this chapter.

13.4.2.3   Texture features from binarized statistical image features

Recent work [27] has proposed another class of texture descriptors leveraging the statistical features of images [13]. Binarized statistical image features (BSIF) are used to extract the distinctive texture information from image, and the responses obtained for each image is used along with the SVM to determine the liveness of the presented biometric sample. The key advantage of using BSIF is that they provide a set of linear filters, which can be combined in different ways to gather unique information from Bona Fide and artefact images. Motivated by this, the authors [27] have indicated the preference to use multi-scale BSIF in their earlier work. Thus, we employ the MBSIF–SVM [27] to evaluate the suitability for ocular biometrics on smartphones.

13.4.2.4   Laplacian pyramid frequency response features

Another recent approach to address the presentation attacks detection for ocular biometrics on smartphones leverages on the space and frequency features in Laplacian space [14]. The Laplacian pyramid frequency response (LPFR)-feature-based PAD algorithm proposed in our earlier work [14] is depicted in Figure 13.5. Given the image/video of the subject in a verification scenario, we first decompose each image/frames into Laplacian pyramids of multiple scales. Each of the resulting images at the specific scale is used to obtain a short-term Fourier transform (STFT) response at four different orientations. The response corresponding to four different orientations is encoded as a single response image, and the features are obtained by taking the histogram as described in this section. The features are used to classify the presentation category as Bona Fide or attack presentation using an SVM classifier. The motivation and intuition behind the LPFR technique are thoroughly discussed in this section.

Laplacian pyramids

Laplacian pyramid decomposition of the image was initially developed with the idea of encoding the image using local operators at many scales with identical basis functions [1]. The significance of the Laplacian pyramid decomposition comes from the fact that the elements of an image are localized both in space and frequency domain.

Further, the Laplacian pyramid can be used effectively to represent images as a series of band-pass filtered images that are sampled successively at sparser representations [1]. Although, the frequency content of the image is well localized using Laplacian pyramids, the orientation information of each frequency content is not obtained.

Algorithm for PAD

In the case of any natural image or video frames, there exists a substantial amount of edge information that contributes to the frequency information of that image or frame [37]. However, when the same images are printed using low-high-resolution printers or when the same images are displayed on an electronic screen, the images present frequency information that is different from the original frequency distribution of the image. This additional frequency information is inherently present in the artefacts generated by printing the live samples or replaying the live samples in the context of biometric samples. Intuitively, localizing this frequency makes the separation of normal presentation versus attack presentation. In order to localize this frequency, we employ Laplacian pyramids at five different scales with binomial filter kernel of size 9 Laplacian pyramid decomposition separates the lower and higher frequency in a well-defined components. We have employed a scale of $n = 5$ in this work. The difference of the obtained low-pass filter and high-pass filter at each scale is used to localize the frequency information further by analysing STFT response corresponding to four different orientations $\phi = \{0°, 45°, 90°, 135°\}$.

If an image at a particular scale s of the Laplacian pyramid is represented by I_s, we obtain the STFT response of the image. The STFT of the image at scale s, which is represented by F_s is the image resulting to response of frequency components in four different orientations such that $\phi = \{0°, 45°, 90°, 135°\}$. The filter response obtained from each orientation is separated for real and complex values subsequently. Each of the responses denoted by b is finally encoded to form the final response map as given by FR_s, where i corresponds to different orientation angles given by $\phi = \{0°, 45°, 90°, 135°\}$:

$$FR_s = \text{Re}\left(\sum_{i=1}^{4} (b_i) * (2^{(i-1)})\right) + \text{Im}\left(\sum_{i=1}^{4} (b_i) * (2^{(i-1)})\right) \tag{13.2}$$

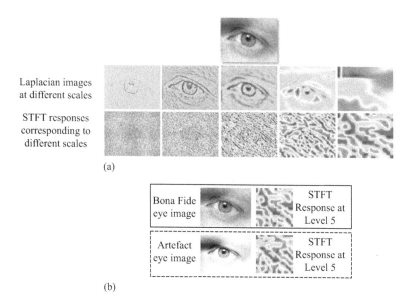

Laplacian images at different scales

STFT responses corresponding to different scales

(a)

Bona Fide eye image

STFT Response at Level 5

Artefact eye image

STFT Response at Level 5

(b)

Figure 13.6   Illustration of features from Bona Fide and artefact presentations using Laplacian decomposed frequency response feature. (a) Decomposition of ocular-image into Laplacian pyramids of scale 5 and corresponding STFT response maps in each scale. (b) The difference in STFT response maps at level 5 for Bona Fide and artefact image. Note: Images from all scales are resized to uniform size for the purpose of illustration only

The feature vector FV, of the image at a particular scale s is formed by obtaining the histogram of the response map at scale FR_s:

$$FV_s = \sum_{i=0}^{255} \{FR_s\}_i \tag{13.3}$$

The final feature vector for the frame or image is formed by concatenating the feature vectors of images from scale 1 to n and orientation $\phi = \{0°, 45°, 90°, 135°\}$. The final feature vector FV_f can be represented as:

$$FV_f = \{FV_{s=1,\phi=0°}, FV_{s=1,\phi=45°},$$
$$FV_{s=1,\phi=90°}, FV_{s=1,\phi=135°},$$
$$\ldots, FV_{s=n,\phi=0°}, FV_{s=n,\phi=45°},$$
$$FV_{s=n,\phi=90°}, FV_{s=n,\phi=135°}\} \tag{13.4}$$

The final feature vector given by (13.4) is used to represent the image for classification purposes. Figure 13.6(a) presents the Laplacian pyramid decomposition at five

Table 13.3   Performance of the various schemes on MobILive 2014
dataset

Techniques (teams)	APCER	BPCER	Mean error rate (or ACER)
HH [36]	29.25	7.00	18.13
IrisKent [36]	0.25	3.75	2.00
Liv-IC-INICAMP [36]	0.50	2.00	1.25
Federico II [36]	1.25	0.00	0.63
GUC [36]	0.75	0.00	0.38
IIT Indore [36]	0.50	0.00	0.25
MBSIF–SVM [27]	**0.00**	**0.00**	**0.00**
LPFR–SVM [14]	**0.00**	**0.00**	**0.00**

The bold values signify best performance of the algorithm in detecting presentation attack.

different scales and its corresponding STFT response maps for a sample frame from Bona Fide video. Figure 13.6(b) illustrates differences in obtained STFT response maps for Bona Fide and artefact at 5th level of the Laplacian Pyramid. It can be observed from the figure that subtle changes in the frequency information along various orientations of the frame from the replay attack video can be highly enhanced by analyzing STFT response maps.

13.5   Experiments and results

This section provides the details on the detection performance of various algorithms on two different databases captured using smartphones.

13.5.1   Results on MobiLive 2014 database

The MobILive 2014 database provides a disjoint/independent training set and testing set which consist of 50 subjects in each test. Different teams have submitted their algorithm on this database in the MobiLive 2014 IJCB competition [36] which have used different features. Further, BSIF–SVM [26] indicated the performance of $ACER = 0\%$. The results obtained on the same dataset using the LPFR–SVM algorithm [14] performed with the same $ACER$ of 0. The complete set of results are presented in the Table 13.3.

13.5.2   Results on the PAVID database

In order to effectively evaluate the different set of algorithms for PAD with the PAVID database, the whole database of 152 unique eye patterns (i.e., instances) obtained using a particular smartphone is divided into three sets: training set, development set and testing set. The training set comprises of 50 unique eye patterns that were used only for training the SVM classifier. The development dataset comprises of 20 unique eye patterns that are used to tune any parameters associated with the PAD

Table 13.4   Division of PAVID database for experiments using ocular videos acquired from each smartphone

	Smartphone	
	Nokia Lumia 1020	**iPhone 5S**
Bona Fide iris/ocular videos		
Development	20	20
Training	50	50
Testing	82	82
Artefact iris/ocular videos for each attack		
Development	40 (20×2)	40 (20×2)
Training	40 (20×2)	40 (20×2)
Testing	224 (112×2)	224 (112×2)

algorithms. The development set if further used to determine the filter kernel for the Laplacian pyramid, the size of the window and scales for the pyramid. The testing dataset comprises 82 unique eye patterns that are solely used to evaluate the PAD algorithms evaluated in this work. The detailed division is provided in Table 13.4.

Table 13.5 presents results obtained on the PAVID database using various state-of-the-art methods such as IQM–SVM [7], LBP–SVM [22] and BSIF–SVM [27]. It can be observed in Table 13.5 that the LPFR method has emerged as the best technique for PAD in the benchmark of all state-of-the-art techniques. The best ACER is obtained consistently across all different attacks from the LPFR technique. The best ACER of 1.49% is obtained when the system employing iPhone as the primary sensor is attacked using enrolment videos captured using iPhone. Similarly, an ACER of 0.64% is obtained when enrolment video captured using Nokia phone is used to attack the ocular recognition system employing iPhone as capture device. The obtained results support the applicability of the LPFR approach for detecting presentation attacks in real life verification scenarios when adapted to video-based ocular recognition systems.

13.6   Discussions and conclusion

Presentation attacks pose a high level of threats to existing biometric systems and have been well illustrated. In this chapter, we have depicted the vulnerability of ocular biometric systems towards presentation attacks. The advancement in types of attacks has evolved from simple print attacks to advanced electronic screen attacks. Although, simple print attacks can be detected using features employing quality metrics or simple texture features, electronic screen attacks present challenges in a higher magnitude. In this chapter, we have explored presentation attacks in ocular biometric system on smartphones in the visible spectrum. We have discussed both kind of attacks – print attacks and electronic screen attacks. We have further thrown light on both image-based and video-based artefact presentations.

Table 13.5 Classification error rates obtained using various schemes for PAVID database

Reference video	Presentation attack video	IQM–SVM [7]			LBP–SVM [22]			MBSIF–SVM [27]			LPFR–SVM [14]		
		BPCER	APCER	ACER	BPCER	APCER	ACER	BPCER	APCER	ACER	BPCER	APCER	ACER
iPhone	iPhone	57.31	11.6	34.45	4.87	0.89	2.88	6.09	9.82	7.955	1.21	1.78	**1.49**
	Nokia	76.92	10.71	43.81	3.84	3.54	3.69	2.56	8.92	5.74	1.28	0	**0.64**
Nokia	iPhone	76.92	4.5	40.71	3.84	4.51	4.175	2.56	10.81	6.68	1.28	4.46	**2.87**
	Nokia	57.31	3.57	30.44	4.87	2.67	3.77	6.09	0.89	3.49	1.21	2.68	**1.95**

In the due course of this chapter, we have employed two publicly available databases that correspond to large-scale image-based artefact and video-based artefacts. For image-based artefacts, we have employed MobiLive 2014 dataset [36], whereas for video-based artefacts, we have used PAVID dataset [14]. PAVID presents a large-scale video iris/ocular database consisting of 152 unique patterns acquired using two different smartphones – iPhone 5S and Nokia Lumia 1020. Furthermore, we have systematically demonstrated the vulnerability of such ocular biometric systems using video replay artefacts. We have evaluated the state-of-art techniques in detecting video replay attacks using features obtained from both quality and texture features.

From the set of experiments, we can conclude that feature vectors representing spatial and frequency features from Laplacian images have proven to be robust in detecting attack presentations. The features are well classified with the use of a popular SVM classifier along with majority voting. The LPFR–SVM method has provided the best ACER of 0.64% for video replay attacks on ocular recognition systems employing iPhone as a capture sensor. While the lowest performance of ACER is observed as 2.87%, the average ACER results in less than 2% for the best performing methods. The obtained results for various evaluation indicate the necessity to continued research to detect artefacts robustly to make smartphone-based ocular biometrics highly secure.

Acknowledgments

The authors thanks to Safran Identity and Security for supporting this work, and in particular to Research & Technology team for the fruitful technical and scientific exchanges related to this particular work. This work was also partially funded by Research Council of Norway (Grant No. IKTPLUSS 248030/O70).

References

[1]   P. J. Burt and E. H. Adelson. The Laplacian pyramid as a compact image code. *IEEE Transactions on Communications*, 31(4):532–540, 1983.

[2]   C.-C. Chang and C.-J. Lin. LIBSVM: A library for support vector machines. *ACM Transactions on Intelligent Systems and Technology*, 2:27:1–27:27, 2011. Software available at http://www.csie.ntu.edu.tw/~cjlin/libsvm.

[3]   A. Czajka. Database of iris printouts and its application: Development of liveness detection method for iris recognition. In *2013 18th International Conference on Methods and Models in Automation and Robotics (MMAR)*, pages 28–33, Aug 2013.

[4]   J. Daugman. How iris recognition works. *IEEE Transactions on Circuits and Systems for Video Technology*, 14(1):21–30, 2004.

[5]   J. Daugman. Iris recognition and anti-spoofing countermeasures. In *7th International Biometrics Conference*, 2004.

[6]   M. De Marsico, C. Galdi, M. Nappi, and D. Riccio. Firme: Face and iris recognition for mobile engagement. *Image and Vision Computing*, 32(12):1161–1172, 2014.

[7]   J. Galbally, S. Marcel, and J. Fierrez. Image quality assessment for fake bio-metric detection: Application to iris, fingerprint, and face recognition. *IEEE Transactions on Image Processing*, 23(2):710–724, 2014.

[8]   Government of India. Aadhaar – a unique identification number. *Research Cell: An International Journal of Engineering Science*, 4(2):169–176, 2011.

[9]   X. He, Y. Lu, and P. Shi. A fake iris detection method based on fft and quality assessment. In *Chinese Conference on Pattern Recognition, 2008. CCPR'08*, pages 1–4. IEEE, 2008.

[10] K. Hughes and K. W. Bowyer. Detection of contact-lens-based iris biometric spoofs using stereo imaging. In *2013 46th Hawaii International Conference on System Sciences (HICSS)*, pages 1763–1772. IEEE, 2013.

[11] ISO/IEC TC JTC1 SC37 Biometrics. *ISO/IEC FDIS 30107-3. Information Technology – Biometric presentation attack detection – Part 3: Testing and reporting*. International Organization for Standardization and International Electrotechnical Committee, 2017.

[12] A. Jain, P. Flynn, and A. A. Ross. *Handbook of Biometrics*. Springer Science & Business Media, 2007.

[13] J. Kannala and E. Rahtu. Bsif: Binarized statistical image features. *Proceedings of the 21st International Conference on Pattern Recognition (ICPR2012)*, Tsukuba: IEEE, 2012, pp. 1363–1366.

[14] Kiran B Raja, R. Raghavendra, and C. Busch. Presentation attack detection using laplacian decomposed frequency response for visible spectrum and near-infra-red iris systems. In *2015 IEEE Seventh International Conference on Biometrics Theory, Applications and Systems (BTAS)*, pages 1–8. Arlington, VA: IEEE, 2015.

[15] Kiran B. Raja., R. Raghavendra, and C. Busch. Video presentation attack detec-tion in visible spectrum iris recognition using magnified phase information. *IEEE Transactions on Information Forensics and Security*, 10(10):2048–2056, Oct. 2015.

[16] Kiran B. Raja, R. Raghavendra, and C. Busch. Collaborative representation of deep sparse filtered feature for robust verification of smartphone periocular images. In *23rd IEEE International Conference on Image Processing (ICIP 2016)*, pages 1–5, 2016.

[17] Kiran B. Raja, R. Raghavendra, and C. Busch. Color adaptive quantized patterns for presentation attack detection in ocular biometric systems. *In Proceedings of the 9th International Conference on Security of Information and Networks (SIN '16)*, pp. 9–15. New York, NY, USA: ACM, 2016.

[18] Kiran B. Raja, R. Raghavendra, M. Stokkenes, and C. Busch. Smartphone authentication system using periocular biometrics. In *2014 International Con-ference on Biometrics Special Interest Group*, pages 1–8. Darmstadt: IEEE, 2014.

[19]   Kiran B. Raja, R. Raghavendra, M. Stokkenes, and C. Busch. Multi-modal authentication system for smartphones using face, iris and periocular. In *IEEE International Conference Biometrics (ICB)*, Phuket, Thailand, 2015.

[20]   Kiran B. Raja, R. Raghavendra, V. K. Vemuri, and C. Busch. Smartphone based visible iris recognition using deep sparse filtering. *Pattern Recognition Letters*, 57(0):33–42, 2015.

[21]   S. J. Lee, K. R. Park, and J. Kim. Robust fake iris detection based on variation of the reflectance ratio between the iris and the sclera. In *2006 Biometrics Symposium: Special Session on Research at the Biometric Consortium Conference*, pages 1–6. Baltimore, MD: IEEE, 2006.

[22]   J. Maatta, A. Hadid, and M. Pietikainen. Face spoofing detection from single images using micro-texture analysis. In *2011 International Joint Conference on Biometrics (IJCB)*, pages 1–7. Washington, DC: IEEE, 2011.

[23]   U. Mahbub, V. M. Patel, D. Chandra, B. Barbello, and R. Chellappa. Partial face detection for continuous authentication. In *2016 IEEE International Conference on Image Processing (ICIP)*, pages 2991–2995, 2016.

[24]   T. Ojala, M. Pietikainen, and T. Maenpaa. Multiresolution gray-scale and rotation invariant texture classification with local binary patterns. *IEEE Transactions on Pattern Analysis and Machine Intelligence*, 24(7):971–987, 2002.

[25]   U. Park, A. Ross, and A. K. Jain. Periocular biometrics in the visible spectrum: A feasibility study. In *Third IEEE International Conference on Biometrics: Theory, Applications, and Systems (BTAS'09)*, pages 1–6, 2009.

[26]   R. Raghavendra and C. Busch. Presentation attack detection algorithm for face and iris biometrics. In *2014 Proceedings of the 22nd European Signal Processing Conference (EUSIPCO)*, pages 1387–1391. Lisbon: IEEE, 2014.

[27]   R. Raghavendra and C. Busch. Robust scheme for iris presentation attack detection using multiscale binarized statistical image features. *IEEE Transactions on Information Forensics and Security*, 10(4):703–715, 2015.

[28]   R. Raghavendra and C. Busch. Learning deeply coupled autoencoders for smartphone based robust periocular verification. In *23rd IEEE International Conference on Image Processing (ICIP 2016)*, pages 1 –5, 2016.

[29]   R. Raghavendra, C. Busch, and B. Yang. Scaling-robust fingerprint verification with smartphone camera in real-life scenarios. In *2013 IEEE Sixth International Conference on Biometrics: Theory, Applications and Systems (BTAS)*, pages 1–8. Arlington, VA: IEEE, 2013.

[30]   R. Raghavendra, Kiran B. Raja, and C. Busch. Ensemble of statistically independent filters for robust contact lens detection in iris images. In *Proceedings of the 2014 Indian Conference on Computer Vision Graphics and Image Processing* (ICVGIP '14). Article 24, 7 pages. New York, NY, USA: ACM, 2014.

[31]   R. Raghavendra, Kiran B. Raja, B. Yang, and C. Busch. Combining iris and periocular recognition using light field camera. In *2013 2nd IAPR Asian Conference on Pattern Recognition (ACPR2013)*. Naha: IEEE, 2013.

[32] R. Raghavendra, K. B. Raja, A. Pflug, B. Yang, and C. Busch. 3D face reconstruction and multimodal person identification from video captured using smartphone camera. In *2013 IEEE International Conference on Technologies for Homeland Security (HST)*, pages 552–557. Waltham, MA: IEEE, 2013.

[33] G. Santos, E. Grancho, M. V. Bernardo, and P. T. Fiadeiro. Fusing iris and periocular information for cross-sensor recognition. *Pattern Recognition Letters*, 57:52–59, 2015.

[34] A. F. Sequeira, J. C. Monteiro, A. Rebelo, and H. P. Oliveira. Mobbio: A multimodal database captured with a portable handheld device. In *2014 International Conference on Computer Vision Theory and Applications (VISAPP)*, volume 3, pages 133–139, Jan 2014.

[35] A. F. Sequeira, J. Murari, and J. S. Cardoso. Iris liveness detection methods in mobile applications. In *2014 International Conference on Computer Vision Theory and Applications (VISAPP)*, volume 3, pages 22–33. Lisbon, Portugal: IEEE, 2014.

[36] A. F. Sequeira, H. P. Oliveira, J. C. Monteiro, J. P. Monteiro, and J. S. Cardoso. Mobilive 2014-mobile iris liveness detection competition. In *2014 IEEE International Joint Conference on Biometrics (IJCB)*, pages 1–6. Clearwater, FL: IEEE, 2014.

[37] E. P. Simoncelli and B. A. Olshausen. Natural image statistics and neural representation. *Annual Review of Neuroscience*, 24(1):1193–1216, 2001.

[38] Q. Tao and R. Veldhuis. Biometric authentication for a mobile personal device. In *2006 3rd Annual International Conference on Mobile and Ubiquitous Systems-Workshops*, pages 1–3. San Jose, CA: IEEE, 2006.

[39] S. Thavalengal, P. Bigioi, and P. Corcoran. Iris authentication in handheld devices-considerations for constraint-free acquisition. *IEEE Transactions on Consumer Electronics*, 61(2):245–253, 2015.

[40] M. Trokielewicz. Iris recognition with a database of iris images obtained in visible light using smartphone camera. In *2016 IEEE International Conference on Identity, Security and Behavior Analysis (ISBA)*, pages 1–6. Sendai: IEEE, 2016.

[41] D. Yambay, J. S. Doyle, K. W. Bowyer, A. Czajka, and S. Schuckers. Livdet-iris 2013-iris liveness detection competition 2013. In *2014 IEEE International Joint Conference on Biometrics (IJCB)*, pages 1–8. Clearwater, FL: IEEE, 2014.

Chapter 14

Countermeasures to face photo spoofing attacks by exploiting structure and texture information from rotated face sequences

Zhen Lei[1], Wang Tao[1], Xiangyu Zhu[1], Tianyu Fu[2], and Stan Z. Li[1]

14.1 Introduction

With the fast development in last decades, face recognition has achieved great success and been widely used in our daily life, such as access control, visual surveillance, remote authentication, and so on [1]. However, most traditional face recognition systems are vulnerable to direct sensory attacks. For example, a simple printed photo or a photo demonstrated on a screen can easily fool the system and an invalid user could gain the access control which may result in severe security vulnerabilities. Face anti-spoofing, which aims to determine the face biometric captured from a genuine person or a fake replica, is becoming a critical technique for face recognition system and has attracted more and more attention in both academic and industry field.

Photo, video, and 3D mask are three common face spoofing attacks in many scenarios (Figure 14.1). In photo-related attacks, a photo from a genuine person is usually printed on a paper or demonstrated on a screen, and then presented in front of the system. In video-related attacks, the movement of face area is recorded and replayed in front of the face recognition system. In 3D mask-related attacks, a 3D mask that mimics a genuine person is produced, which is then presented and try to fool the system. There are, of course, also many other spoofing attacks that we have not imagined.

With the development of face spoofing attacks, the difference between the genuine subject and fake replica is smaller and smaller, making face anti-spoofing a challenging problem. As we know, different types of spoofing attacks mimic different characteristic of genuine faces. For example, photo-related attacks pursue as high resolution as possible face images so that sufficient face texture information is preserved. Video-based spoofing attacks try to mimic the movement of face region and 3D masks counterfeit the texture and shape structure of faces. Because of the complexity of face spoofing, it is difficult and even impossible to use a single generic

[1]Center for Biometrics and Security Research & National Laboratory of Pattern Recognition, Institute of Automation, Chinese Academy of Sciences, China
[2]Authenmetric Co., Ltd., China

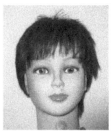

printed photo   video replay   3D model

Figure 14.1   Examples of three face spoofing methods

face anti-spoofing approach to detect all the spoofing successfully, let alone some unseen types. One feasible way is to adopt different strategies to address different spoofing attacks, respectively, and these strategies can then be combined to solve the face anti-spoofing problem [2].

This work focuses on the most common and cheapest face spoofing methods, i.e., photo attacks (including the printed photo on a paper or a photo demonstrated on an electronic screen). Many previous works [3–6] propose to classify genuine and fake samples based on frontal face images and achieve good performance on several face spoofing databases. However, in real applications, the imposter will try his best to fool the system and the texture difference between the genuine and fake samples is usually very small. In order to achieve robust face anti-spoofing performance, other cues like 3D face structure and motion pattern can be incorporated. In this work, we propose to detect spoofing photo attacks based on a sequence of rotated face images. Both the structure and texture information from the rotated face sequence are exploited. In practice, the users are only asked to take simple movement (i.e., rotate their faces). As pointed in [7], this head rotation requirement is much simpler than traditional challenge-response-based face anti-spoofing method, in which a combination of multiple movements is usually necessary. The proposed anti-spoofing method is applicable to face recognition applications such as face access control and remote authentication on mobile devices. The simple head rotation requirement is acceptable in these applications. The contributions of this work can be summarized as follows.

- A sparse 3D structure recovery based on detected landmarks from rotated face sequences is proposed. Based on the significant structure difference between the genuine and spoofing samples, the recovered structure is a useful cue to differentiate genuine and photo replica.
- Unlike the previous texture-based work that is based on frontal face images, we find that the texture-based method with rotated face sequences can also achieve high accuracy and good generalization across sensors, which is very promising in real application.
- By combining the structure and texture information, we propose an accurate and reliable face anti-spoofing method based on rotated face sequences. The good

performance on both intra- and cross-sensor indicates that it is a good solution to face anti-spoofing problem.

• A novel face anti-spoofing database, captured by three different devices, is collected. The face database consists of rotated face sequences and is a good complementary to the existing ones.

This chapter is an extension of our previous conference paper [8]. We have collected a novel face anti-spoofing database with more subjects. The subjects are asked to rotate their faces and movements are required. Therefore, the database contains variations, including face expression, pose, rotation speed, etc. The data is collected under different environments without specific requirement. Besides the structure clue, we also investigate the performance of texture-based method and its combination with structure information on this database. More comparison experiments are conducted to analyze the performance of the proposed method comprehensively.

The rest of the chapter is organized as follows. In Section 14.2, we review related anti-spoofing works. Section 14.3 briefly introduces the pipeline of structure- and texture-based face anti-spoofing method. The sparse 3D structure recovery and the face anti-spoofing classification are detailed in Sections 14.4 and 14.5, respectively. In Section 14.6, extensive experiments and discussions are conducted with different methods and databases, and in Section 14.7, we conclude this work.

14.2 Related works

In general, face anti-spoofing methods can be roughly divided into two categories, interactive and noninteractive. The interactive methods mainly refer to challenge-response-based ones, in which users are asked to respond to specific actions given by the system so that the genuine and fake face can be classified.

Most of the published face anti-spoofing methods fall into the latter category. According to the cues used, these methods can be further categorized into texture-, multispectral-, motion-, and 3D structure-based methods.

Texture-based method is based on difference of texture and contour between the genuine and fake faces. Määttä *et al.* [9] utilize multiscale LBP to model the micro-texture difference between the genuine and fake faces and achieve perfect performance on the first face anti-spoofing competition. Pereira *et al.* [10] propose LBP-TOP to model the appearance difference of face sequences of genuine and fake samples. Yang *et al.* [3] propose component-dependent descriptors to extract micro-texture differences of genuine and fake samples on different face components, respectively. Schwartz *et al.* [11] combine a set of low-level descriptors such as HOG, color frequency (CF), co-occurrence matrix, and histograms of shearlet coefficients to detect spoofing faces. Kim *et al.* [12] combine the texture and frequency information to differentiate the genuine and 2D fake faces, mainly based on the assumption that the illumination component from genuine (3D structure) and fake face (2D photo) is different. Recently, Galbally *et al.* [4] adopt 25 general image quality features to distinguish the genuine and spoofing face images, which demonstrate good generalization on different scenarios. Wen *et al.* [5] propose image distortion analysis (IDA) method to detect spoofing faces. Four different features including specular reflection, blurriness, chromatic moment, and color diversity are extracted. Patel *et al.* [6] propose to use moiré patterns to differentiate live and spoofing faces. The moiré patterns are represented by multiscale Local Binary Patterns (LBP) and Scale-Invariant Feature Transform (SIFT) features. Although the texture-based methods can achieve good performance on a specific scenario (database), its performance is sensitive to environmental variation (especially lighting change), which limits its application in practice. Moreover, with the development of capture devices, more and more high-resolution face photos are available, which increases the difficulty of genuine/fake classification by texture.

Multispectral methods analyze the reflectance properties of human skin under different spectrums to distinguish genuine and fake faces. Pavlidis and Symosek illustrate how to capture face images at two wavelengths in near infrared (NIR), and use a simple thresholding method to classify genuine and fake faces. Zhang *et al.* [14] observe that the energy of genuine and spoofing faces under various spectrums is different, which is adopted as the cue of face liveness detection. Yi *et al.* [15] train two spoofing classifiers for visual and NIR cameras, which are then combined to resist the visual and NIR face photo attacks. Multispectral methods usually achieve better performance than texture-based ones. However, these methods need additional multispectral capture devices, which increases the hardware cost and is not always available in real applications.

Face motion is one of the most significant live signals. Various face anti-spoofing methods based on face motion, such as eye blinking [16] and mouth movement [17–19], have been proposed. Kollreider *et al.* [20] propose structure tensors to extract the motion clues and classify the genuine and fake face given the assumption that a 3D face from a genuine subject and a 2D face from a photo have different motion patterns. André Anjos *et al.* [21] analyze the motion correlation between the face region and background region using optical flow to classify the genuine and fake samples. Yan *et al.* [22] use low rank matrix decomposition approach to extract the nonrigid motion to verify the genuine faces. In ICB 2013, 2D face spoofing competition [23], both of the two winner algorithms incorporated motion information. However, the performance of existing motion-based anti-spoofing methods may not be as good when a segmented photo without background or a warped face photo is used to attack the system.

3D structure-based methods make use of 3D shape information to classify the genuine and fake faces. 3D information can be recovered from a face video or be captured by a depth sensor directly. Obviously, a planar photo gives a flat structure whereas a genuine face yields a quite different structure (e.g., nose is convex compared to cheek). Choudhury *et al.* [24] mention that the depth information can be used for differentiating planar faces and real ones. Maria *et al.* [7] proposes to use 3D projective invariant (cross ratios) in face rotation to distinguish genuine and photo spoofing samples. Limited experiments show that their method can well detect the photo attacks with shift rotation, bending, and zooming. Its effectiveness on face rotation sequence with warped photo is unknown. 3D information can be effectively utilized to distinguish a genuine person and a flat object; however, it is still vulnerable to 3D mask attacks [25,26].

As pointed in [27], no single anti-spoofing technique could detect all spoofing attacks reliably. Researchers have proposed to combine different live cues to enhance the performance of face anti-spoofing. Pan *et al.* [28] combine eye-blinks and scene context information to detect face liveness. In [22], authors combine the facial motion and image quality information to distinguish genuine and fake faces. Komulainen *et al.* [2] fuse motion- and texture-based clues to address face anti-spoofing problem. Reference [29] incorporate the anti-spoofing and face verification tasks to improve the trustworthiness of the system.

There are pros and cons for both interactive and noninteractive methods. Interactive-based methods usually can achieve better and more reliable face anti-spoofing performance than noninteractive ones. However, it is inconvenient to users because the user is asked to take several specific actions which degrades the user experience. Moreover, the commonly used actions such as eye-blinking and mouth movement expose the cues used for face anti-spoofing and are easy to be attacked (e.g., by using photo with eye or mouth cut). For noninteractive methods, although requiring no cooperation from users, many of them, which are texture and motion based, cannot extract sufficient information to classify the genuine person and spoofing face accurately.

In this chapter, we propose a face anti-spoofing method based on a sequence of rotated face images. Face rotation is one of the simplest face movements so that user's inconvenience in face capture process is minimized. Users are asked to rotate their faces without other specific requirements. Various face liveness information, including the face texture variation, face 3D structure, and face motion patterns, can be extracted from the rotated face sequence and it increases the difficulty for intruders to attack. We exploit both structure and texture information from the rotated face images to improve the discriminative and the robustness in different scenarios. Experiments in both intra- and cross-sensor scenarios show that the proposed method has high spoofing detection rate and good generalization against photo attacks.

14.3   Overview of the proposed method

We propose to classify genuine and spoofing photos from a sequence of rotated face images. The structure recovered from the genuine faces usually contain richer 3D structure information than that from, which is usually planar in depth. Moreover, the texture differences between the genuine and spoofing photos in rotated face sequence are also enlarged compared to those from a single frontal image. Therefore, we can exploit the structure and texture information from a rotated face sequence to realize face anti-spoofing accurately.

Figure 14.2 illustrates the pipeline of the proposed method. It mainly consists of structure and texture information extraction modules. For structure extraction, we first detect the face regions and locate the landmarks of faces. A sparse 3D shape recovery algorithm is then developed to recover the 3D sparse structure based on the detected facial landmarks. The genuine subject is supposed to be of rich stereo

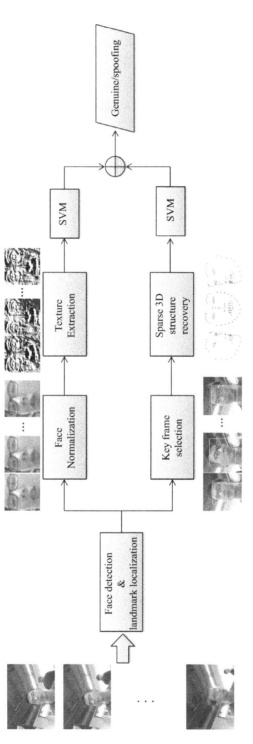

Figure 14.2   Pipeline of the proposed face anti-spoofing method based on a sequence of rotated face images

structure while the fake photo is not, which can be used as a discriminative clue to classify genuine and fake samples. For texture extraction, we apply LBP feature proposed in [9] to describe the appearance variations of rotated face images. Finally, the structure and texture information is combined to classify the genuine and spoofing samples.

14.4   Sparse 3D facial structure recovery

The rotated face image sequence can be captured from a fixed camera with face rotating or vice versa. Without loss of generality, in this work, we assume that the sequence is captured from a fixed camera with face rotating. To recover facial structure, we first use the recent developed landmark detection algorithm [30] to locate the sparse facial landmarks. As shown in Figure 14.3, there are 68 landmarks located at eyes, nose, lip, check, etc.

Next, considering that neighboring frames have almost the same viewpoint, which is least useful for structure recovery, we estimate pose for each frame and select proper frames (called key frames in the following) with specific pose variations for structure recovery.

Once we have obtained two key frames, the camera parameters and initial facial structure are recovered. Notably, general geometry reconstruction algorithm is not applicable to a deformable object, such as a face. However, face deformations are often caused by local expressions, which occurs most likely in the regions of mouth. Among all the landmarks, only a bit of them are located near the mouth. Therefore, we recover facial structures disregarding the deformation, which is proved to be feasible to face liveness problem in our experiments.

Merely two face images are not enough to derive reliable facial structure recovery because of possible inaccurate landmark detection and rough estimation of camera parameters. We further add more key frames into the bundle adjustment to improve the recovered 3D facial structure.

Figure 14.3   68 landmarks used in sparse 3D facial structure recovery

14.4.1   Initial recovery from two images

In this part, we describe our algorithm for initial facial structure recovery based on facial landmarks from two face images with different viewpoints. Let $q = \{q_1, q_2, \ldots, q_N\}$ and $q' = \{q'_1, q'_2, \ldots, q'_N\}$ be landmarks extracted from two face images, respectively, where q_i denotes a 2D point (coordinate) in the image and N is the number of facial landmarks for a face. In our work, we adopt the perspective camera model [31], in which a 2D point q_i projected from a 3D structure point Q_i can be represented as: $q_i = PQ_i$, where P is a 3×4 perspective projection matrix. Note that both q_i and Q_i are in homogeneous coordinates. As we know, camera projection matrix P can be further decomposed as: $P = K[R, t]$, where K is the camera intrinsic matrix and $[R, t]$ is the relative pose between the two given images. The next step is to estimate these parameters so that the structure recovery can be completed by triangulation method [32].

14.4.1.1   Intrinsic parameter estimation

To make the proposed method easily applicable to different cameras, we propose an auto-calibration method for our face anti-spoofing application. We make the common assumption that the intrinsic matrix K is constant over the whole video sequence. In real application, we can first estimate the parameters from a video sequence as an initialization step and fix them in future. Generally, the intrinsic matrix is formulated as a 3×3 upper triangular matrix [31]:

$$
K = \begin{pmatrix} f_x & s & u_x \\ 0 & f_y & u_y \\ 0 & 0 & 1 \end{pmatrix}, \tag{14.1}
$$

where f_x and f_y are the focal length in terms of pixel dimensions in the direction of x-axis and y-axis, respectively. u_x and u_y are the projection of optical center. s is referred to the skew parameter. Suppose that the camera sensor pixels are in the shape of a square, and the projection center is coincide with the image center, we have $f_x = f_y = f$, and $u_x = u_y = 0$ if the origin of the image coordinate is set at the center of image. The skew parameter s is equal to zero for most normal cameras. As a result, the intrinsic matrix can be simplified with only one parameter f, that is, $K = diag(f, f, 1)$.

In this work, we propose to estimate K by fitting the detected landmarks q from the input face images to a predefined normal (upright and frontal) 3D facial structure $Q^{model} = \{Q_1^{model}, Q_2^{model}, \ldots, Q_N^{model}\}$. The normal (upright and frontal) 3D facial structure Q^{model} can be easily obtained from a general 3D face database and is scaled to the physical size of a face in our work (i.e., about 10 cm between the center of two eyes). The intrinsic parameter K thus can be roughly estimated as

$$
K^* = \underset{K}{\arg\min} \sum_{i=1}^{N} d\left(q_i, K[R_m, t_m]_{3 \times 4} Q_i^{model}\right), \tag{14.2}
$$

where $[R_m, t_m]_{3 \times 4}$ represents the rigid transformation (rotation matrix $R_{3 \times 3}$ and translation $t_{3 \times 1}$) on Q^{model} and $d(\cdot)$ denotes the Euclidean distance between two points represented in homogeneous coordinates. Because simultaneously estimating $\{K, R_m, t_m\}$ is intractable, in this part, we alternatively to traverse all possible values of (R_m, t_m) to find the best matching. Let, ω_x, ω_y, ω_z be the rotation angles of Q^{model}, whose 2D projection is similar to q, in three axes, respectively. First, we rotate the landmarks around z-axis (perpendicular to image plane) to make the pair of eyes horizontal. This rotation can only get an upright face projection ($\omega_z = 0$), whereas ω_x and ω_y may be nonzero for deflecting viewpoint. Therefore, we uniformly discretize the rotations around x-axis and y-axis to n_x and n_y states, respectively. Then, based on least square error (LSE) algorithm, we find the optimal K with minimum residual error by traversing all the $n_x \times n_y$ possible ω_x and ω_y (in this chapter, $n_x = n_y = 20$). It is worth noting that the above procedure only gives a rough K, which will be further refined in the following step. Clearly, if the camera has been calibrated accurately, we can skip above procedure. However, knowing exact intrinsic matrix is not feasible in many practical cases.

14.4.1.2 Facial structure reconstruction

As stated in [31], the relative pose $[R, t]$ can then be obtained by estimating the essential matrix ε. The essential matrix ε has the form of $\varepsilon = [t_\times]R$, where $[t_\times]$ is the skew-symmetric cross-product matrix of $t = (t_x, t_y, t_z)^T$:

$$[t_\times] = \begin{pmatrix} 0 & -t_z & t_y \\ t_z & 0 & -t_x \\ -t_y & t_x & 0 \end{pmatrix}. \tag{14.3}$$

Once the essential matrix is estimated, camera relative pose can be extracted straightforwardly. Given the landmark correspondences $\{(q_1, q_1'), (q_2, q_2'), \ldots, (q_N, q_N')\}$, the essential matrix ε can be estimated using the epipolar constraint $q_i^T \varepsilon q_i' = 0$. The general linear algorithm [31] needs eight or more point correspondences and minimize epipolar distance over all points. In our case, considering that the facial landmarks may be detected with noises, we adopt a random sample consensus scheme (RANSAC) [33] to estimate the essential matrix ε robustly. For each RANSAC loop, a candidate essential matrix is obtained using an efficient nonlinear minimal algorithm [34] which needs at least five correspondences. Upon the completion of RANSAC, we choose optimal ε with minimum average residual error.

After getting camera intrinsic matrix K and relative pose $[R, t]$, the projection matrixes of two images are obtained by the following equation:

$$\begin{aligned} P &= K[I, 0], \\ P' &= K[R, t], \end{aligned} \tag{14.4}$$

where I is a 3×3 identity matrix. Given P and P', we implement a triangulation algorithm [32]:

$$\min \sum_{i=1}^{N} [d(q_i, \hat{q}_i) + d(q_i', \hat{q}_i')], \tag{14.5}$$

where $\hat{q}_i$ and $\hat{q}_i'$ are the reprojections of the recovered structure $\hat{Q} = \{\hat{Q}_1, \hat{Q}_2, \ldots, \hat{Q}_N\}$ on two given images.

14.4.2   *Facial structure refinement*

As aforementioned, the facial structure can be recovered based on two images from different viewpoints. However, such recovered results may not be accurate due to inaccurate detection of landmarks or rough estimation of K. In this subsection, we describe the facial structure refinement step to obtain a more accurate facial structure, where new key frames are added into the refinement one by one. Given the projection matrixes $\mathbf{P} = \{P_1, P_2, \ldots, P_M, P_{new}\}$ of refined M key frames and a new key frame, our goal of refinement is to minimize the reprojection error between predicted points and detected landmarks over all key frames:

$$\min \sum_{j=1}^{M+1} \sum_{i=1}^{N} d\left(\hat{q}_{i,j}, q_{i,j}\right), \tag{14.6}$$

where $\hat{q}_{i,j} = P_j \hat{Q}_i$ is the ith predicted point on image j and $q_{i,j}$ is the ith landmark on image j.

First, the pose (external parameters) $[R_{new}, t_{new}]$ of the new key frame is estimated using Grunert's algorithm [35], rather than the method described in the subsection above. According to the estimated $[R, t]$, the distance error between predicted points $\hat{q}$ and landmarks q can be computed. For those frames which have large error, we regard them as useless key frames and skip to refine the next key frame. Next, we group the parameters which will be refined together: $\Theta = [K, T, \hat{Q}]$, where $T = \{[R_1, t_1], [R_2, t_2], \ldots, [R_M, t_M], [R_{new}, t_{new}]\}$ is the external parameters for every frame. Subsequently, we use a sparse bundle adjustment algorithm [36] to optimize (14.6) with the input of Θ, which solves by a fast Levenberg–Marquardt (LM) optimization algorithm.

In implementation, we notice that when the viewpoints of two images are too similar, the triangulated 3D points will be far from the real structure, especially in the depth (z-axis). To address this problem, we add an extra soft constraint on the triangulation. Equation (14.6) can be rewritten as

$$\min \sum_{j=1}^{M+1} \sum_{i=1}^{N} d\left(\hat{q}_{i,j}, q_{i,j}\right) + \omega\left(\theta\right) \sum_{i=1}^{N} d\left(\hat{Q}_i, Q_i^{model}\right), \tag{14.7}$$

where ω is the weight function to the soft constraint, which is defined as $\omega(\theta) = \exp(-15\theta/\pi)$ in this work. θ is the maximum angle difference of $M + 1$ faces. This constraint can be regarded as a prior on the facial structure. The recovered structure, which is different from a natural face much, is less likely to be a correct estimation.

With the increasing number of key frames, we can recover a more and more accurate facial structure. Considering the tradeoff between the accuracy and the computational cost, we at most select 10 key frames for facial structure recovery as stated in Section 14.4.3. Figure 14.4 shows the recovered examples of genuine faces and spoofing faces with this method, from which one can see that the genuine and spoofing photos can be classified using proper structure information.

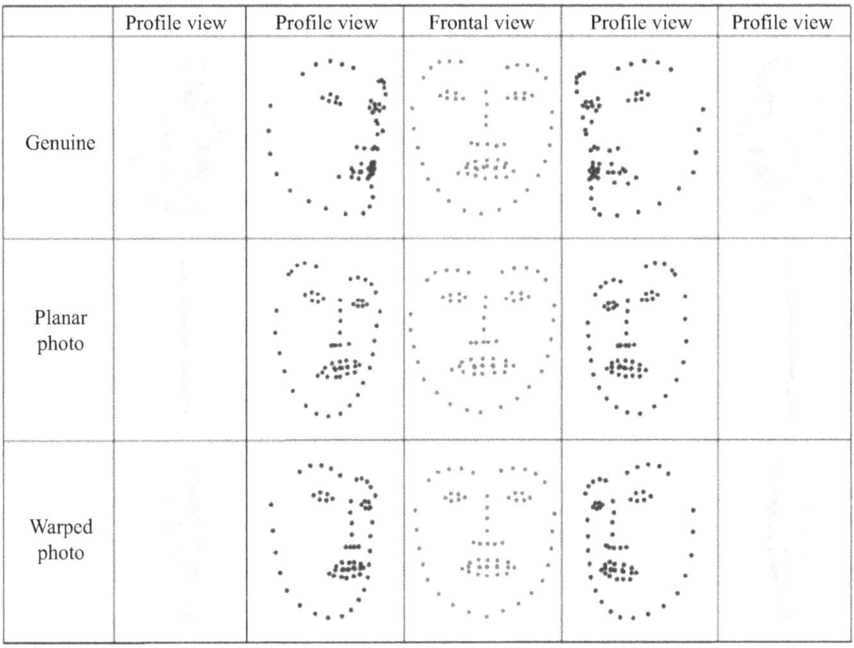

	Profile view	Profile view	Frontal view	Profile view	Profile view
Genuine					
Planar photo					
Warped photo					

Figure 14.4   Examples of recovered facial structure from genuine and spoofing photos. Reprinted with permission from Tao Wang, Jianwei Yang, Zhen Lei, Shengcai Liao, Stan Z. Li. "Face Liveness Detection Using 3D Structure Recovered from a Single Camera." In Proceedings of the Sixth IAPR International Conference on Biometrics

14.4.3   Key frame selection

As we know, images with similar poses provide limited information for 3D structure recovery. In this work, we first estimate the pose of each input frame and select the images with large pose variations to finish the recovery task. We roughly estimate angles of a face by aligning the landmark positions in image plane of input face to an average 3D face model. Let's denote the 2D position (X, Y) of landmarks in image plane as $X = [x_1, x_2, \ldots, x_N]$ and $Y = [y_1, y_2, \ldots, y_N]$, where N is the number of landmarks. The 3D coordinates of corresponding landmarks $(\hat{X}, \hat{Y}, \hat{Z})$ on an average 3D model is denoted as $\hat{X} = [\hat{x}_1, \hat{x}_2, \ldots, \hat{x}_N]$, $\hat{Y} = [\hat{y}_1, \hat{y}_2, \ldots, \hat{y}_N]$, and $\hat{Z} = [\hat{z}_1, \hat{z}_2, \ldots, \hat{z}_N]$. Following weak perspective projection, there is formulation between $\{X, Y, Z\}$ and $\{X, Y\}$ as follows:

$$[X^T, Y^T]^T = s[\mathrm{R}, t][\hat{X}^T, \hat{Y}^T, \hat{Z}^T, 1^T]^T, \tag{14.8}$$

where s, R, and t are scale parameter, rotation, and translation matrix, respectively. The least square solutions to (14.8) can be obtained by using the method in [37]. After getting the rotation matrix R, one can compute the three angles (pitch, yaw, and roll) consequently [38].

For simplicity and no loss for generality, we only use yaw angle to select face images of different poses. Particularly, we divided the angle ($[-\pi/2, \pi/2]$) into 10 nonoverlapping sets and for each interval, at most one image is selected. Therefore, we select at most 10 key frames with different poses for sparse face shape recovery.

14.5 Face anti-spoofing classification

14.5.1 Structure-based anti-spoofing classifier

After the sparse 3D structure recovery, we first align these structures and then the feature vector is formed by concatenating the 3D coordinates of the structure for classification. In this work, we align all the structure of samples $Q = \{Q_i\}$ to a reference (average) 3D model $Q^{model} = \{Q_i^{model}\}$. The goal is to estimate the rigid transformation (scale s, rotation R, and translation t) to minimize the sum of distances between these two sets of 3D points. This can be realized by minimizing the following objects using least squares method:

$$\{s_0, \mathrm{R}_0, t_0\} = arg \min_{\{s,\mathrm{R},t\}} \sum_{i=1}^{N} d(s[\mathrm{R}, t]Q_i, Q_i^{model}). \tag{14.9}$$

The above coarse alignment assigns each 3D point an equal weight in the minimization procedure. In real application, due to the inaccurate face landmark detection, there may exist some outliers in the recovered facial structure. To improve the robustness of the alignment, we propose a coarse to fine strategy in this part. With the coarse initialization of the least squares solution, we further refine the solution $\{s, \mathrm{R}, t\}$ iteratively. In each iteration, we compute the distances of each point pair $(s[\mathrm{R}, t]Q_i, Q_i^{model})$ with the current estimation of $\{s, \mathrm{R}, t\}$. After that, we select five-point pairs with the smallest distances, which are regarded as the most reliable correspondences among all the pairs. These five pairs are utilized to estimate new $\{s, \mathrm{R}, t\}$. The iterative process proceeds until the five-point pairs with smallest distances do not change or the overall iteration times exceed a predetermined threshold ($T = 10$).

In classification, we concatenate the 3D coordinates of recovered points (68 points in this work) as the structure feature vector and train a linear SVM classifier to distinguish genuine and spoofing photos.

14.5.2 Texture-based anti-spoofing classifier

Most texture-based face anti-spoofing methods are based on frontal face images. In this work, we apply the similar method to rotated face sequences. Multiscale LBP descriptor is implemented to exploit the texture cue (Figure 14.5). Given a face video sample, we detect the face region in a frame-wise way. Each detected face region is then cropped to 64×64 in size according to automatically detected eye positions. As proposed in [9], for each cropped face image, we apply $LBP_{8,1}^{u2}$, $LBP_{8,2}^{u2}$, and $LBP_{16,2}^{u2}$ to extract multiscale features. In histogram feature extraction, for $LBP_{8,1}^{u2}$,

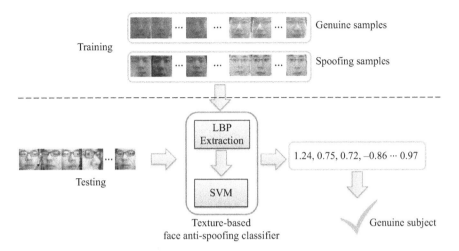

Figure 14.5 Texture-based face anti-spoofing classifier

the face image is divided into 3×3 overlapping blocks with an overlap of 14 pixels. Therefore, the feature dimension for $LBP_{8,1}^{u2}$ is $59 \times 9 = 531$. For $LBP_{8,2}^{u2}$ and $LBP_{16,2}^{u2}$, the histogram feature is computed from the whole face image, with the dimension of 59 and 243, respectively. These three histogram-based features are finally combined to form a 833-dimensional feature to represent each face image. In training phase, we use the collected genuine and spoofing photo samples to train an SVM classifier. In testing phase, given a rotated face sequence, each frame is fed into the texture extraction and classification engine to output the confidence of genuine/spoofing image independently. The confident scores are finally fused using sum rule to predict the label of the input rotated face sequence.

14.6 Experiments

Recently, there are several publicly available face databases (e.g., CASIA database [39], NUAA database [40], and Idiap database [41]) for face spoofing detection. However, it is necessary to collect multiple face images with different poses for the proposed method. Therefore, most of the available face anti-spoofing databases, which contain only frontal face images, cannot meet the requirements of the proposed method. In this work, we collect a new face anti-spoofing database including various face poses to examine the performance of proposed method.

14.6.1 Database description

Digital camera (DC), mobile phone (MP), and web camera (WC) are three popular devices to capture face images in many real applications. In this part, we collect genuine faces and photo attacks using these three devices, with the resolution of 1920×1080, 1280×720, and 640×480, respectively. For genuine faces, the subjects are asked to rotate their faces naturally without specific speed requirement. Figure 14.6(a) shows an example of rotated face sequence. For photo attacks, two attack types, the planar photo and the warped photo, are collected (Figure 14.6(b)). We also try to rotate the photos to simulate the genuine rotated face sequence. We record video clip with about 20 s for each subject. For genuine faces, we collect 56, 68, and 33 video clips (one video clip for each subject) using DC, MP, and WC, respectively. For both two types of spoofing samples, we collect 65 video clips from 65 subjects using three devices, respectively. There are lighting, pose, background, and motion variations in these videos. In the following experiments, we split each video into five clips, each of which is of about 4 s. In this way, there are in total 280, 340, and 165 video clips for genuine faces and 325 video clips for planar/warped photos collected by DC, MP, and WC, respectively. Table 14.1 illustrates the sample

(a) Rotated face sequence from a genuine subject.

(b) Planar photo attack (the first row) and the warped photos (the second row).

Figure 14.6   *Example of genuine rotated face sequence and two types of spoofing photo attacks*

Table 14.1   *The number of video samples with three devices*

Database	Genuine	Spoofing photos	
		Planar	Warped
DC	280	325	325
MP	340	325	325
WC	165	325	325

distribution of the database. In following experiments, we examine the face anti-spoofing performance by using planar photo, warped photo, and both planar and warped photo attacks, which are denoted as *genuine vs. planar (G vs. P)*, *genuine vs. warped (G vs. W)*, and *genuine vs. spoofing (G vs. S)*, respectively.

14.6.2 Evaluation protocols

The generalization of face anti-spoofing method is one of the most important problems in real applications. Most previous work report face anti-spoofing performance based on the database collected by the same sensor. In this work, we adopt two protocols, namely intra-sensor and cross-sensor to examine the performance of face anti-spoofing extensively. In intra-sensor evaluation, a fivefold cross-validation is adopted. For each database (DC, MP, and WC), the video samples are randomly splitted into five subsets in terms of subjects. That is, there is no intersection of subjects among the five subsets. In each iteration, four of five subsets are used for training and the other is used for testing. The mean classification accuracy of five results is reported. In cross-sensor evaluation, we train the classification model on the database collected by one of three devices and test it on other two databases collected by different devices. This scenario is more close to real applications. The classification accuracy and ROC performance (true positive rate (TPR), false accept rate (FAR), and equal error rate (EER)) are reported. In the following, when report the classification accuracy, unless with explicit explanation, the threshold of SVM classifier is set to 0, which is the default value determined on training set.

14.6.3 Results of structure-based method

14.6.3.1 Intra-sensor evaluation

In this part, we conduct experiments on DC, MP, and WC databases following the intra-sensor protocol. As aforementioned, the proposed structure recovery relies on the facial landmark detection (face alignment). Face alignment algorithm usually achieves higher accuracy on landmarks in the inner part of face region (e.g., landmarks around the eye, nose, and mouth areas), which is called inner structure. Due to vague definition and large appearance variation of landmarks on contour, their location accuracy is usually worse than those in inner part. In this part, we report the results based on the structure recovered from the inner points (51 points) and overall points (68 points) to examine the effect of contour landmarks on face anti-spoofing performance.

Table 14.2 illustrates the performance of inner and overall structure on DC, MP, and WC databases following intra-sensor protocol. From the results, the inner structure achieves slightly better results than overall structure in the case of G vs. P photo, whereas in the case of G vs. W photo and G vs. W spoofing photo, the results of overall structure are better than those of inner structure. The results indicate that although contour landmarks detection may be less accurate than inner points, the utilization of contour points (shape information) is still helpful to improve face spoofing detection accuracy. Therefore, in the following experiments, we use all the landmarks (68 points) to exploit the structure information for photo spoof detection. The results presented in Table 14.2 validate that structure information is useful to distinguish

Table 14.2   Spoofing classification accuracy ((mean ± SD)%) following intra-sensor protocol using structure information

Database	Method	G vs. P	G vs. W	G vs. S
DC	Inner Structure	97.51 ± 2.83	94.91 ± 3.88	95.62 ± 3.31
	Overall structure	97.67 ± 2.08	95.55 ± 2.65	95.48 ± 4.22
MP	Inner structure	96.43 ± 2.46	93.08 ± 3.33	93.57 ± 2.27
	Overall structure	96.14 ± 2.40	94.55 ± 2.96	94.29 ± 1.58
WC	Inner structure	98.16 ± 0.48	94.29 ± 1.30	94.42 ± 5.00
	Overall structure	97.98 ± 0.97	94.51 ± 0.73	95.83 ± 3.87

Table 14.3   Spoofing classification accuracy (%) following cross-sensor protocol using structure information

Database	G vs. P	G vs. W	G vs. S
DC→MP	94.89	88.42	91.21
DC→WC	96.94	91.84	94.60
MP→DC	98.18	96.36	96.45
MP→WC	98.37	93.67	95.46
WC→DC	96.86	92.23	92.47
WC→MP	96.24	87.07	90.61

genuine and spoofing photos. It achieves higher than 96% in case of G vs. P photo on both three databases, indicating that structure information is effective to detect planar spoofing photo. Regarding warped photo attacks, because of richer stereo structure information contained in warped photos, which is easily confused with genuine subject, its spoofing classification performance is slightly worse than that of G vs. P photo. For the G vs. S photo, where the planar and warped photos are merged, the overall structure-based method achieves higher than 94% accuracy in all databases.

14.6.3.2   Cross-sensor evaluation

Tables 14.3 and 14.4 illustrate the performance of structure-based face anti-spoofing following cross-sensor protocol and Figure 14.7 shows the corresponding ROC curves of G vs. S photo. The FAR and TPR corresponding to zero threshold is illustrated in Table 14.5. Notation A→B means that the classifier is trained on database A and tested on database B. Comparing the results of cross-sensor and intra-sensor, one can see that:

- For the G vs. P photo, the face spoofing detection performance of cross-sensor is close to that of intra-sensor, indicating the structure-based face anti-spoofing method has good generalization performance in this case.

Table 14.4   ROC performance (%) following cross-sensor protocol using structure
information

Database	G vs. P			G vs. W			G vs. S		
	TPR@ FAR = 0.01	TPR@ FAR = 0.1	EER	TPR@ FAR = 0.01	TPR@ FAR = 0.1	EER	TPR@ FAR = 0.01	TPR@ FAR = 0.1	EER
DC→MP	97.35	97.35	1.32	79.12	93.24	6.09	80.29	97.35	6.09
DC→WC	100.00	100.00	0.31	76.36	92.12	7.78	76.97	90.91	7.15
MP→DC	98.57	98.57	0.71	92.86	98.57	3.73	92.14	97.50	3.86
MP→WC	99.39	99.39	0.61	57.58	92.73	7.00	78.79	93.94	6.88
WC→DC	99.64	99.64	0.18	80.36	98.57	3.32	94.64	97.86	2.61
WC→MP	97.35	97.35	1.48	31.18	95.88	6.17	74.71	97.06	4.98

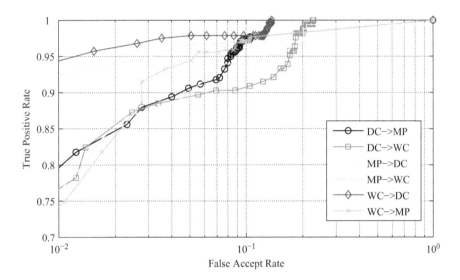

Figure 14.7   ROC curves of G vs. S photo across different sensors using structure
information

Table 14.5   FAR and TPR (%) of genuine vs. spoofing
corresponding to zero threshold
following cross-sensor protocol using
structure information

Database	FAR	TPR
DC→MP	0.00	86.76
DC→WC	0.00	60.61
MP→DC	0.15	98.57
MP→WC	0.15	96.97
WC→DC	0.00	91.43
WC→MP	0.15	83.53

- In contrast, the performance of face spoofing detection on G vs. W photo of cross-sensor declines more significantly compared to the results of intra-sensor. On one hand, landmark detection on warped photo is more difficult because of the severe image distortion, which affects the facial structure recovery result. On the other hand, richer stereo structure from warped photo reduces the discriminative gap between the genuine and warped photo, which also decreases face spoofing detection performance.
- Overall speaking, regarding the G vs. S photo, the spoofing detection accuracy across sensor exceeds 90% and all the EERs in ROC curves are below 8%, which is very promising and also indicates that there is space to improve spoofing detection performance by integrating other cues.

14.6.3.3   False classified sample analysis

Most false classified samples are caused by the inaccurate landmark location. Figure 14.8 shows some inaccurate face alignment results from genuine and warped photos. For genuine subject, the facial landmark engine may fail due to large pose variations, so that the facial structure recovery could fail. For the attack photos, because of

(a) Genuine subject

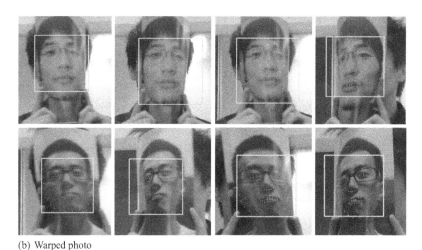

(b) Warped photo

Figure 14.8   Example of inaccurate landmark detection results

distortion of face images, the facial landmarks cannot always be detected accurately. This problem is especially serious in warped photos. From Figure 14.8(b), one can see that there is strong facial structure of genuine subject in the detected landmarks because of the priori contained in the landmark detection model, thus leading to the genuine-like facial structure recovery. It unfortunately increases the difficulty of distinguishing genuine and spoofing photo, and therefore, the classification performance between the genuine and warped photo is decreased.

14.6.4 Results of texture-based method

We implement the multiscale LBP-based face anti-spoofing method [9]. For each video sample, we apply a face detection engine [42] to automatically detect the face region in a frame-wise way. The detected faces are then cropped into 64 × 64 size according to the located eye positions (Figure 14.9). In training phase, we use all the cropped face images from genuine and spoofing video samples to learn a linear SVM classifier. In testing phase, all the detected face images in each video sample are classified by the learned SVM model. The label of testing video sample is finally predicted by fusing the confident score of each frame using a sum rule. The threshold of SVM classifier is set to 0 when report the classification accuracy.

Tables 14.6 and 14.7 illustrate face spoofing detection performance of texture-based method based on rotated face sequences following intra-sensor and cross-sensor, respectively. Figure 14.10 and Table 14.8 show the corresponding ROC curves and results across different sensors. The FAR and TPR corresponding to zero threshold are listed in Table 14.9. From results, one can see that the generalization performance between high-resolution images (i.e., between DC and MP) is good and much better

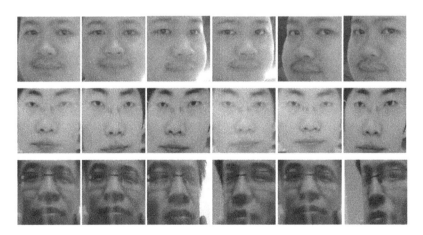

Figure 14.9 Examples of cropped face images. The first row is the genuine subject. The second and the third rows represent planar photo and warped photo, respectively

*Table 14.6   Spoofing classification accuracy ((mean ±
            SD)%) using LBP and SVM classifier following
            intra-sensor protocol*

Database	G vs. P	G vs. W	G vs. S
DC	93.90 ± 6.73	96.67 ± 2.77	98.41 ± 0.96
MP	96.44 ± 3.12	98.19 ± 1.72	96.73 ± 3.24
WC	96.98 ± 4.23	100.00 ± 0.00	97.69 ± 3.49

*Table 14.7   Spoofing classification accuracy
            (%) using LBP and SVM classifier
            following cross-sensor protocol*

Database	G vs. P	G vs. W	G vs. S
DC→MP	97.44	98.95	95.69
DC→WC	95.71	91.43	90.80
MP→DC	99.83	100.00	99.57
MP→WC	95.71	98.98	96.81
WC→DC	87.77	95.37	84.30
WC→MP	92.33	90.08	86.97

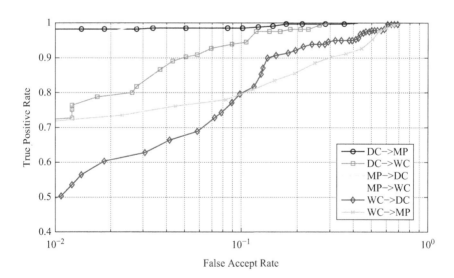

*Figure 14.10   ROC curves of G vs. S photo across different sensors using texture
            information*

Table 14.8   ROC performance (%) following cross-sensor protocol using texture information

Database	G vs. P			G vs. W			G vs. S		
	TPR@ FAR = 0.01	TPR@ FAR = 0.1	EER	TPR@ FAR = 0.01	TPR@ FAR = 0.1	EER	TPR@ FAR = 0.01	TPR@ FAR = 0.1	EER
DC→MP	97.94	99.71	1.36	99.71	100.00	0.45	98.24	98.53	1.27
DC→WC	90.91	97.58	3.96	94.55	100.00	1.08	72.12	93.94	6.49
MP→DC	100.00	100.00	0.18	100.00	100.00	0.00	99.64	100.00	0.54
MP→WC	96.36	100.00	1.38	100.00	100.00	0.46	93.94	100.00	1.69
WC→DC	80.71	91.43	8.96	91.79	95.36	8.34	47.86	79.64	11.85
WC→MP	86.76	90.29	6.93	83.82	90.88	8.34	72.65	80.00	14.00

Table 14.9   FAR and TPR (%) of G vs. S corresponding to zero threshold following cross-sensor protocol using texture information

Database	FAR	TPR
DC→MP	0.00	88.53
DC→WC	0.00	54.55
MP→DC	0.15	98.93
MP→WC	3.85	99.39
WC→DC	19.23	92.14
WC→MP	8.31	77.94

than that between high (DC, MP) and standard resolution (WC), indicating texture-based method is sensitive to image resolution. Comparing the results of texture with structure, one can see that the variation of texture performance on different databases is a little larger than structure performance, meaning that texture information is not as stable as structure. Intuitively, there are complementary cues in structure and texture information. There is space to improve the performance by combining texture and structure information further.

14.6.5   Combination of structure and texture clues

In this part, we examine the spoofing classification performance by combining structure and texture. In testing phase, after obtaining the confident scores using structure and texture SVM classifiers, respectively, we then fuse them using sum rule. The video sample is classified as a genuine sample if the fused score is larger than 0 and a spoofing one if otherwise. Figures 14.11 and 14.12 illustrate the face

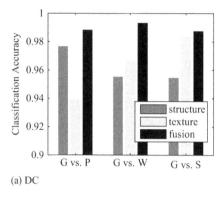

(a) DC

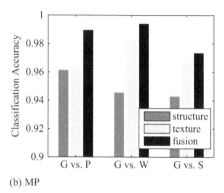

(b) MP

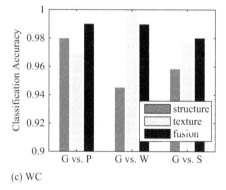

(c) WC

Figure 14.11 *Spoofing classification accuracy by combining structure and texture*
information on "DC," "MP," "WC" database following intra-sensor
protocol

spoofing detection accuracy. Figure 14.13 and Table 14.10 show the ROC performance. Table 14.11 lists the FAR and TPR corresponding to zero threshold. In
most cases, the combination of structure and texture does improve the face spoofing
detection performance compared to the single clue, though not significant. Another
observation of the fusion is that its performance is more stable in different cases. The
(*best*, *worst*) EER on cross sensors in the case of G vs. S for the fusion-based method
is (0.59%, 2.66%), while that for structure- and texture-based method are (2.61%,
6.88%) and (0.54%, 14.00%), respectively. In all cross-sensor experiments, the EER
of fusion is below 3%, which means that least 97% samples are correctly classified,
validating that the proposed structure- and texture-based face anti-spoofing based on
rotated face sequence is an effective method.

We compare the performance of proposed method with recently developed (IDA)-
based face spoof detection method [5]. We also report the performance of MsLBP [9]
as a baseline. Table 14.12 illustrates the results across different sensors. It is shown that

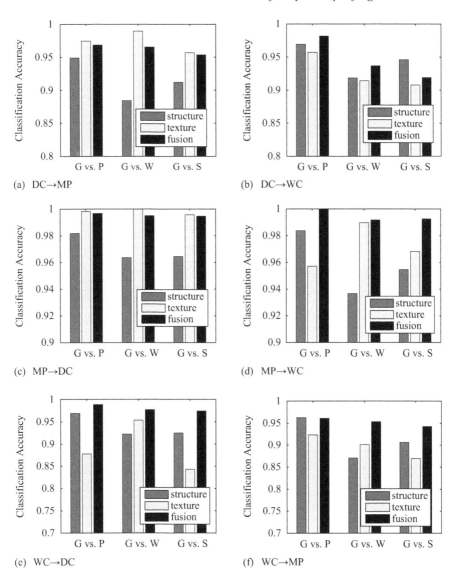

Figure 14.12   Spoofing classification accuracy by combining structure and texture information following cross-sensor protocol

the proposed method achieves competitive performance with state-of-the-art methods and is an effective method to address face spoof detection problem. It outperforms texture-based methods on the databases across "DC" and "WC" sensors.

The IDA does not perform as well as in [5]. The color information is sensitive to lighting change. The diverse lighting environments in this database affect the color analysis in IDA. Moreover, the face rotation may bring about image blur and

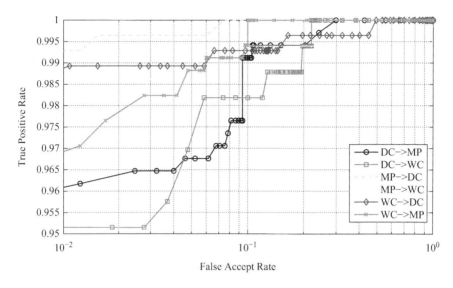

Figure 14.13   ROC curves of G vs. S photo across different sensors by combining structure and texture information

Table 14.10   ROC performance (%) following cross-sensor protocol by fusing structure and texture information

Database	G vs. P			G vs. W			G vs. S		
	TPR@ FAR = 0.01	TPR@ FAR = 0.1	EER	TPR@ FAR = 0.01	TPR@ FAR = 0.1	EER	TPR@ FAR = 0.01	TPR@ FAR = 0.1	EER
DC→MP	99.41	99.71	0.45	96.76	97.35	1.93	97.06	99.41	1.85
DC→WC	100.00	100.00	0.31	96.36	98.79	1.97	95.15	98.18	2.66
MP→DC	100.00	100.00	0.00	99.29	99.64	0.36	99.29	100.00	0.59
MP→WC	100.00	100.00	0.00	98.79	99.39	0.76	98.18	99.39	1.06
WC→DC	99.64	99.64	0.18	96.43	99.29	1.92	98.93	99.29	0.84
WC→MP	99.71	100.00	0.30	95.88	99.41	1.80	97.06	99.41	1.85

distortion, which affects the blurriness and other distortion evaluation in IDA. These factors degrade the performance of IDA. The unsatisfied performance of texture-based method also indicates that the texture cue is not robust enough in this scenario to classify genuine and fake samples accurately.

14.6.6   Computational cost analysis

The proposed face anti-spoofing system mainly consists of face detection, landmark location, sparse 3D structure recovery, texture extraction, and SVM classification.

Table 14.11　*FAR and TPR (%) of G vs. S corresponding to zero threshold following cross-sensor protocol by fusing structure and texture information*

Database	FAR	TPR
DC→MP	0.00	88.53
DC→WC	0.00	54.55
MP→DC	0.15	98.93
MP→WC	3.85	99.39
WC→DC	19.23	92.14
WC→MP	8.31	77.94

Table 14.12　*Classification accuracy (%) of G vs. S following cross-sensor protocol*

Database	MsLBP [9]	IDA [5]	Proposed method
DC→MP	**95.96**	91.01	95.35
DC→WC	90.80	85.40	**91.90**
MP→DC	**99.57**	94.95	99.46
MP→WC	96.81	94.85	**99.26**
WC→DC	84.30	85.59	**97.42**
WC→MP	86.97	83.84	**94.24**

Bold values mean the best result among the three methods in each row.

The computational costs of face detection and landmark location have been greatly improved recently [43]. The computational cost of linear SVM in testing phase can be ignored compared to the sparse 3D structure recovery and texture extraction. We evaluate the efficiency of proposed algorithm with unoptimized MATLAB® code on a PC with 3.20GHZ i5 CPU and 8G RAM. For the sparse 3D structure recovery, each update of sparse 3D structure recovery with a new input key frame takes about 17 ms and for the texture extraction, the computational cost of multiscale LBP filtering and histogram extraction on each cropped face image is about 7.4 ms. It is feasible to transplant the proposed algorithm on a mobile device.

14.7　Conclusion

In this chapter, a sparse 3D structure-based face anti-spoofing method on rotated face sequence is proposed. Based on the facial landmarks from images with different poses, a sparse 3D facial structure can be successfully recovered, with which the genuine and spoofing photo (planar and warped) can be distinguished. Moreover, we integrate the texture information as a complementary cue to improve face anti-spoofing performance. To examine the performance of the proposed method, a novel face database consisting of rotated face sequence is collected using three devices. Experiments on different databases validate the effectiveness of the proposed method. The proposed method is efficient and can be implemented on mobile devices.

References

[1]   S. Z. Li and A. K. Jain (eds.), *Handbook of Face Recognition, 2nd ed.* New York: Springer-Verlag, 2011.

[2]   J. Komulainen, A. Hadid, M. Pietikäinen, A. Anjos, and S. Marcel, "Complementary countermeasures for detecting scenic face spoofing attacks," in *ICB*, 2013, pp. 1–7.

[3]   J. Yang, Z. Lei, S. Liao, and S. Z. Li, "Face liveness detection with component dependent descriptor," in *ICB*, 2013, pp. 1–6.

[4]   J. Galbally, S. Marcel, and J. Fiérrez, "Image quality assessment for fake biometric detection: Application to iris, fingerprint, and face recognition," *IEEE Transactions on Image Processing*, vol. 23, no. 2, pp. 710–724, 2014.

[5]   D. Wen, H. Han, and A. K. Jain, "Face spoof detection with image distortion analysis," *IEEE Transactions on Information Forensics and Security*, vol. 10, no. 4, pp. 746–761, 2015.

[6]   K. Patel, H. Han, A. K. Jain, and G. Ott, "Live face video vs. spoof face video: Use of moiré patterns to detect replay video attacks," in *ICB*, 2015, pp. 98–105.

[7]   M. D. Marsico, M. Nappi, D. Riccio, and J. Dugelay, "Moving face spoofing detection via 3D projective invariants," in *ICB*, 2012, pp. 73–78.

[8]   T. Wang, J. Yang, Z. Lei, S. Liao, and S. Z. Li, "Face liveness detection using 3D structure recovered from a single camera," in *ICB*, 2013, pp. 1–6.

[9]   J. Määttä, A. Hadid, and M. Pietikäinen, "Face spoofing detection from single images using micro-texture analysis," in *IJCB*, 2011, pp. 1–7.

[10]   T. de Freitas Pereira, A. Anjos, J. M. D. Martino, and S. Marcel, "Lbp – top based countermeasure against face spoofing attacks," in *ACCV Workshops (1)*, 2012, pp. 121–132.

[11]   W. R. Schwartz, A. Rocha, and H. Pedrini, "Face spoofing detection through partial least squares and low-level descriptors," in *IJCB*, 2011, pp. 1–8.

[12]   G. Kim, S. Eum, J. K. Suhr, I.-D. Kim, K. R. Park, and J. Kim, "Face liveness detection based on texture and frequency analyses," in *ICB*, 2012, pp. 67–72.

[13]   I. Pavlidis and P. Symosek, "The imaging issue in an automatic face/disguise detection system," in *IEEE Workshop on Computer Vision Beyond the Visible Spectrum: Methods and Applications*, 2000, pp. 15–24.

[14]   Z. Zhang, D. Yi, Z. Lei, and S. Z. Li, "Face liveness detection by learning multispectral reflectance distributions," in *FG*, 2011, pp. 436–441.

[15]   S. Marcel, M. S. Nixon, and S. Z. Li (eds.), *Handbook of Biometric Anti-Spoofing*. London: Springer, 2014.

[16]   G. Pan, L. Sun, Z. Wu, and S. Lao, "Eyeblink-based anti-spoofing in face recognition from a generic webcamera," in *ICCV*, 2007, pp. 1–8.

[17]   M. Wagner and G. Chetty, 'Liveness' verification in audio–video authentication," in *INTERSPEECH*, 2004.

[18]   M. I. Faraj and J. Bigün, "Audio-visual person authentication using lip-motion from orientation maps," *Pattern Recognition Letters*, vol. 28, no. 11, pp. 1368–1382, 2007.

[19]   K. Kollreider, H. Fronthaler, M. I. Faraj, and J. Bigün, "Real-time face detection and motion analysis with application in 'liveness' assessment," *IEEE Transactions on Information Forensics and Security*, vol. 2, no. 3, pp. 548–558, 2007.

[20]   K. Kollreider, H. Fronthaler, and J. Bigün, "Evaluating liveness by face images and the structure tensor," in *AutoID*, 2005, pp. 75–80.

[21]   A. Anjos, M. M. Chakka, and S. Marcel, "Motion-based counter-measures to photo attacks in face recognition," *IET Biometrics*, 2013.

[22]   J. Yan, Z. Zhang, Z. Lei, D. Yi, and S. Z. Li, "Face liveness detection by exploring multiple scenic clues," in *ICARCV*, 2012, pp. 188–193.

[23]   I. Chingovska, J. Yang, Z. Lei, *et al.* "The 2nd competition on counter measures to 2D face spoofing attacks," in *ICB*, 2013, pp. 1–6.

[24]   T. Choudhury, B. Clarkson, T. Jebara, and A. Pentland, "Multimodal person recognition using unconstrained audio and video," in *Proceedings of International Conference on Audio- and Video-Based Person Authentication*, 1999, pp. 176–181.

[25]   N. Erdogmus and S. Marcel, "Spoofing 2D face recognition systems with 3D masks," in *BIOSIG*, 2013, pp. 209–216.

[26]   N. Erdogmus and S. Marcel, "Spoofing face recognition with 3D masks," *IEEE Transactions on Information Forensics and Security*, vol. 9, no. 7, pp. 1084–1097, 2014.

[27]   J. Komulainen, A. Hadid, and M. Pietikäinen, "Context based face anti-spoofing," in *BTAS*, 2013, pp. 1–8.

[28]   G. Pan, L. Sun, Z. Wu, and Y. Wang, "Monocular camera-based face liveness detection by combining eyeblink and scene context," *Telecommunication Systems*, vol. 47, no. 3–4, pp. 215–225, 2011.

[29]   I. Chingovska, A. Anjos, and S. Marcel, "Anti-spoofing in action: joint operation with a verification system," in *CVPR Workshops*, 2013, pp. 98–104.

[30]   X. Xiong and F. D. la Torre, "Supervised descent method and its applications to face alignment," in *CVPR*, 2013, pp. 532–539.

[31]   R. Hartley and A. Zisserman, *Multiple View Geometry in Computer Vision*, 2nd edition. New York: Cambridge University Press, 2000, vol. 2.

[32]   R. Hartley and P. Sturm, "Triangulation," *Computer vision and image understanding*, vol. 68, no. 2, pp. 146–157, 1997.

[33]   M. Fischler and R. Bolles, "Random sample consensus: a paradigm for model fitting with applications to image analysis and automated cartography," *Communications of the ACM*, vol. 24, no. 6, pp. 381–395, 1981.

[34] D. Nistér, "An efficient solution to the five-point relative pose problem," *IEEE Transactions on Pattern Analysis and Machine Intelligence*, vol. 26, no. 6, pp. 756–777, 2004.

[35] B. Haralick, C. Lee, K. Ottenberg, and M. Nölle, "Review and analysis of solutions of the three point perspective pose estimation problem," *International Journal of Computer Vision*, vol. 13, no. 3, pp. 331–356, 1994.

[36] M. Lourakis and A. Argyros, "The design and implementation of a generic sparse bundle adjustment software package based on the levenberg-marquardt algorithm," Technical Report 340, Institute of Computer Science-FORTH, Heraklion, Crete, Greece, Tech. Rep., 2004.

[37] A. M. Bruckstein, R. J. Holt, T. S. Huang, and A. N. Netravali, "Optimum fiducials under weak perspective projection," *International Journal of Computer Vision*, vol. 35, no. 3, pp. 223–244, 1999.

[38] D. A. Forsyth and J. Ponce, *Computer Vision – A Modern Approach, 2nd ed.* Prentice Hall, USA: Pitman, 2012.

[39] Z. Zhang, J. Yan, S. Liu, Z. Lei, D. Yi, and S. Z. Li, "A face antispoofing database with diverse attacks," in *ICB*, 2012, pp. 26–31.

[40] X. Tan, Y. Li, J. Liu, and L. Jiang, "Face liveness detection from a single image with sparse low rank bilinear discriminative model," in *ECCV (6)*, 2010, pp. 504–517.

[41] A. Anjos and S. Marcel, "Counter-measures to photo attacks in face recognition: a public database and a baseline," in *IJCB*, 2011, pp. 1–7.

[42] B. Yang, J. Yan, Z. Lei, and S. Z. Li, "Aggregate channel features for multi-view face detection," *CoRR*, vol. abs/1407.4023, 2014.

[43] S. Ren, X. Cao, Y. Wei, and J. Sun, "Face alignment at 3000 fps via regressing local binary features," in *CVPR*, 2014.

Chapter 15

Biometric antispoofing on mobile devices

Zahid Akhtar[1], Christian Micheloni[2],
and Gian Luca Foresti[2]

15.1 Introduction

Nowadays, mobile devices with built-in sensors (e.g., camera) have become ubiquitous, which are indeed widely being used worldwide not only for basic communications but also as a tool to store e-mail, personal photos, online history, passwords and even payment information. Due to the sensitive nature of these devices, biometric-based access control is a very common feature available almost on all mobile devices. Biometrics is used as a natural alternative to passwords that utilizes biological or behavioral characteristics of the users to grant access to data and/or use of the device. For example, iPhone 6s, Motorola Atrix, Samsung Galaxy S5 and newer Fujitsu F505i, Lenovo ThinkPad come with touch sensors that automatically unlock the phone and laptop, respectively, using fingerprint. Similarly, Android KitKat mobile OS and Lenovo, Asus and Toshiba laptops come with in-built biometric systems that authenticate users by their faces. Likewise, Fujitsu, UMI, SecuriMetrics and Siam, Amazon produce smart mobile devices that employ iris and ear to recognize individuals, respectively. Acceptance of biometrics on mobile devices by individuals is growing year after year. In fact, a recent industry survey estimates that almost 3.4 billion users will have biometrics on mobile devices by 2018 [1].

In spite of many advantages, biometric systems like any other security applications are vulnerable to a wide range of attacks. Adversary attacks exploit the system vulnerabilities generally at one or more modules or interfaces. There are eight possible different points where security of biometric systems can be compromised [2,3] as shown in Figure 15.1. These attacks fall broadly into two categories: *direct* (spoofing attacks) and *indirect*.

Direct attacks are defined as "presentation of an artifact or human characteristic to the biometric capture subsystem with the goal to interfere the operation of biometric system" [2–5]. They occur externally at the sensor level (point 1 in Figure 15.1). Thus, digital protection techniques, such as hashing, encryption, and digital signatures,

[1]INRS-EMT, University of Quebec, Canada
[2]Department of Mathematics, Computer Science and Physics, University of Udine, Italy

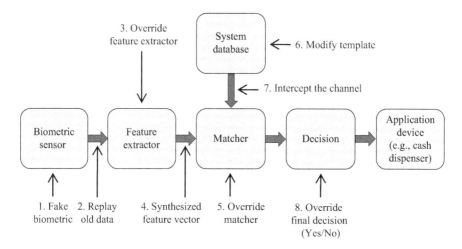

Figure 15.1   Eight points of attack on a generic biometric system

are not applicable to them. On the other hand, intruders (e.g., cybercriminals or hackers) perform indirect attacks inside the system by bypassing the feature extractor or matcher (points 3 and 5 in Figure 15.1, respectively), manipulating templates in the database (point 6 in Figure 15.1), or exploiting possible weak points in the communication channels (points 2, 4, 7 and 8 in Figure 15.1). Clearly, indirect attacks require advanced programming skills.

   Among the potential attacks discussed in the literature, the one with greatest practical relevance is "*spoofing attack*," which consists in submitting a synthetically replicated artifact[1] (e.g., gummy finger, printed iris image or face mask) or mimicking the behavior of a genuine user to the sensor in order to fool the biometric system, and thereby gaining illegitimate access and advantages [2,5,6]. Spoofing attack is also known as *direct-* or *presentation-attack* as it is carried out directly on the biometric sensor by presenting artificial traits. It is worth mentioning that spoofing attacks does not require advanced technical skills, therefore the potential number of attackers is very high.

   While many efforts are being focused on the performance improvement of biometrics on mobile platforms, the issues related to their vulnerabilities to spoofing attacks are generally overlooked [2,7]. For instance, within 2 days of its release, German Hackers group Chaos Computer Club using a counterfeited fingerprint fooled iPhone5S [3,4,8], whereas Samsung Galaxy S5 also got hacked using fake fingerprints [9]. Similarly, a team from University of Hanoi (Vietnam) has demonstrated

---

[1]In this chapter, we will use both "fake" and "spoofed" terms interchangeably, to indicate an artificial replica of genuine user's biometric trait.

*Table 15.1　The categorization of attacks and countermeasures for biometric
systems. For each attack technique, the attack location (targeted
component) and the attack point(s), according to Figure 15.1, are
also reported*

Attack technique	Attack point(s)	Attack location	Defense
Spoofing	1	Sensor	Liveness detection, multibiometrics
Replay	2,4,7	Interfaces/channels	Encrypted channel, time stamp, challenge-response, physical isolation
Hill climbing	2,4	Interfaces/channels	Encrypted channel, time stamp, challenge-response, physical isolation, score quantization
Malware infection	3,5,8	Modules/algorithms	Secure code, specialized hardware, algorithm integrity
Template theft, substitution and deletion	6	Template database	Template encryption, cancellable/ revocable templates

that how to trick Lenovo, Asus and Toshiba laptops' Face Recognition using genuine user's photographs [10].

The aforementioned cases and other analogue research studies have shown the importance of improving system's security as well as the necessity of devising specific protection methods against attacks (spoofing attacks specifically) in order to bring rapidly emerging mobile biometric technology into more practical use. Researchers thus have developed peculiar counteragents, which empower biometric systems to detect spoofing attacks and successfully rejecting them. The eight possible points of attacks as shown in Figure 15.1 and their techniques with corresponding countermeasures proposed so far are summarized in Table 15.1.

Besides multibiometrics as antispoofing approach, researches had been particularly focused on *liveness detection* (*presentation attack* detection) techniques [2,4,11]. This quintessential countermeasure to spoofing attacks aims at detecting whether the submitted biometric trait is a live or an artificial one by observing physiological signs of life, such as eye blinking, precipitation, etc. [2,4,6]. Antispoofing or liveness detection can be done in four different modes: (i) with available sensors detecting a pattern characteristic of attack in the signal; (ii) with dedicated sensors detecting an evidence of genuineness, which is not always feasible to deploy; (iii) with a challenge-response technique where an attack can be detected by asking user to interact with the system or (iv) with recognition techniques intrinsically robust against presentation attacks, if any.

Over the last decade, a great amount of works on biometric antispoofing [2,4–7,11–13] has been done to advance the state-of-the-art. However, simultaneously attacking techniques have also evolved to fabricate more and more sophisticated spoofing attacks. Consequently, many of existing biometric systems are still vulnerable to spoofing attempts using novel spoofing samples. A systematic analysis of state-of-the-art antispoofing techniques revealed that there are still big challenges to be faced in detecting biometric direct attacks [2,4,8,14], e.g., (i) lack of generalization—most of the existing liveness detection approaches are spoof material- and trait-dependent, such that feature descriptors proposed for face spoofing may not function effectively if employed for iris or fingerprint spoofing and vice versa; (ii) unuseable on mobile platform—due to the use of complex features and/or high computational cost, namely most of the existing antispoofing methods are not applicable for mobile applications; (iii) high error rates—yet none of the methods have shown to reach a very low acceptable error rates.

In the present chapter, after a thorough review of state-of-the-art in biometric antispoofing, we present a software-based spoof detection prototype for mobile devices, named **MoBio_LivDet** (**Mo**bile **Bio**metric **Liv**eness **Det**ection) that can be used in multiple biometric systems. MoBio_LivDet analyzes local features and global structures of face, iris and fingerprint biometric images using a set of low-level feature descriptors and decision-level fusion. In particular, we propose to use image descriptor classification algorithms Locally Uniform Comparison Image Descriptor (LUCID) [15], CENsus TRansform hISTogram (CENTRIST) [16] and Patterns of Oriented Edge Magnitudes (POEM) [17] for face, iris and fingerprint spoof detection. The proposed system allows user to choose "Security Level" (SL) against spoofing, between "low," "medium" and "high." Depending on SL, the system selects unit-descriptor or multidescriptors-fusion-based liveness detection. These descriptors are computationally inexpensive, fast and novel approach to real-time image description, which are desirable requisites for mobile processors. Experiments on publicly available data sets containing several real and spoofed faces, irises and fingerprints show promising results.

The rest of the chapter is structured as follows. An exhaustive review of relevant related works in the field of face, fingerprint and iris antispoofing is given is Section 15.2. A case study based on MoBio_LivDet system design is presented in Section 15.3. To conclude, the summary and discussion with some insight into future research directions are given in Sections 15.4 and 15.5.

15.2 Biometric antispoofing

Spoofing attacks are defined by the standardization project (ISO/IEC 30107 [18]) as "presentation of an artifact or human characteristic to the biometric capture subsystem with the goal of interfering with the operation of the biometric system." The spoofing attacks are also known as presentation attacks, direct attacks or nonzero-effort attacks. As illustrated in Figure 15.2, face, iris and fingerprint images captured from spoofing attacks can look very similar to images captured from real ones. The countermeasure to spoofing is equipping the systems with a presentation attack detection (PAD) method. The standardization project (ISO/IEC 30107 [18]) defines PAD as "detection of fake, altered biometric characteristics, coercion, nonconformity and obscuration."

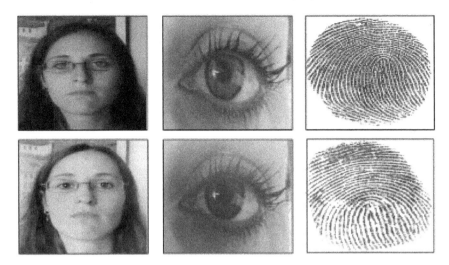

Figure 15.2   Example of images captured from real face, iris and fingerprint (upper row) and from spoofing attacks (lower row): face photo attack, iris photo attack and fingerprint latex attack. The appearance similarity explains the difficulty of spoof detection

Thus, *antispoofing* (also referred as *liveness detection* or *vitality detection*)[2] belongs to the group of PAD methods and works under the assumption that attacker is trying to intrude the system pretending as a genuine user by felonious biometric means (i.e., synthetically produced artifacts imitating the genuine trait) [2].

Developing a liveness detection technique is challenging, since it has to meet certain demanding requirements, such as fast (i.e., outputs have to be produced in a shortest interval of time possible so that user does not have to interact with the sensor long), noninvasive (i.e., not harmful to the user), performance (i.e., low false rejection together with high detection rate), user-friendly (i.e., easy to use to make user less reluctant), and low cost (in order to get adopted on mass level).

From a general perspective, antispoofing techniques may be categorized into three groups: hardware-based, software-based and hybrid-based methods. Hardware-based techniques, also referred as sensor-level techniques, add some extra hardware or device to the system in order to detect particular properties of a living biometric trait (e.g., blood pressure). Hardware-based methods typically measure any combination of three characteristics: *intrinsic properties* of living human body such as physical, electrical, spectral or visual properties; *involuntary signals* of living human body such as perspiration or brain wave signals (EEG); and *reactions to external stimuli*, also called challenge-response approach, such as voluntary or involuntary responses to lighting.

---

[2]In this chapter, we will use "anti-spoofing", "PAD", "liveness detection" and "vitality detection" terms interchangeably, unless explicitly stated otherwise.

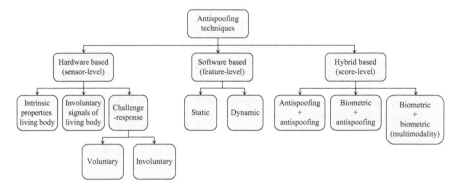

Figure 15.3   The three main categories of antispoofing outlined in this chapter

Software-based techniques, also mentioned as feature-level methods, exploit signal processing procedures on the biometric sample, which has been acquired with a standard sensor. Namely, features extracted from the captured biometric trait are used to distinguish real samples from fake ones. Software-based systems can be further bifurcated into static and dynamic antispoofing methods, depending on whether they operate with only one instance of biometric sample, or with a sequence of samples captured over time [4].

Hybrid-based techniques, also referred as score-level methods, either integrate software- and hardware-based procedures or use multimodal biometrics to increase resistance against spoofing attacks. This type of antispoofing is less common, and usually is used as supplementary measures to above mentioned two categories due to their inherent performance limitations. The source of information in hybrid-based techniques—in the form of any potential combination of image (sensor-level), feature and score—may come from: (i) unimodal biometric modules and antispoofing techniques; (ii) only different antispoofing modules or (iii) only, number of uni-modal biometric modules. It is worth mentioning that multibiometric antispoofing is based on the hypothesis that circumventing multibiometric system requires breaking all unimodal modules [3]. However, this hypothesis has recently been shown to be untrue [19].

All in all, hardware- and software-based methods are mostly adopted among above-mentioned three categories, since they generally provide complementary information thus leading to increased security of biometric systems. Nevertheless, software-based solutions are the most interesting and challenging ones, since they do not employ any additional and possibly invasive measurements such as blood pressure, etc. A graphical representation of the categorization outlined above is shown in Figure 15.3. In the following, we present an overview of spoofing and antispoofing techniques proposed in the literature for the most commonly used biometric traits, viz., face, fingerprint and iris.

15.2.1   State-of-the-art in face antispoofing

Most of the existing academic and commercial facial recognition systems may be spoofed by: (i) a photo of a genuine user; (ii) a video of a genuine user; (iii) a 3D face model (mask) of a genuine user; (iv) a reverse-engineered face image from the template of a genuine user; (v) a sketch of a genuine user; (vi) an impostor wearing specific make-up to look like a genuine user; (vii) a photo or a video, generated using computer graphics, of a genuine user; (viii) an impostor who underwent plastic surgery to look like a genuine user.

The published face antispoofing methods can be coarsely classified, based on the cues used for spoof attacks detection, into four categories: (i) *motion-analysis-*, (ii) *texture-analysis-*, (iii) *image-quality-analysis-* and (iv) *hardware* based.

(i)   *Motion-analysis*-based methods: These feature-level methods broadly try to detect spontaneous movement clues (dynamic features) generated when counterfeits are presented to the camera of the system. For instance, Pan *et al.* [20] proposed eye-blink-based liveness detection using an undirected conditional random field framework. The method likely fails for higher quality video spoof samples. Similarly, Google's Android 4.1 Jelly Bean use eye-blinking model to detect face spoofing on mobile devices [21]. It is evident that real human faces (which are 3D objects) will move significantly differently from planer objects, and such deformation patterns can be employed for liveness detection. For example, Tan *et al.* [22] considered Lambertian reflectance model with difference-of-Gaussians (DoG) to derive differences of motion deformation patterns between 2D face photos presented during spoofing attacks and 3D live faces. Likewise, Maria *et al.* [23] used 3D geometric invariant motion features and implemented it on mobile devices for face liveness detection. Moreover, several commercial vendors are marketing their products with in-built majorly motion-based face liveness detection modules such as Lenovo's Veriface is using head motion and expressions to detect spoofs [24], BioID is using head motion or voice [25], Jumio's Netverify is using facial movements [26] and KeyLemon's OASIS is using eye blink and head, mouth, background motion [21]. However, methods in [23] and Google's Android 4.1 Jelly Bean including above mentioned other commercial vendor's solutions are mostly prone to 3D mask, videos and image manipulations attacks. Kollreider *et al.* [13] proposed a liveness detection approach based on a short sequence of images using a binary detector, which captures and tracts the subtle movements of different selected facial parts using a simplified optical flow analysis followed by a heuristic classifier. In the similar fashion, Bao *et al.* [27] also used optical-flow to estimate motion for detecting attacks produced with planar media such as prints or screens. Also, dynamic-face antispoofing techniques have been proposed to detect video-based attacks, e.g., exploiting the 3D structure of the face through the analysis of several 2D images with different head poses [28]. Since the frequency of facial motion is restricted by the human physiological rhythm, thus motion-based methods take a relatively long time (usually >3 s) to accumulate stable vitality features for face spoof detection. Moreover, they may be circumvented or confused by other motions, e.g., background motion in the video attacks. However, motion-analysis-based algorithms usually present high-performance rates, because they exploit both spatial and temporal information, which unfortunately leads to one main limitation, i.e., they cannot be used in systems where only a single face image of the user is available.

(ii) *Texture-analysis*-based methods: The need for a temporal sequences of a certain duration in motion-analysis-based method motivated texture-analysis-based methods, which examine the skin properties (e.g., skin texture/reflectance) in a single static image rather than video data under the assumption that surface properties (e.g., pigments) of real faces and prints are different. Examples of detectable texture patterns (static features) due to artifacts are printing failures or blurring. Li *et al.* [29] described an antispoofing method using 2D Fourier spectra of images. The method only works well for down-sampled photos of the attacked identity, but likely fails for higher-quality samples. In [10,30], authors developed microtexture-analysis (e.g., LBP) based methods to detect printed photo-attacks and even 3D spoof mask. One limitation of presented methods is the requirement of reasonably sharp input image. Very recently face antispoofing techniques specifically targeted on mobile devices using local descriptors were proposed in [7,31,32]. The static-features-based antispoofing approaches may be applied to video data as well, e.g., fusing the outcomes of frame-by-frame analysis to attain single final decision [2]. Compared to other techniques, texture-analysis-based algorithms are generally faster to classify a spoof attack. Nevertheless, they could be easily overfitted to one particular illumination and imagery condition leading them to have poor generalization ability [2,4].

(iii) *Image-quality-analysis*-based methods: Recently, Galbally *et al.* [33] designed a face spoof detection scheme based on 25 different image quality measures such as: 21 full-reference measures and 4 nonreference measures. However, all 25 image quality measures are required to get good results, and no face-specific information has been considered in designing informative features for face spoof detection. On the contrary, four face-specific features were designed using image distortion analysis in [34]. This method compared to procedure in [33] aims to improve generalization ability under crossdatabase scenarios. Some of the above mentioned feature-level methods been also fused successfully to increase accuracy compared to individual parameters [4]. In fact, the feature-level fusion of static and dynamic approaches (i.e., motion-, texture- and image-quality-based methods) attains best performance [35]. A comparative study of various dynamic and static approaches may be found in the 2011 and 2013 Competitions on Countermeasures to 2D facial spoofing attacks [35,36].

(iv) *Hardware*-based methods: Few interesting sensor-level face antispoofing techniques have been proposed so far based on imaging technology outside the visual spectrum, such as 3D depth [28], complementary infrared (CIR) or near infrared (NIR) images [37] by comparing the reflectance information of real faces and spoof materials using a specific set-up of LEDs and photo-diodes at two different wavelengths. Preliminary efforts on thermal imaging for face liveness detection have also been exploited, including the acquisition of large database of thermal face images for real and spoofed access attempts [38]. Besides, numbers of researchers have explored multimodality as face antispoofing techniques as well. They have mainly considered combination of face and voice by utilizing the correlation between the lips movement and the speech being produced [39]. Similarly, challenge-response strategy consider-ing voluntary eye blinking and mouth movement by following a request from the system has been studied in [13]. Though, hardware-based methods pro-vide better results and performances, they require extra piece of hardware that increases the cost of the system. A summary with relevant features of the most representative works in face antispoofing is presented in Table 15.2.

Presently, there are seven significant face-spoofing databases publicly available: NUAA PI DB [22], YALE-Recaptured DB [40], PRINT-ATTACK DB [35], CASIA FAS DB [41], REPLAY-ATTACK DB [36], 3D MASK-ATTACK DB [30] and MSU MFS DB [34].

The NUAA PI DB (NUAA Photograph Imposter database [22]) is the one of the first public-domain face-spoofing database, which was released in 2010. It contains genuine and attack attempts of 15 users captured using cheap generic webcam. This database only contains printed photo-attack.

The YALE-Recaptured DB [40] is composed of 640 real faces and 1920 LCD spoofs, by displaying the images from the Yale Face Database B on 3 LCD monitors, from 10 users under 64 different illumination conditions. The spoofing attempts were captured using Kodak C813 and Samsung Omnia i900 cameras.

The PRINT-ATTACK DB [35] consists of 200 videos of real access and 200 videos of print-attack attempts from 50 users under different lighting conditions captured using MacBook laptop.

The CASIA FAS DB (CASIA Face Anti-Spoofing Database (FASD) [41]) con-tains short videos of both real and attack attempts of 50 different identities. The CASIA database contains more diverse samples in terms of the acquisition devices (high-resolution Sony NEX-5 camera and low-, normal-quality USB cameras), face variations (pose and expression variations) and attack attempts (warp photo, cut photo, and HD displayed video, paper, mobile phones and tablets).

The REPLAY-ATTACK DB [36] consists of 1300 video recordings of both real-access and attack attempts of 50 different subjects acquired from MacBook laptop under controlled and adverse illumination conditions. The face spoof attacks were gen-erated via printed photos, displayed photos/videos on iPhone's screen and displayed photos/videos on iPad's screen.

The 3D MASK-ATTACK DB [30] is the first public database that considered face-mask attacks, which also provides depth information as well. It compromises genuine and attack samples captured using Microsoft Kinect from 17 different users. In particular, single operator wearing the life-size 3D masks of genuine subjects performed the attacks.

Table 15.2 A summary and comparison of different face spoof detection categories

Method	Feature and technique used	Year	Pros	Cons
Motion-analysis-based methods	• Motion detection using retinotopic grid [13] • Eye blink detection using Conditional Random Fields (CRF) [20] • Face motion detection using Optical Flow Lines (OFL) [27] • Context-based using correlation between face motion vs. background motion [36]	2009 2007 2010 2013	• Good generalization capability	• High computational cost since mostly image registration is needed • Slow response (>3 s) • Low robustness because they can be easily circumvented by fake motions
Texture-analysis-based methods	• Face texture using Lambertian model [22] • Face texture using LBPs [10] • Multiscale LBP for 3D mask [30] • Deep convolutional networks [42]	2010 2013 2014 2015	• Fast response (<1 s) • Low computational cost	• Poor generalization capability since they are vulnerable to the variations in acquisition conditions
Image-quality-analysis-based methods	• 25 general image quality measures [33] • Image-distortion-based quality measures [34]	2014 2015	• Fast response (<1 s) • Good generalization capability • Low computational cost	• Different classifiers needed for different spoofing attacks
Hardware-based methods	• Multimodality: Face and Voice [39] • Thermal Images [38] • Reflectance in 3D [43]	2005 2013 2014	• High robustness • High generalization capability	• Slow response (>3 s) • Extra hardware requirement • High system cost

The MSU MFS DB (MSU Mobile Face Spoof Database [34]) is the first public mobile face spoof dataset which is composed of 440 videos of photo and video attack attempts of 55 subjects. The samples were recorded using MacBook laptop camera and Google Nexus 5 Android phone. Table 15.3 provides a summary of the above seven databases in terms of sample size, acquisition devices, attack types and so on.

15.2.2 State-of-the-art in fingerprint antispoofing

Fingerprint spoofing is an old practice [2]. Fingerprint recognition systems can be fooled by: (i) a 2D (flat) fake fingerprint of a genuine user; (ii) a synthesized 3D fake fingerprint of a genuine user; (iii) a reverse-engineered fingerprint image from template of a genuine user; (iv) a cadaver fingerprint of a genuine user; (v) a finger cut from a genuine user.

Spoof fingerprints can be fabricated either by "consensual/cooperative/direct casts" or "nonconsensual/noncooperative/indirect casts" method by using easily available materials like latex, etc. [2,4]. Till now, 57 materials and variants have been recognized for spoof fingerprints. In consensual method, the fake fingerprints are created directly from real fingers with person's consent, while in nonconsensual method fake fingerprints are fabricated from latent fingermarks on daily use product or sensors; hence cooperation of the user is not required.

Existing fingerprint antispoofing techniques can be clustered in seven groups as: (i) *perspiration*-, (ii) *skin-deformation*-, (iii) *pore-detection*-, (iv) *image-quality*-, (v) *texture*-, (vi) *hybrid*- and (vii) *hardware* based.

(i) *Perspiration*-based methods: One of the very first efforts to counteract fingerprint spoofing was using feature-level techniques based on dynamic analysis of perspiration pattern present in the fingerprint. Schuckers *et al.* [6] did pioneer study and used static patterns and dynamic changes in the moisture structure of skin around sweat pores caused by perspiration. In later works, they also used a wavelet-based approach. In [44] authors stated that perspiration-based approaches are time-consuming, since the user is required to present his finger twice thus inapplicable in real-time applications.

(ii) *Skin-deformation*-based methods: They are also dynamic feature-level schemes exploiting flexibility properties of skin (skin elasticity) and different fingerprint distortion models. For instance, Zhang *et al.* [44] developed a novel method to capture finger distortion using Thin-plate Spline model at different rotated angles. The main drawback of skin deformation-based schemes is that users need to have special training, e.g., they have to move their finger while pressing it against the scanner surface, thereby deliberately exaggerating the skin distortion. This is because the movement of real finger on scanner surface produces a significant amount of distortion that is generally quite different from the one produced by fake fingers that are usually more rigid than skin. Moreover, in this kind of methods the scanners require to be capable of delivering frames at a proper rate.

(iii) *Pore-detection*-based methods: Pore-detection-based methods try to overcome some limitations of skin-deformation and perspiration techniques using level-3 fingerprint features. Espinoza *et al.* [45] proved that pores can be used as

Table 15.3 A summary of seven face spoof databases in the public-domain

Database	Year of release	No. of subjects	No. of samples	Sample types	Acquisition devices	Attack types
NUAA PI DB [22]	2010	15 15	• 5105 genuine • 7509 spoof	Images	• Web-cam (640 × 480)	• Printed photo
YALE-Recaptured DB [40]	2011	10 10	• 640 genuine • 1920 spoof	Images	• Kodak C813 camera (96 × 96) • Samsung Omnia i900 camera (96 × 96)	• Display photo (HD)
PRINT-ATTACK [35]	2011	50 50	• 200 genuine • 200 spoof	Videos	• MacBook camera (320 × 240)	• Printed photo
CASIA FAS DB [41]	2012	50 50	• 150 genuine • 450 spoof	Videos	• Low-quality camera (640 × 480) • Normal-quality camera (480 × 640) • High-quality (Sony NEX-5) camera (1280 × 720)	• Printed photo • Eye region cut photo • Replayed video (HD)
REPLAY-ATTACK DB [36]	2012	50 50	• 200 genuine • 1000 spoof	Videos	• MacBook camera (320 × 240)	• Printed photo • Display photo (mobile/HD) • Replayed video (mobile/HD)
3D MASK-ATTACK DB [30]	2013	17 17	• 170 genuine • 85 spoof	Videos	• Microsoft Kinect for Xbox 360 camera (640 × 480)	• Life-size 3D mask
MSU MFS DB [34]	2014	55 55	• 110 genuine • 330 spoof	Videos	• MacBook Air camera (640 × 480) • Google Nexus 5 camera (720 × 480)	• Printed photo • Replayed video (mobile/HD)

a liveness sign and suggested to use pore quantity differences between query image (real or fake) and a reference image. Manivanan *et al.* [46] evolved a method using two filtering techniques: high pass and correlation. The former method extracts active sweat pore and latter one locates position of pores. Nevertheless, using novel spoof materials and systems such as laser technology, one can produce fairly accurate pore information in the fake fingers to fool the pore-detection-based methods.

(iv) *Image-quality*-based methods: To overcome the shortcomings of dynamic methods, Galbally *et al.* [11] formulated a static scheme based on 10 different fingerprint specific quality measures such as: ridge strength, ridge continuity and ridge clarity. Likewise, in [33] authors devised a technique that uses 25 different general-purpose image quality measures. Unfortunately, this scheme needs all 10/25 quality measures to attain better accuracy. Very recently, NexID Biometrics has produced Live Finger Detection (LFD) module exploiting image quality differences measurements, which is usable on smartphones [47].

(v) *Texture*-based methods: This static-based algorithms attempt to curb the short-comings of above outlined categories by using image or signal processing tools to analyze the fingerprint texture and extract distinctive features to detect fake fingerprints. Nikam *et al.* [48] utilized ridgelet transform with ensemble classi-fier to capture texture information from fingerprint image. But, this technique is dataset dependent and performs poor under different acquisition conditions. While Ghiani *et al.* [49] exploited local phase quantization (LPQ) features for liveness detection. Fingerprint antispoofing techniques specifically targeted on mobile devices using local descriptors, such as LUCID [15], CENTRIST [16] and POEM [17], were proposed in [31,32]. Nevertheless, texture-based algorithms have very low sensor and material interoperability capability.

(vi) *Hybrid*-based methods: Various researchers combined static and dynamic fea-tures including local and global at features level to improve significantly the results of individual technique. For instance, Jai *et al.* [4] applied skin elas-ticity and perspiration pattern via five features from a sequence of fingerprint images. Also, novel protection techniques focused on the vulnerabilities of bio-metric systems to spoofing, and consequently devising more robust matchers and fusion strategies to attain high robustness against spoofing attempts is a growing trend [50].

(vii) *Hardware*-based methods: Majority of the sensor-based methods requires specific sensors to measure intrinsic properties of the living body (e.g., sub-cutaneous features) or involuntary signals produced by real fingerprints (e.g., blood flow, skin electric properties, odor or oximetry). Such fingerprint anti-spoofings are based on multispectral imaging [4], blood flow detection [2], ultrasound [11] or optical coherence tomography [51] and so on. It is worth mentioning that commercial vendors such as Lumidigm [52] and M2SYS [53] use multispectral imaging and finger vein sensor, respectively, to detect spoof-ing on mobile devices. Comparative results for some of the existing fingerprint antispoofing could be found in the 2009, 2011 and 2013 and 2015 fingerprint liveness detection competitions [49,54,55,56]. Moreover, a summary with relevant features of the most representative works in fingerprint antispoofing is presented in Table 15.4.

At present, five largest fingerprint-spoofing databases are publicly available for research purposes: ATVS-FFp DB [11], LivDet 2009 DB [54], LivDet 2011 DB [49], LivDet 2013 DB [55] and LivDet 2015 DB [56]. Unfortunately, till date no fingerprint-spoofing database collected using mobile devices is made publicly available.

The ATVS-FFp DB (ATVS-Fake Fingerprint Database [11]) comprises the index and middle fingers of both hands of 17 users ($17 \times 4 = 68$ different fingers). For each real finger, two spoofs were generated using silicon with two procedures (i.e., with and without the user's cooperation). Four samples of each fingerprint (real and spoofed) were captured in one acquisition session with three sensors (i.e., Biometrika FX2000, Precise SC100 and Yubee). Thus, the database comprises 68 fingers $\times$ 4 samples $\times$ 3 sensors $= 816$ real image samples and as many spoofed images for each procedure.

The LivDet 2009 DB (Fingerprint Liveness Detection Competition 2009 [54]) is made up of three datasets of real and spoofed fingerprints acquired with three different sensors (Biometrika, CrossMatch and Identix). The gummy fingers were generated using three different materials (gelatin, playdoh and silicone) using a consensual.

The LivDet 2011 DB (Fingerprint Liveness Detection Competition 2011 [49]) incorporates fingerprint images captured by four different sensors: Biometrika, Italdata, Digital Persona, Sagem. For Biometrika and Italdata, spoof fingerprints were fabricated using five materials: gelatin, latex, ecoflex, silicone and woodglue, while using gelatin, latex, playdoh, silicone, woodglue for Digital Persona and Sagem.

The LivDet 2013 DB (Fingerprint Liveness Detection Competition 2013 [55]) consists of images from four sensors: Biometrika, Italdata, Crossmatch, Swipe. For Biometrika and Italdata spoofs were fabricated using five materials (ecoflex, gelatin, latex, modasil and woodglue) under noncooperative manner, while for Crossmatch and Swipe spoofs were fabricated using four materials (bodydouble, latex, playdoh and woodglue) under cooperative manner.

The LivDet 2015 DB (Fingerprint Liveness Detection Competition 2015 [56]) contains images from four different optical devices: Green Bit, Biometrika, Digital Persona and Crossmatch. Live images were acquired in a variety of ways in order to mimic real scenarios: normal mode, with wet and dry fingers and with high and low pressure. Spoofs were fabricated using cooperative method. For Green Bit, Biometrika and Digital Persona sensors, Ecoflex, gelatin, latex, woodglue, a liquid Ecoflex and RTV (a two-component silicone rubber) spoof materials were used, and for Crossmatch sensor, Playdoh, Body Double, Ecoflex, OOMOO (a silicone rubber) and a novel form of Gelatin were utilized. Contrary to previous editions of LivDet competitions, the testing sets included spoof images of unknown materials, i.e., materials that were not included in the training set, to test the generalization capability of the systems. Table 15.5 shows a comparative summary of above-mentioned five databases considering most important features.

Table 15.4 A summary and comparison of different fingerprint spoof detection categories

Method	Feature and technique used	Year	Pros	Cons
Perspiration-based methods	• Periodicity of sweat [57] • Sweat diffusion pattern [6] • Wavelet-based approach [4]	2003 2005 2008	• Practicable performances	• Time consuming (requires two image samples) • Sensitive to pressure of finger, environment, user and time interval
Skin-deformation-based methods	• Thin-Plate Spline model [44] • Distortion model [58]	2007 2001	• Not privacy invasive	• Slow response • User needs training • Sensor needs to be capable to deliver frames at proper rate • Poor performance under spoof materials similar to live skin
Pore-detection-based methods	• Number and quality of pores [45] • High pass filter + correlation [46]	2011 2010	• Same pore related features useful both for liveness detection and fingerprint recognition	• Requirement of high-resolution images • Low generalization
Image-quality-based methods	• Fingerprint specific 10 quality measures [11] • General purpose twenty-five quality measures [33]	2012 2014	• Fast response • Convenient to use • Low computational complexity • Easy to implement	• Low robustness to various spoof materials, i.e., low generalization
Texture-based methods	• Local binary pattern [48] • Local phase quantization [49] • Deep convolutional networks [42]	2008 2012 2015	• Fast response • Does not require user cooperation • Mostly needs single image	• Vulnerable to low textural attacks • Poor interoperability to datasets, materials and sensors
Hybrid-based methods	• Skin elasticity + Perspiration [4] • Matching score + Liveness score [50]	2010 2012	• High performance	• Requirement of different classifiers for different attacks
Hardware-based methods	• Multispectral imaging [4] • Blood flow detection [4] • Ultrasound [11] • Optical coherence tomography [51]	2008 2008 2012 2006	• High robustness • High generalization capability	• Slow response • Extra hardware requirement • High system cost • Mostly privacy invasive

Table 15.5 A summary of five fingerprint spoof databases in the public domain

Database	Year of release	No. of subjects	No. of samples	Acquisition devices	Spoof Fabrication materials	Spoof fabrication methods
ATVS FFp DB [8]	2006	• 68 • 68	• 816 live • 816 spoof	• Biometrika (312×372) • Precise (300×300) • Yibee (233×412)	• Silicone	• Cooperative • Noncooperative
LivDet 2009 [44]	2009	• 100 • 35	• 5500 live • 5500 spoof	• Biometrika (312×372) • CrossMatch (480×640) • Identix (720×720)	• Gelatin • Playdoh • Silicone	• Cooperative
LivDet 2011 [41]	2011	• 200 • 50	• 8000 live • 8000 spoof	• Eiometrika (312×372) • Italdata (640×480) • Sagem (352×384) • Digital Persona (355×391)	• Gelatin • Latex • Ecoflex • Silicone • Woodglue • Playdoh	• Cooperative
LivDet 2013 [45]	2013	• 500 • 128	• 9000 live • 8000 spoof	• Eiometrika (312×372) • Italdata (640×480) • CrossMatch (640×480) • Swipe (208×1500)	• Gelatin • Latex • Ecoflex • Modasil • Woodglue • Playdoh • Bodydouble	• Cooperative • Noncooperative
LivDet 2015 [46]	2015	• 510 • 100	• 9000 live • 8000 spoof	• Eiometrika (312×372) • Digital Persona (355×391) • CrossMatch (640×480) • Green Bit (500×500)	• Gelatin • Latex • Ecoflex • Liquid Ecoflex • Woodglue • Playdoh • Bodydouble • RTV • OOMOO	• Cooperative

15.2.3 State-of-the-art in iris antispoofing

Iris recognition is generally conceded to be the most accurate person identification method [2,4]. However, iris recognition systems could be deceived by: (i) an iris photo of a genuine user; (ii) an iris video of a genuine user; (iii) a printed contact lens of a genuine user; (iv) an artificial eye made of glass or plastic of a genuine user; (v) a reverse-engineered iris image from the template of a genuine user; (vi) a iris photo or a video, generated using computer graphics, of a genuine user; (vii) a natural eye removed from the body of a genuine user.

Last decade has seen various notable efforts to enhance the security of iris recognition systems against spoofing [2,4]. Existing iris antispoofing (liveness detection) schemes can be categorized in six categories: (i) *frequency-spectrum-analysis-*, (ii) *reflectance-analysis-*, (iii) *dynamics-analysis-*, (iv) *image-quality-analysis-*, (v) *texture-analysis*-based and (vi) *hardware* based.

(i) *Frequency-spectrum-analysis*-based methods: Daugman *et al.* [4] indicated that iris spoof fabrication methods can leave detectable traces on spoofing artifacts, which could be find out by simple frequency spectrum information analysis. Czajka *et al.* [59] designed a method to analyze statistical properties of 2D Fourier spectra together with iris image quality assessment. This category's methods are able to detect spoof attacks using glass eyes, photographs or dead tissues but fails for attacks using replayed videos.

(ii) *Reflectance-analysis*-based methods: These approaches are mainly based on "red-eye effect." Lee *et al.* [60] proposed to detect live and fake irises based on reflectance attributes using the theoretical reflectance model. Such liveness detection methods are efficient against print attacks and artificial-eyes attacks but perform poorly under contact lenses or iris printouts (where the pupil has been cut out) attacks.

(iii) *Dynamics-analysis*-based methods: Dynamics-analysis-based methods acquire several images while manipulating the illumination level to observe dynamics in pupil and/or entire eyeball. The primary theory of these methods is very similar to previous methods (i.e., the analysis of the pupil dynamics) together with challenge-response approaches. Moreover, last group of method is based on passive measurement of dynamic object, while this one is based on active measurement of dynamic object. Pacut *et al.* [2,59] introduced a composite iris liveness detection method that analyzes behavioral eye features such as pupil dynamics, image frequency spectrum and controlled light reflection from the cornea. However, this method does not work properly with replay video attacks. Generally, dynamic-analysis-based algorithms entail a high level of discomfort for the user due to sudden lighting changes.

(iv) *Image-quality-analysis*-based methods: To overcome the limitations of abovementioned three categories of iris antispoofing, Galbally *et al.* [33,61] adopted motion features (i.e., spontaneous dynamic features) with image quality properties caused by either motion of the iris or of the sensor. Nevertheless, these methods need all quality measures to attain lower error rates, thus increasing computational cost.

(v) *Texture-analysis*-based methods: Texture-analysis-based methods aim at developing user-friendly, easy and computationally inexpensive iris antispoofing. Many local image texture-analysis-based methods have been proposed [12,31,32,62] to study the antispoofing on mobile applications. For example, especially, Gragnaniello *et al.* [12] evaluated the effectiveness of the various local texture descriptors on data set collected from mobile devices. They found that simple local binary pattern (LBP) descriptor could provide a satisfactory performance. The comparative performances of various iris antispoofing techniques could be found in Mobile Iris Liveness Detection Competition (MobILive 2014) [5] and Liveness Detection-Iris Competition 2013 [63].

(vi) *Hardware*-based methods: Few sensor-based methods have been investigated using either multispectral imaging technologies or NIR illumination at different wavebands and positions in order to detect the reflection properties of the different parts of the eye. For instance, Lee *et al.* [64] proposed a method for fake iris detection via calculating the theoretical positions and distances between the Purkinje images obtained from collimated IR-LED and human eye models. However, this method fails for contact lenses. A general overview with relevant features of the most representative works in iris antispoofing is presented in Table 15.6.

At present, three iris-spoofing databases are publicly available: MobBIO*fake* DB [5], LivDet-Iris 2013 DB [63] and ATVS-FIr DB [61]. It is worth noting that only MobBIO*fake* DB is collected using mobile device.

The MobBIO*fake* DB [5] is comprised of 800 iris images and its corresponding fake printed copies, captured using back camera (TF300T-000128) of Asus Transformer Pad TF 300T with Android version 4.1.1. The live and spoofed irises were captured under two different lighting conditions, in a room with both natural and artificial sources of light, with variable eye orientations and occlusion levels, so as to comprise a larger variability of unconstrained scenarios.

The LivDet-Iris 2013 DB [63] was used to evaluate antispoofing algorithms in first iris liveness detection held in 2013 [26]. It comprises three subsets named as Notre Dame, Warsaw and Clarkson. The Notre Dame subset was collected using IrisAccess LG4000 sensor. While Warsaw and Clarkson subsets were acquired by EyeGuard AD100 and Genie TS-Teledyne Dalsa sensors, respectively. The Spoofed irises were fabricated using printed attacks in Warsaw portion, while textured contact lenses and pattern contact lenses were used in Notre Dame and Clarkson, respectively.

The ATVS-FIr DB [61] is composed of 50 subjects $\times$ 2 eyes $\times$ 4 images $\times$ 2 sessions $= 800$ spoofed iris images and its corresponding original samples acquired using LG IrisAccess EOU3000. The spoofing attacks were generated using high quality commercial printer. Table 15.7 presents a comparative summary of above-mentioned three databases considering most important features.

Table 15.6 A summary and comparison of different iris spoof detection categories

Method	Feature and technique used	Year	Pros	Cons
Frequency-spectrum-analysis-based methods	• Fast Fourier Transform [59]	2006	• Low computational complexity	• Low robustness (can be circumvented by high-resolution images and videos)
Reflectance-analysis-based methods	• Theoretical eye reflectance model [60]	2005	• Comparative performances	• Poor generalization capability (vulnerable to contact lenses and pupil cutout in printed irises)
Dynamic-analysis-based methods	• Pupil behavior [4] • Pupil Challenge-response [65]	2006 2009	• Good generalization capability	• Not user-friendly (due to sudden light change) • High computational complexity • Slow response
Image-quality-analysis-based methods	• Iris specific 16 quality measures [61] • General purpose 25 quality measures [33]	2012 2014	• Fast response • Convenient to use • Easy to implement	• High computational complexity (many quality measures and feature selection needed) • Different classifier needed for different spoof attacks
Texture-based methods	• Local binary pattern (LBP) [12] • Gray level co-occurrence matrix (GLCM) [62] • Deep convolutional networks [42]	2015 2014 2015	• Fast response • Does not require user cooperation • Mostly needs single image	• Poor interoperability to datasets, materials and sensors • Poor generalization ability (vulnerable to unseen spoofing attacks and acquisition conditions)
Hardware-based methods	• Purkinje images [64] • Brightness between two LED lighting [66]	2005 2007	• High robustness • High-generalization capability	• Slow response • Additional sensing or processing unit needed • High system cost

Table 15.7 A summary of three iris spoof databases in the public domain

Database	Year of release	No. of subjects	No. of samples	Acquisition devices	Spoof fabrication materials
ATVS-FIr [61]	2012	• 100 • 100	• 8000 live • 8000 spoof	• IrisAccess LG EOU3000 (640 × 480)	• Printed Photo
LivDet-Iris 2013 [63]	2013	• 324 • 216	• 1726 live • 2600 spoof	• IrisAccess LG4000 (640 × 480) • EyeGuard AD100 (640 × 480) • Genie TS-Teledyne Dalsa (780 × 600)	• Printed Photo • Textured Contact Lens • Pattern Contact Lens
MobBIO*fake* [5]	2014	• 100 • 100	• 800 live • 800 spoof	• Asus Transformer Pad TF 300T (250 × 200)	• Printed Photo

15.3   Case study: MoBio_LivDet system

The above literature survey states that most of the existing face, iris and finger-print liveness detections are either very complex (and thus not very practical for mobile applications that require fast processing) or using nonconventional imaging method/device (e.g., thermal cameras). Moreover, existing schemes function typically well only against individual trait's attacks for which they are designed, but not for other traits' attacks. Namely, image descriptor proposed for iris spoofing may not necessarily work if employed for face or fingerprint spoofing and vice versa. Hence, we proposed three individual and their combination approaches for face, iris and fingerprint liveness detection on mobile devices, which are very simple, computationally fast and uses conventional images plus requires no user-cooperation.

Biometric liveness detection can be seen as a two-class classification problem, where the input biometric sample has to be flagged as either live or spoof. The keynote of the process is attaining a discriminant features set along with an appropriate classifier that gives the probability of the image realism. In particular, face, iris and fingerprint images captured from spoof attacks may visually look very similar to images captured from a live person (see Figure 15.2), but a close look reveals that spoof attack images contain some specific artifacts. Thus, motivated by characterization of artefacts, we developed a prototype face, iris and fingerprint representation (or a feature space) that is capable of capturing distinctive attributes of real and spoofed face, iris and fingerprint images. The antispoofing approach described in this section is a step toward trait-independent MoBio_LivDet, which was first presented and validated in [32].

The MoBio_LivDet is a novel software-based multibiometric system that can be used for mobile and real-time applications. The presented system learns the fine differences between images of real and spoofed faces, irises and fingerprints via LUCID [15], CENTRIST [16] and POEM [17].

The proposed system as shown in Figure 15.4 enables user to control SL against spoofing, between "low," "medium" and "high." Thus, allowing user to balance security (robustness against spoofing) and convenience that they want. "Low" be the SL, easier to fool the system, higher the success rate of spoof attacks and lower the computation time; "High" be the SL, harder to fool the system, lower the success rate of spoof attacks and higher the computation time. The keynote of our scheme is to intensify local and global structure differences in the feature space. Thus, depending on SL, the system selects unidescriptor- or multidescriptors-fusion-based liveness detection, thereby adopting complementary properties of image descriptors, since LUCID encodes the local information and CENTRIST global information. Moreover, POEM provides additional local and global structure information.

The proposed method encodes image patterns into enhanced feature vectors using LUCID, CENTRIST and POEM. The results are then fed to Support Vector Machine (SVM) classifiers. Dependent on SL, the final decision, whether input biometric is coming from a live person or not, is determined by either single SVM or decision-level

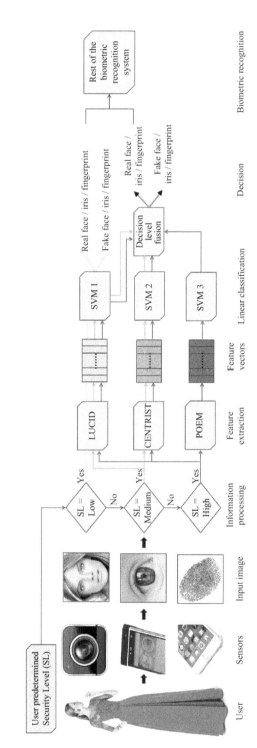

Figure 15.4 Block diagram of the MoBio_LivDet approach

fusion of individual SVM outputs using AND rule. We describe below MoBio_LivDet approach in detail.

When SL is set "Low," the system adopts only LUCID descriptor to analyze local features of face, fingerprint and iris input images and encodes local patterns into enhanced feature vector. The resulting vectors are then fed to SVM classifier, which determines whether feature description corresponds to a live person, or not.

LUCID is a novel approach to feature description based on order permutations, that is computable in linear time with respect to number of pixels and does not require floating point computation, beside the fact that typical mobile devices perform poorly for floating point applications. LUCID is simple, invariant to photometric transformations and noise; and implicitly encapsulate all possible intensity comparisons in a local area of an image. Let p be $n \times n$ image patch with c color channels. We can compute LUCID descriptor for the patch in one line of MATLAB® as shown in Figure 15.5. Here desc is the order permutation representations for p. Let $m = cn^2$, then a stable comparison-free linear time sort-based native implementation takes $O(m)$ time and space. LUCID has three parameters; blur kernel width, image patch size and the option to use color or gray-scale images. Before LUCID construction an averaging blur is applied to remove noise that may perturb order permutation.

When SL is set "Medium," in addition to LUCID-based analysis, system also uses CENTRIST features to capture the global representation of face, fingerprint and iris images. Hence, the key idea is to emphasize complementary local and global information differences in the feature space. Each resulting representation is then fed to respective SVM classifiers and decision-level fusion using AND rule is applied to combine the outputs of individual SVMs for spoof detection.

CENTRIST is a holistic representation that captures structural properties of an image by modeling distribution of local structures. CENTRIST is fast, easy to implement and robust to illumination and gamma variations. Its computational cost is linear in number of pixels of the interested region. CENTRIST is based on Census Transform (CT), which compares the intensity value of a pixel with its eight neighboring pixels. If the center pixel is bigger than (or equal to) one of its neighbors, a bit 1 is set in the corresponding location. Otherwise, a bit 0 is set. The bit stream resulting from the eight comparisons for each individual pixel is then converted into a base-10 number in [0 255], as illustrated in Figure 15.5. Once all CT values are calculated, one can easily transform them into a histogram with 256 bins which results into so called CENTRIST descriptor.

When SL is set "High," system employs POEM descriptor in conjunction with LUCID and CENTRIST to additionally enhance both local and global representations. Thus, image description consists of three enhanced feature vectors, which encode the local information and global structures of the biometric images. Feature vectors are then fed to respective SVM classifiers and outputs of individual SVMs are fused using decision level and fusion rule to ascertain whether the input image corresponds to a live person or not.

POEM is an oriented spatial multiresolution descriptor that captures rich local and global information (self-similarity structure) by the distribution of local intensity gradients or edge directions. POEM is robust to local image transformations due to

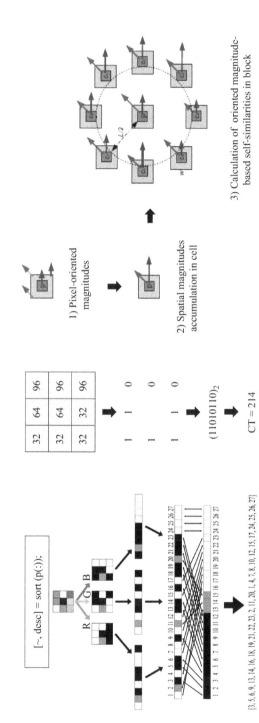

Figure 15.5 *Left: LUCID feature construction method in 1 line of MATLAB (Top). An illustration of an image patch split into its RGB color channels, vectorized and then sorted; inducing a permutation on the indices (Bottom). Middle: Calculation of CT value in CENTRIST. Right: Main steps of POEM feature extraction*

variations of pose, lighting, expression and occlusion. Moreover, POEM is of low complexity and is therefore practical for real-time applications.

The POEM feature extraction consists of following three steps as also shown in Figure 15.5 (see [16] for further details). First, the gradient image is computed, and then orientation of each pixel is discretized over $0-\pi$ (unsigned) or $0-2\pi$ (signed). The second step is to incorporate information from neighboring pixels by computing a local histogram of gradient orientations over all pixels within a cell. In the final step, the accumulated magnitudes are encoded using the self-similarity LBP-based operator within a larger image patch. The final POEM feature set at each pixel is the concatenation of all unidirectional POEMs.

15.3.1 Experiments

In the experimental validation, we have used three publicly available datasets for face, fingerprint and iris spoofing introduced above in Sections 15.2.1, 15.2.2 and 15.2.3, i.e., PRINT-ATTACK DB [35], ATVS-FFp DB [11] and ATVS-FIr DB [61].

For each above-mentioned data set, we used randomly selected 40% of the users (real and spoofed) as training set, whereas the remaining 60% users (real and spoofed) were used to build the testing set as in [11,31,33,61]. We run the above procedure five times. Reported results are average values over the 5 runs with standard deviations. We used LUCID-24-RGB for face, LUCID-24-Gray for iris and fingerprint, which respectively are LUCID on image patches of size 24×24 in RGB color and gray-scale. Before LUCID construction a 5×5 averaging blur is applied. POEM parameters were set as unsigned representation with 3 bins, built on 8×8 pixel blocks and 7×7 pixel cells. The experiments were conducted on Asus K52F laptop with 2.13 GHz Intel dual-core CPU with 3 GB of RAM using unoptimized MATLAB code. The average time required for feature extraction is reported in Table 15.8, which also explains the additional rationale behind the use of specific features in a definite fashion, besides the use of local and global information as primary motive. A liveness detection method is subject to two types of errors, either the real access is rejected (false rejection) or a spoofing attack is accepted (false acceptance). We evaluated performance of liveness detection systems as in [10,11,30,31,33–36] using Half Total Error Rate (HTER), that combines the False Rejection Rate (FRR) and the False Acceptance Rate (FAR) as: $HTER(\tau, D) = [FAR(\tau, D) + FRR(\tau, D)]/2$ [%]; where D denotes the used data set. FAR and FRR are strongly interrelated since both depends on the same threshold τ. We have chosen threshold τ on the equal error rate (EER) at training set and reported HTER using test set data.

In Table 15.8, we report the performance comparison between the proposed approach and individual LUCID, CENTRIST and POEM descriptors along with corresponding existing methods in discriminating live face, iris and fingerprint images from fake ones. Several observations may be extracted from Table 15.8: (i) the proposed method presents a very high potential as a simple, fast and novel method to detect spoofing attacks on mobile devices, which reaches a great classification accuracy for different biometric traits ("multibiometric"); (ii) error rates goes on decreasing as SL goes from low to high; (iii) it is possible to detect face/iris/fingerprint spoofing from single image; (iv) unlike state-of-the-art, presented individual image feature descriptors, i.e., LUCID/CENTRIST/POEM, seems to deliver encouraging liveness detection

Table 15.8   *Performance [Half Total Error Rate (HTER)-%] comparison between the proposed approach and individual LUCID, CENTRIST and POEM feature descriptors along with existing iris [9], face [45] and fingerprint [57] methods in discriminating live face, iris and fingerprint images from fake ones. The average time required for feature extraction is reported in the last column*

Method	Iris (ATVS)	Face (print attack)	Face (NUAA)	Fingerprint (ATVS)	Time (s)
LUCID (SL = Low)	1.03 ± 0.34	2.88 ± 0.88	1.54 ± 0.16	7.17 ± 1.97	0.003
CENTRIST	0.29 ± 0.15	6.84 ± 1.12	3.83 ± 0.30	3.84 ± 1.76	0.004
POEM	0.19 ± 0.17	0.43 ± 0.25	0.13 ± 0.04	2.20 ± 2.05	0.042
SL = Medium	0.05 ± 0.02	0.40 ± 0.20	0.08 ± 0.05	0.98 ± 0.24	0.007
SL = High	0.01 ± 0.01	0.10 ± 0.04	0.06 ± 0.03	0.25 ± 0.35	0.049
Exiting method	4.66 ± 1.15	4.54 ± 1.35	0.54 ± 0.10	14.22 ± 4.10	0.081

performance for all three modalities, namely face, iris and fingerprint; (v) even when "SL = Low," the proposed method performed better than existing methods on all data sets (except NUAA); (vii) The computational time required by proposed method is much lower than existing methods. In particular, when "SL = Low," it is comparable with others while performing better, which is vital for mobile applications.

First, we perform experiments to analyze discrete performances of LUCID, CEN-TRIST and POEM image representation. We can notice in Table 15.8 that LUCID descriptor performed best for iris liveness detection (HTER = 1.03%), though the spoofing are of high-quality iris printed images. While the method in [61], which combines two focus (IQF15 and IQF16), two occlusion (IQF3 and IQF19) and one pupil dilation (IQF22) features, achieved 4.66% HTER. It is also notable that in proposed approach contrary to existing methods, iris detection and segmentation is not required for liveness detection, thus making the process more faster. The HTERs of face spoofing detection on Print Attack and NUAA data sets are 2.88% and 1.54%, respectively, whereas the reflectance-analysis-based approach in [43] attained 4.54% and 0.54% HTER on Print Attacks and NUAA, respectively. We can also observe that LUCID and LPQ-based method in [49] attained 7.17% and 14.22% HTER for fingerprint spoofing, respectively. The considerable error rate may be occurring due to less quality difference in spoofed and real fingerprint images. LUCID and LPQ features thus may contain noisy and redundant information that leads to the overlap between spoofed and live classes and hence their bad performance.

It is evident that CENTRIST performs better than LUCID. The only exception is face spoofing. Our conjecture is that, this behavior is due to the fact that the employed face data sets contain variations due to spoof production procedure and CENTRIST is not robust to affine transformations [16]. Moreover, holistic representations are generally known to perform poor for faces. On the other hand, POEM proved to be significantly better than others for all traits, due to use of both local and global information.

Results clearly show that the LUCID, CENTRIST and POEM descriptions are able to provide complementary information to enhance the performances. For instance, when "SL = Low" (LUCID only) the HTER for fingerprint liveness detection reaches to 7.04%, while it decreases to 0.25% in case of "SL = High" (LUCID + CENTRIST + POEM).

All in all, user can get noteworthy security and convenience against spoofing even when he/she sets SL low. The proposed technique is simple, fast and effective, which makes it highly suitable for real-time or mobile devices. As the method does not deploy any trait-specific property (e.g., minutiae points, face detection or iris position), thus the computational load is minimized. Moreover, contrary to the existing approaches, we propose to employ only one image descriptor (LUCID/CENTRIST/POEM) to gain notable accuracy for three biometric traits, viz., face, iris and fingerprint, liveness detection. In addition, our method uses only one image for liveness detection, which can also be used for biometric recognition.

15.4   Research opportunities

15.4.1   Mobile liveness detection

A recent industry survey estimated that by 2018, almost 3.4 billion users would access biometrics on their devices [1]. However, very little research on mobile biometric antispoofing has been conducted. Moreover, existing liveness detection methods are unsuited for mobile applications because of the complex features they analyze or high computational cost. So, to make such applications more practical, researchers must address the issue of presentation attacks on mobile devices.

15.4.2   Mobile biometric spoofing databases

The Lack of public databases containing spoofing attacks on mobile devices has further stymied research on mobile antispoofing. Many commercial vendors have incorporated biometrics authentication in mobile devices, which have been proved to be vulnerable to spoofing. However, there exist very few datasets containing spoofing attacks. In particular, still no fingerprint spoofing dataset using mobile devices is available publicly. Likewise, existing datasets are quite small and do not cover different spoofing methods and scenarios. There is vital need to collect and release in public domain several challenging and large mobile biometric spoofing datasets with diverse spoofing scenarios, techniques and sensors.

15.4.3   Generalization to unknown attacks

Reported studies [8,14] suggest a threefold increase in the error rate of the antispoofing methods when spoofs using new materials (not used during the training stage) are encountered during the testing or operational stage. This means that the generalization capability (i.e., interoperability) of existing antispoofing methods is limited across materials, sensors and datasets. A particular emphasis should be put on generalized antispoofing methodologies that are capable of detecting varying or previously unseen spoofing attacks.

15.4.4   Randomizing input biometric data

Another defensive measure to spoofing is where users are required to enroll multiple biometric samples, such as several fingerprints or face and fingerprint. Verification will then randomize the sample requested thus adding complexity to any attempt to circumvent the biometric authentication. This procedure would be a variation of challenge-response strategy to counteract the spoofing attacks. How and where to utilize this strategy has been mostly ignored and has not been much explored. The developed methods should be user-friendly and less computationally expensive. This kind of defensive measures will also assist in defeating attempts to reuse, e.g., latent fingerprints on the fingerprint reader.

15.4.5 Fusion of biometric system and countermeasures

Most works on antispoofing tend to focus only on spoof detection part, hence omitting to integrate the counter measure into a recognition system that consequently damages the accuracy of the biometric verification systems. The essential challenge is devising techniques for combining antispoofing and biometric recognition system without compromising the robustness to spoofing attacks and recognition accuracy, which is relatively under explored subject.

15.5 Conclusion

In spite of big advances in mobile biometrics, even commercial products developed by the most advanced technological vendors fail mostly to withstand the problem posed by spoofing attacks. Various countermeasures to spoofing attacks have been devised, nevertheless formulating efficient protection methods against this threat has proven to be a challenging task yet. In this context, we have presented a comprehensive overview of the state-of-the-art in mobile biometric spoofing and antispoofing with special attention to three largely deployed modalities (i.e., fingerprint, face and iris).

After the general overview of the progress in the field of mobile biometric antispoofing, a multibiometric prototype named **MoBio_LivDet**, which unlike the existing methods is trait-independent and can be used for face, iris and fingerprint spoofing attack detection on mobile devices, is presented. The proposed approach allows one to balance security (robustness against spoofing) and convenience that the user wants. Moreover, it is simple, fast, effective and does not require user-cooperation, making it hence ideal for mobile platforms. We believe that it may serve as a stimulus for the development of generalized mobile liveness detection. Lastly, we have outlined few of the potential research directions in mobile biometric antispoofing to evolve the state-of-the-art.

References

[1] Goode, A.: 'Mobile Biometric Security Market Forecasts 2013–2018'. *Goode Intelligence*, October 2013.
[2] Akhtar, Z., Micheloni, C., and Foresti, G. F.: 'Biometric Liveness Detection: Challenges and Research Opportunities'. *IEEE Security & Privacy*, 2015; **13**(5):63–72.
[3] Akhtar, Z.: 'Security of Multimodal Biometric Systems against Spoof Attacks'. PhD thesis, University of Cagliari, Italy, 2012.

[4] Marcel, S., Nixon, M.S., and Li, S.Z. (eds.): *Handbook of Biometric Anti-Spoofing*. Berlin: Springer, 2014.

[5] Sequeira, A.F., Oliveira, H.P., Monteiro, J.C., Monteiro, J.P., and Cardoso, J.S.: 'MobILive 2014 – Mobile Iris Liveness Detection Competition'. *Proc. Int. Joint Conf. on Biometrics (IJCB)*, 2014, pp. 1–6.

[6] Parthasaradhi, S.T.V., Derakhshani, R., Hornak, L.A., and Schuckers, S.A.C.: 'Time-series detection of perspiration as a liveness test in fingerprint devices'. *IEEE Trans. in Systems, Man, and Cybernetics, Part C: Applications and Reviews*, 2005:**35**(3):335–343.

[7] Patel, K., Han, H., Jain, A.K., and Ott, G.: 'Live face video vs. spoof face video: Use of moiré patterns to detect replay video attacks'. *Proc. Int. Conf. in Biometrics*, 2015, pp. 98–105.

[8] Akhtar, Z., Micheloni, C., and Foresti, G.L.: 'Correlation based fingerprint liveness detection' *Proc. International Conference on Biometrics (ICB)*, 2015, pp. 305–310.

[9] Tait, B.: 'Secure cloud based biometric signatures utilizing smart devices'. *Proc. International Conference on Cloud Security Management*, 2014, pp. 109–119.

[10] Matta, J., Hadid, A., and Pietikäinen, M.: 'Face spoofing detection from single images using texture and local shape analysis', *IET Biometrics*, 2012:1(1): 3–10.

[11] Galbally, J., Alonso-Fernandez, F., Fierrez, J., and Ortega-Garcia, J.: 'A high performance fingerprint liveness detection method based on quality related features'. *Future Generation Computer Systems*, 2012:**28**(1):311–321.

[12] Gragnaniello, D., Sansone, C., and Verdoliva, L.: 'Iris liveness detection for mobile devices based on local descriptors'. *Pattern Recognition Letters*, Volume, 2015:**57**(1):81–87.

[13] Kollreider, K., Fronthaler, H., and Bigun, J.: 'Non-intrusive liveness detection by face images'. *Image and Vision Computing*, 2009:**27**(2):233–244.

[14] Rattani, A., Scheirer, W.J., and Ross, A.: 'Open set fingerprint spoof detection across novel fabrication materials'. *IEEE Transactions on Information Forensics and Security*, 2015:(**10**):11:2447–2460.

[15] Ziegler, A., Christiansen, E., Kriegman, D., and Belongie, S.: 'Locally uniform comparison image descriptor'. *Proc. Advances in Neural Information Processing Systems*, 2012, 1–9.

[16] Wu, J., and Rehg, J.M.: 'CENTRIST: A visual descriptor for scene categorization'. *IEEE Transactions on Pattern Analysis and Machine Intelligence*, 2011:**33**(8):1489–1501.

[17] Vu, N., and Caplier, A.: 'Face recognition with patterns of oriented edge magnitudes'. *Proc. European Conference on Computer Vision*, 2010, pp. 313–326.

[18] Information Technology—Biometric Presentation Attack Detection—Part 1: Framework, ISO/IEC DIS 30107-1, *Int'l Org. for Standardization*, 2014.

[19] Akhtar, Z., Fumera, G., Marcialis, G.L., and Roli, F.: 'Evaluation of serial and parallel multibiometric systems under spoofing attacks'. *Proc. IEEE Int. Conf. on Biometrics: Theory, Applications and Systems*, 2012, pp. 283–288.

[20]   Pan, G., Lin, S., Zhaohui, W., and Shihong, L.: 'Eyeblink-based anti-spoofing in face recognition from a generic webcamera'. *Proc. Int. Conf. on Computer Vision*, 2007, pp. 1–8.

[21]   Omar, L., and Ivrissimtzis, I.: 'Evaluating the resilience of face recognition systems against malicious attacks'. *Proc. Computer Vision Student Workshop (BMVW)*, 2015, 51–59.

[22]   Tan, X., Li, Y., Liu, J., and Jiang, L.: 'Face liveness detection from a single image with sparse low rank bilinear discriminative model'. *Proc. European Conference on Computer Vision (ECCV)*, 2010, pp. 504–517.

[23]   Maria, D.M., Chiara, G., Michele, N., and Daniel, R.: 'FIRME: face and iris recognition for mobile engagement'. *Image and Vision Computing*, 2014:**32**(12):1161–1172.

[24]   Li, Y., Li, Y., Yan, Q., Kong, H., and Deng, R.H.: 'Seeing your face is not enough: an inertial sensor-based liveness detection for face authentication'. *Proc. ACM Conf. on Comp. and Comm. Security (CCS)*, 2015, pp. 1–8.

[25]   Frischholz, R.W., and Dieckmann, U.: 'BioID: a multimodal biometric identification system. *IEEE Computer*, 2000:**33**(2):64–68 [https://www.bioid.com/Blog/2015/6/30/guarding-against-biometric-fraud-with-liveness-detection].

[26]   White paper: 'Fastfill and netverify mobile implementation guide for Android'. *Jumio*, 2014 [https://www.jumio.com/2015/02/jumio-unveils-new-id-scan-image-capturing-technology-in-latest-version-of-award-winning-netverify/].

[27]   Bao, W., Li, H., Li, N., and Jiang, W.: 'A liveness detection method for face recognition based on optical-flow field'. *Proc. IEEE Int'l Conf. on Image Analysis and Signal Processing*, 2009, pp. 233–236.

[28]   Wang, T., Yang, J., Lei, Z., Liao, S., and Li, S.Z.: 'Face liveness detection using 3d structure recovered from a single camera'. *Proc. IEEE/IAPR Int. Conf. on Biometrics* (ICB), 2013, pp. 1–6.

[29]   Li, J., Wang, Y., Tan, T., and Jain, A. K.: 'Live face detection based on the analysis of Fourier spectra'. *Proc. SPIE Biometric Tech. for Human Identification*, 2004, pp. 296–303.

[30]   Erdogmus, N., and Marcel, S.: 'Spoofing face recognition with 3D masks'. *IEEE Transactions on Information Forensics and Security*, 2014:**9**(7): 1084–1097.

[31]   Akhtar, Z., Michelon, C., and Foresti, G.L., 'Liveness detection for biometric authentication in mobile applications'. *Proc. Int. Conference on Security Technology (ICCST)*, 2014, pp. 1–6.

[32]   Akhtar, Z., Micheloni, C., Piciarelli, C., and Foresti, G.L.: 'MoBio_LivDet: mobile biometric liveness detection'. *Proc. IEEE Int. Conf. on Advanced Video and Signal Based Surveillance*, 2014, pp. 187–192.

[33]   Galbally, J., Marcel, S., and Fierrez, J.: 'Image quality assessment for fake biometric detection: application to iris, fingerprint, and face recognition'. *IEEE Transactions on Image Processing,* 2014: **23**(2):710–724.

[34]   Wen, D., Han, H., and Jain, A.K.: 'Face spoof detection with image distortion analysis'. *IEEE Transactions on Information Forensics and Security*, 2015:**10**(4), pp. 746–761.

[35]   Chakka, M.M., Anjos, A., Marcel, S., *et al.*: 'Competition on countermeasures to 2-d facial spoofing attacks'. *Proc. IEEE Int. Joint Conf. on Biometrics (IJCB)*, 2011, pp. 1–6.

[36]   Chingovska, I., Yang, J., Lei, Z., *et al.*: 'The 2nd competition on counter measures to 2d face spoofing attacks'. *Proc. IAPR Int. Conf. on Biometrics (ICB)*, 2013, pp. 1–6.

[37]   Zhang, Z., Yi, D., Lei, Z., and Li, S.Z.: 'Face liveness detection by learning multispectral reflectance distributions'. *Proc. Conf. Face and Gesture*, 2011, pp. 436–441.

[38]   Dhamecha, T. I., Nigam, A., Singh, R., and Vatsa, M.: 'Disguise detection and face recognition in visible and thermal spectrums'. *Proc. IEEE Int. Conf. on Biometrics (ICB)*, 2013, pp. 1–6.

[39]   Chetty, G., and Wagner, M.: 'Liveness detection using cross-modal correlations in face-voice person authentication'. *Proc. Int. Conf. on Speech Communication Association (Interspeech)*, 2005, pp. 2181–2184.

[40]   Peixoto, B., Michelassi, C., and Rocha, A.: 'Face liveness detection under bad illumination conditions'. *Proc. IEEE Int. Conf. on Image Processing (ICIP)*, 2011, pp. 3557–3560.

[41]   Zhang, Z., Yan, J., Liu, S., Lei, Z., Yi, D., and Li, S. Z.: 'A face anti-spoofing database with diverse attacks'. *Proc. Int. Conference on Biometrics (ICB)*, 2012, pp. 26–31.

[42]   Menotti, D., Chiachia, G., Pinto, A., *et al.*: 'Deep representations for iris, face, and fingerprint spoofing detection'. *IEEE Transactions on Information Forensics and Security*, 2015:**10**(4):864–879.

[43]   Kose, N., and Dugelay, J.L.: 'Mask spoofing in face recognition and countermeasures'. *Image and Vision Computing*, 2014:**32**(10):779–789.

[44]   Zhang, Y., Tian, X., Chen, X., Yang, X., and Shi, P.: 'Fake finger detection based on thin-plate spline distortion model'. *Advances in Biometrics Lecture Notes in Computer Science*, 2007, pp. 742–749.

[45]   Espinoza, M., and Champod, C.: 'Using the number of pores on fingerprint images to detect spoofing attacks'. *Proc. International Conference on Hand-Based Biometrics*, 2011, pp. 1–5.

[46]   Manivanan, N., Memon, S., and Balachandran, W.: 'Automatic detection of active sweat pores of fingerprint using highpass and correlation filtering'. *Electronic Letters*, 2010(**46**):18:1268–1269.

[47]   Caldwell, T.: 'NexID biometrics brings liveness detection to Android'. *Biometric Technology Today*, 2014:(6):3–4.

[48]   Nikam, S., and Agarwal, S.: 'Ridgelet-based fake fingerprint detection'. *Neurocomputing*, 2008:(72):2491–2506.

[49]   Ghiani, L., Marcialis, G.L., and Roli, F.: 'Fingerprint liveness detection by local phase quantization'. *Proc. International Conference on Pattern Recognition (ICPR)*, 2012, pp. 537–540.

[50]   Marasco, E., Ding, Y., and Ross, A.: 'Combining match scores with liveness values in a fingerprint verification system'. *IEEE International Conference on Biometrics: Theory, Applications and Systems*, 2012, pp. 418–425.

[51] Avanaki, N., Reza, M., Meadway, A., *et al.*: 'Anti-spoof reliable biometry of fingerprints using en-face optical coherence tomography'. *Optics and Photonics Journal*, 2011:(**01**):3:91–96.

[52] Li, S.Z., and Jain, A.K.: *Encyclopaedia of Biometrics*. Berlin: Springer, 2015.

[53] Schuckers, S., Tan, A., Lewicke, A., *et al.*: 'Evaluation of liveness or anti-spoofing in biometric systems'. *Proc. Int'l Biometric Performance Testing Conference*, 2010, pp. 1–48.

[54] Marcialis, G.L., Lewicke, A., Tan, B., *et al.*: 'First international fingerprint liveness detection competition—LivDet 2009'. *Proc. Int. Conf. Image Analysis and Processing*, 2009, pp. 1–6.

[55] Ghiani, L., Yambay, D., Mura, V., *et al.*: 'LivDet 2013 fingerprint liveness detection competition 2013.' *Proc. Int. Conf. on Biometrics (ICB)*, 2013, pp. 1–6.

[56] Mura, V., Yambay, D., Ghiani, L., Marcialis, G.L., Schuckers, S., and Roli, F.: 'LivDet 2015 – fingerprint liveness detection competition 2015'. *Proc. Int. Conf. on Biometrics: Technology, Applications and Systems*, 2015, pp. 1–6.

[57] Derakhshani, R., Schuckers, S., Hornak, L.A., and O'Gorman L.: 'Determination of vitality from a non-invasive biomedical measurement for use in fingerprint scanners', *Pattern Recognition*, 2003(36):2:383–396.

[58] Cappelli, R., Maio, D., and Maltoni, D.: 'Modelling plastic distortion in fingerprint images'. *Proc. International Conference on Advances in Pattern Recognition*, 2001, pp. 371–378.

[59] Pacut, A., and Czajka, A.: 'Aliveness detection for iris biometrics'. *Proc. IEEE International Carnahan Conferences Security Technology*, 2006, pp. 122–129.

[60] Lee, S., Park, K., and Kim, J.: 'A study on fake iris detection based on the reflectance of the iris to the sclera for iris recognition'. *Proc. ITC-CSCC*, 2005, pp. 397–403

[61] Galbally, J., Ortiz-Lopez, J., Fierrez, J., and Ortega-Garcia, J.: 'Iris liveness detection based on quality related features'. *Proc. Intl. Conference on Biometrics (ICB)*, 2012, pp. 271–276.

[62] Sequeira, A.F., Murari, J., and Cardoso, J.S.: 'Iris liveness detection methods in mobile applications'. *Proc. Intl. Conference on Computer Vision Theory and Applications (VISAPP)*, 2014, pp. 22–33.

[63] Yambay, D., Doyle, J.S., Bowyer, K.W., Czajka, A., and Schuckers, S.: 'LivDet-iris 2013-iris liveness detection competition 2013'. *Proc. IEEE Int. Joint. Conf. on Biometrics (IJCB)*, 2014, pp. 1–8.

[64] Lee, E.C., Park, K.R., and Kim, J.: 'Fake iris detection by using Purkinje image'. *Proc. ICB*, 2005, pp. 397–403.

[65] Bodade, R., and Talbar, S.: 'Dynamic iris localisation: a novel approach suitable for fake iris detection'. *Proc. Intl. Conf. on Ultra Modern Telecommunications & Workshops*, 2009, pp. 1–5.

[66] Kanematsu, M., Takano, H., and Nakamura, K.: 'Highly reliable liveness detection method for iris recognition'. *Proc. Annual Conference on SICE*, 2007, pp. 361–364.

Chapter 16
Biometric open protocol standard
Scott Streit[1] and Hector Hoyos[2]

16.1 Introduction

Convenience drives consumers toward the biometrics-based access management solutions, say studies from Ericsson, PayPal, IBM®, and Microsoft®.

According to the Ericsson study "Your body is the new password," 52% of smartphone users want to use their fingerprints instead of a password, a further 61% want to use fingerprints to unlock their phones, and 48% want to use eye recognition.

The study conducted by PayPal says that consumers approve biometrics for access management. In terms of readiness to switch from traditional password protection to the new technology, 53% of the surveyed population would be comfortable replacing passwords with fingerprints, and 45% would choose a "retinal scan," which is presumably an iris scan (the misplaced terminology points to the lack of consumer education).

IBM Fellow and Speech Chief Technology Officer David Nahamoo states that, over the next 5 years, your unique biological identity and biometric data—facial definitions, iris scans, voice files, even your DNA—will become the key to safeguarding your personal identity and information and will replace the current user-ID-and-password system.

A Microsoft Research-funded study titled "The Quest to Replace Passwords: A Framework for Comparative Evaluation of Web Authentication Schemes," concluded that the vast password-replacement transition should conform to the following criteria: nothing to carry, efficient to use, and easy recovery from a loss. The Microsoft study goes as far as concluding that such criteria could be achieved mostly through the biometric schemes.

Biometric technologies provide consumers with a long-awaited convenience to securely enter cyberspace on the front end. The biometric open protocol standard (BOPS), developed by Hoyos Labs, protects digital assets and digital identities on the backend.

BOPS is a biometrics-agnostic standard that opens an application programing interface (API) for registered developers. Entering as a game-changer, BOPS communication architecture enables two-way secure sockets layer (SSL) or transport

[1]Computer Science Innovations, LLC, USA
[2]Hoyos Labs, USA

layer security (TLS) connection over the encryption mechanism to the server, which employs an intrusion detection system (IDS). The IDS is an external system responsible for blacklisting devices that are violating the replay portion of this specification.

Identity assertion, role gathering, multilevel access control, assurance, and auditing are provided by the BOPS. The BOPS implementation includes software running on a client device (smartphone or mobile device), a trusted BOPS server, and an IDS. The BOPS implementation allows pluggable components to replace existing components' functionality, accepting integration into current operating environments in a short period of time. The BOPS implementation provides continuous protection to the resources and assurance of the placement and viability of adjudication and other key features. Accountability is the mechanism that proves a service-level guarantee of security. The BOPS implementation allows the systems to meet security needs by using the API. The BOPS implementation need not know whether the underlying system is a relational database management system (RDBMS) or a search engine. The BOPS implementation functionality offers a "point-and-cut" mechanism to add the appropriate security to the production systems as well as to the systems in development. The architecture is language neutral, allowing representational state transfer (REST), JavaScript object notation (JSON), and SSL implemented as TLS to provide the communication interface. The architecture is built on the servlet specification, Open SSLs, Java, JSON, REST, and an open persistent store. All tools adhere to open standards, allowing maximum interoperability.

16.2   Overview

16.2.1   Scope

The BOPS provides identity assertion, role gathering, multilevel access control, assurance, and auditing. The BOPS implementation includes software running on a client device (e.g., smartphone or mobile device), a trusted BOPS Server, and an IDS. The BOPS implementation allows pluggable components to replace existing components' functionality, accepting integration into the current operating environments in a short period of time. The BOPS implementation adheres to the principle of continuous protection in adjudicating access to resources. Accountability is the mechanism that proves a service-level guarantee of security. The BOPS implementation allows the systems to meet security needs by using the API as a RESTFul interface that used 2-way Secure Socket Layers (SSL). The BOPS implementation need not know whether the underlying system is a RDBMS or a search engine. The BOPS implementation functionality offers a "point-and-cut" mechanism to add the appropriate security to the production systems as well as to the systems in development.

16.2.2   Purpose

This standard provides a biometric agnostic, multilevel security protocol.

16.2.3 Intended audience

The intended audience of this document includes security evaluators, system underwriters, developers, and systems engineers. The BOPS is subject to changes and updates.

16.3 Definitions, acronyms, and abbreviations

16.3.1 Definitions

For the purposes of this document, the following terms and definitions apply:

Account: A user account that was validated (against an external system or by an email validation mechanism). It can have associated one or multiple mobile devices. The enrollment process ends by creating a client certificate for the device that shall be used for subsequent calls to authenticate against BOPS.

Bell-LaPadula: The multilevel model that was proposed by Bell and LaPadula for enforcing access control in government and military applications. A subject can only access objects at certain levels determined by their security level.

JUnit: A testing framework for Java programing language.

RESTful: Refers to REST, which is a software architecture style.

SHA512: Secure hash algorithm 512, providing one way encryption.

16.3.2 Acronyms and abbreviations

admin	administrator
AOP	aspect-oriented programing
API	application programing interface
app	a mobile client application
BOPS	biometric open protocol standard
CPU	central processing unit
CSRF	crosssite request forgery
ID	identifier
IDS	intrusion detection system
IP	internet protocol
JSON	JavaScript object notation
LDAP	lightweight directory access protocol
MAC	mandatory access control
NSA	National Security Agency (US)
QR	code quick response code
RDBMS	relational database management system
REST	representational state transfer
SSL	secure socket layer
TCSEC	trusted computer system evaluation criteria
TLS	transport layer security
URL	uniform resource identifier
XNTP	extended network time protocol

16.4   Conformance

The BOPS comprises the rules governing secure communication between a variety of client devices and the trusted server. This standard is based on the tested computer-based implementation of the Trusted Computer System Evaluation Criteria (TCSEC).

BOPS conforms to the TCSEC, which is the US Department of Defense Standard that sets basic requirements for assessing the effectiveness of computer security controls built into a computer system. TCSEC was created by the National Computer Security Center, an arm of the National Security Agency (NSA) and is also frequently referred as "Orange Book, section B1." BOPS also conforms to the Director of the Central Intelligence Directive 6/3 protection level 3, level 4, and level 5 (PL3, PL4, and PL5), and to the standards of the Multiple Independent Levels of Security/Safety (MILS) architecture.

16.5   Security considerations

BOPS largely considers the server side component to an end to end biometric solution. This solution is recorded as IEEE standard 2410–2015 and describes all the components necessary for the server side of end-to-end biometric security. The standard lists requirements for the client component to comply with the Server Side Solution. The standard describes a set of new technologies including but not limited to liveness. Liveness is a term that means insuring the biometric is from a live source, such as a human being and not from an imitation such as a wax replica, a picture, or a video.

This technology may be implemented on a public cloud such as Amazon Web Services or any private cloud. The specification makes no assumptions of the "how" of a solution. The specification describes the "what" and the overall goals of the solution.

Users no longer have to remember their password, or worse yet, having their passwords end up in the wrong hands. BOPS provides a framework for safety and convenience. Biometrics is part of us, unique to us, and excellent for leveraging in a secure solution.

Using multiple rounds of penetration tests, the IEEE vetting process, industry and client critiques, BOPS has been validated. This mature and secure framework provides an end-to-end solution without the compromise of privacy.

16.5.1   Background

One of the BOPS's main functions is to provide authentication instead of authorization in a way that the server does not retain the client information but rather recognizes one client from another. The key components of security considerations include identity assertion, role gathering, access control, auditing, and assurance.

BOPS has some basic rules which include:

(1)   All passivated biometric data is encrypted.
(2)   Compromising the server yields no useful information mitigating insider risk.

(3)   Compromising any client yields no useful information mitigating local hacks.
(4)   White hat hackers may have the client source code and still not penetrate the system.

Passivated data, or data at rest is the subject of incredible debate. Data stored in a relational database is plaintext. This plaintext storage allowed automated access, but is susceptible to inside attacks. The internal "badactor" garners access to plaintext and compromises privacy. On the other hand, encrypted passivated data protects internal compromises, but makes the use of automated querying nearly impossible. The reasonable compromise for BOPS is linkages, including biometrics, or partial biometrics is encrypted. Supplemental profile data is plaintext, but with no reference to an identity is rendered useless. Section 16.5.3 describes the approach in detail.

Points 2 and 3 occur by encrypting and splitting the Initial Biometric Vector.

White hat hackers having the client source code highlights the lofty goals of BOPS. Even with source code, the system is impervious to attack.

Authentication is who am I. Authorization is what am I allowed to do.

To do authentication, we have to determine the original identity. This prevents the case of enrolling as the wrong person. We also have a level of assurance on the original identity. To do this, we will examine components that make up authorization, Genesis, Enrollment, and Match.

To do Authorization, we require a variety of things. First of all, we must gather the roles, dictating what an identified user can do or data than can see. Then we have access control requirements, including the optional labeling of data and the storage of the actual data.

We may store linkages to externally stored data, or stored the labeled data internally to BOPS. In either case, BOPS provides access control by implementing a role-based adjudication model.

For authentication we assert an identity. For authorization we ask, may this identity perform the following operation. They are vastly different questions.

When we enroll, we perform a variety of operations after Genesis, including the gathering of a certificate, the creation of an Initial Biometric Vector and the encryption and decryption of data necessary for the handshake.

When we authorize we compare the roles associated with the Identity, with the roles associated with an action or the access of data. If the roles of the user dominate the roles necessary to perform the action or access the data, the request is allowed.

16.5.2 Genesis

A key component of the End-to-End solution is the decoupling of Genesis and Enrollment. Genesis is how we determine the identity of a subject without concern for the biometric, certificate, or any other automated processing.

BOPS solution is open and it can enable any customization for Genesis flow. Some examples of Genesis could include a username and password access to Active Directory, a validating email or text message, or an officer of an organization physically verifying the identity.

Preregistration of user account may occur in batch is based on business requirements. Ultimately, Genesis forms a full dependency on risk management and may determine downstream processing.

Postgenesis, the user enrolls his or her biometric(s). The enrollment process includes a unique client certificate issued for the enrolled device. In addition, a One-Time Password (seed) is established between Mobile and Server and additional seed value for replay attack prevention.

A user may have many devices, a device may have many users. A device may have many biometrics. This many-to-many relationship occurs through the separation of Genesis and Enrollment. The identified subject through Genesis may enroll many times with many biometrics.

The enrollment process uses a 2-way SSL certificate which is server generated. This generation occurs after Genesis guaranteeing that the certificate is for the well-defined subject.

BOPS platform has provision in this direction and provides flexible mechanism to impose different security levels. For example, the highest level of Genesis may be the user validated in front of an officer. The lowest level may be a username and password in conjunction with a validating email. At the end of the day, the levels of Genesis, and the verification process is a business decision frequently unique to each and every organization.

Subsequent processing may change based on the Genesis level. For example, the systems may allow a one hundred thousand dollar transfer for a high level of Genesis, but only a one hundred dollar transfer for a lower level of Genesis.

The key component here is Genesis impacts authorization. This is an area where Identity and Authorization meet. The definition of these Genesis-oriented transactional rules occurs in the Administration Console.

16.5.3   Enrollment

BOPS addresses the speed of biometric authentication transaction and solves the problem of a virtualized threat on a mobile device. Such a threat assumes that an intruder decompiles the code on a copied virtual image of a mobile device, uses this source code to stop authentication calls, and attempts to get a control of a server that authenticates and grants permissions.

To mitigate these risks, BOPS encrypts the initial biometric value (IBV) without the encryption key, then stores a half of the IBV on the client device and the other half on the server. The biometric matching occurs on the server.

In this way, stolen device cannot bypass authentication, hence compromised device or server renders no useful information to an attacker.

These assumptions are essential for establishing a processing agreement to biometrics authentication in BOPS.

The biometric vector is split between the client and server.

The approach to authentication is biometric agnostic.

For face recognition, the size of the initial biometric vector is roughly 20k, which could be minimized by the up/down of an http-request and http-response, and therefore is accepted.

The splitting algorithm for face is: zero bit is the white and one bit the black. BOPS corresponds to visual cryptography.

An example of Visual Cryptography is below. It is important to note that Visual Cryptography works nicely with encryption, splitting the IBV and the reconstruction of the IBV without key management.

Example

Bit	Probability	Shares #1	Shares #2	Superposition of the two shares	
	$p = 0.5$				Zero Bit
	$p = 0.5$				
	$p = 0.5$				One Bit
	$p = 0.5$				

Black is 1, white is 0
For this example, assume the IBV is 0-0-1-0-0-1-1-0.

We prefer XOR reconstruction because the solution is pure Boolean.

Original	0	0	1	0	0	1	1	0
Share 1	01	10	10	01	10	10	01	01
Share 2	01	10	01	01	10	01	10	01
OR-reconstruction	01	10	11	01	10	11	11	01
XOR-reconstruction	00	00	11	00	00	11	11	00

The original biometric vector encryption occurs using visual cryptography, and the results of this are two vectors noted as sheets, which contain only white noise. The mobile storage contains one of the sheets and the server contains the other.

The verification process combines the two sheets using a simple Boolean operation which results in the original biometric vector fully reconstructed.

Figure 16.1 shows the (2,2) visual cryptography (VCS) where each bit encrypts into shares. Note that the choice of shares for a zero and one bit is a random process. When encoding a zero or one bit, take a value from the table for one share and the adjacent value in the table for the other share. At the end of the process, neither shares provide any clue about the original bit. Superimposing the two shares (using OR or XOR) determines the value of the original bit.

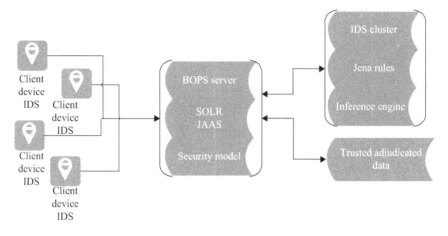

Figure 16.1   BOPS system diagram

This is an example for (2,2) VCS. VCS may extend to more than two shares by changing the random process probability.

Changing the probability of the random process from 0.5 to 0.25 results in the shares having 4 bits instead of the 2 bits present in the 0.5 example. In addition, changing the probability of the random process to 0.125 results in an encryption of 8 bits for each input bit.

16.5.4   Matching agreement

For a match, BOPS employs multiple initial biometric vectors. On the API side, there are two RESTful web services calls that communicate via TLS.

One call includes all IBVs (halves) in addition to the current (to the authentication session) biometric and returns a floating point value signifying the strength of the match.

Another call offers one IBV (half) at a time and the current biometric and returns a floating point value signifying the strength of the match. For the second call, it is assumed that there may be several consecutive calls: one IBV at a time to determine a match.

The sizing calculations per a matching agreement, as follows:

20 kb per face vector, 5 frames per second
for 10 s = 50 vectors
50 × 20 kb = 1,000 kb

The matching logistics could be described, as follows:

Sending 1,000 kb to the server to perform the matching. Matching one by one, once authentications session is matched, the server stops matching and sends the result.

Sending 100 kb to the server to perform the matching. If authentications session is not a match, ask for the second 100 kb, and so on, until a floating point value is determined.

According to the matching algorithm, the current frame requires 200 ms plus a 125-ms up/down time to the server. Hence, the frame transmission brings the transaction speed to 325 ms per frame plus the match. The match is upper bounded at 100 ms, then the frame transmission is roughly at 425 ms for the most common case.

If the most common case fails, then we send a batch of frames (e.g., five at a time), and attempt to match again. In the best case, BOPS conducts matching under a second, in the worst possible case, match happens in 10 s.

16.5.4.1   Identity assertion

The BOPS implementation helps provide continuous protection to resources and assurance of the placement and viability of adjudication and other key features. Accountability is the mechanism that can help provide a service-level guarantee of security.

The BOPS implementation identity assertion helps confirm that named users are who they claim to be. The identity assertion implies reliance on human biometrics; however, the BOPS is an interoperable standard and can incorporate any identity asserter or a number of different asserters associated with the same named user. The application of the IDS provides active monitoring to help prevent spoofing of the credentials set and to blacklist a subject or device that makes malicious attempts.

16.5.5   Role gathering

Role gathering is focused on the data confidentiality and privileged access that is based on the rules enforced by a known system. To determine whether a specific access mode is allowed, the privilege of a role is compared to the classification of the group to determine if the subject is authorized for a confidential access. The object's structure is defined by the access control. Role gathering occurs on the system level or through the client/server call. The BOPS implementation server stores role gathering information to associate a unique user with a unique device.

16.5.6   Access control

16.5.6.1   General

Access control asks whether a given subject (person, device, or program) may read, write, execute, or delete a given object. The community further divides access control into discretionary access control and a more granular form of access control called mandatory access control (MAC).

16.5.6.2   Discretionary access control

The BOPS implementation supports access control between the named users and the named objects (e.g., files and programs). The adjudication mechanism is role based and allows users and administrators to specify and control sharing of those objects by named individuals and defined groups of individuals. The discretionary access control mechanism provides that objects are protected from unauthorized access. Discretionary access control provides protection at the group or individual level across a single or group of objects. The granularity ranges from individual to group.

16.5.6.3   Mandatory access control

The BOPS implementation shall enforce a MAC policy over all subjects and storage objects under its control (e.g., processes, files, segments, devices). These subjects and objects are assigned sensitivity labels, which are a combination of hierarchical classification levels and nonhierarchical categories. The labels are used in the adjudication as the basis for MAC decisions. The client software shall maintain labels or have the BOPS server maintain the data in order to force adherence to labeling of the subject and objects. The BOPS implementation server maintains a trusted store as a component of BOPS.

The following requirements hold all accesses between subjects and objects controlled by the BOPS:

(a)   A subject can read an object only if the hierarchical classification in the subject's security level is greater than or equal to the hierarchical classification in the object's security level.
(b)   The nonhierarchical categories in the subject's security level include all the nonhierarchical categories in the object's security level.
(c)   A subject can write an object only if the hierarchical classification in the subject's security level is less than or equal to the hierarchical classification in the object's security level and all the nonhierarchical categories in the subject's security level are included in the nonhierarchical categories in the object's security level.
(d)   Identification and authentication data should be used by the BOPS to authenticate the user's identity and to maximize that the security level and authorization of subjects external to the BOPS that may be created to act on behalf of the individual user are dominated by the clearance and authorization of that user.

16.5.7   Auditing and assurance

16.5.7.1   Background

The worst possible case is when a system is compromised and the compromise goes undetected. To prevent this case, the aforementioned specifications require auditing and proof of the security model, which is called assurance.

16.5.7.2   Auditing

The BOPS implementation supports all auditing requests at the subject/object level or at the group level. The BOPS implementation uses aspect-oriented programing to maximize the likelihood that all calls are safely written to an audit trail. An interface of RESTful web services and JSON provides a mechanism to read the audit trail. Auditing may occur at the subject per action, the object per action, or the group per action. For example, a group of users called "accounting" may audit all writes to general ledger, or the chief financial officer may have audits for reads of the income statement.

16.5.7.3   System integrity

JUnit tests exist for all boundary conditions of the BOPS. The suite of tests includes testing all boundary components and conditions of the system.

16.6   BOPS interoperability

The BOPS implementation allows the systems to meet security needs by using the API. The BOPS implementation does not need to know whether the underlying system is an RDBMS or a search engine. The BOPS implementation functionality offers a "point-and-cut" mechanism to add the appropriate security to the production systems as well as to the systems in development. The architecture is language neutral, allowing REST, JSON, and SSL/TLS to provide the communication interface. The architecture is built on the servlet specification, open SSL, Java, JSON, REST, and a persistent store. All tools adhere to open standards, allowing maximum interoperability as shown in Figure 16.1.

16.6.1   Application

The entire BOPS suite could be used through the access control or simply be added to the identity assertion of an already existing framework. The BOPS implementation enables trusted processing by performing the minimum actions in the production environment and, in most cases, does not require the change of any application software.

Two-way SSL/TLS, which is built on top of one-way SSL/TLS, provides communication starting at the client. The enrollment communication establishes the origin of the client's identity and passes the BOPS-compliant two-way certificate that the client uses for subsequent communication in conjunction with the session-oriented identity assertion. It is important to note that the client application shall have a preloaded two-way SSL/TLS key that allows subsequent identityGenesis.

The BOPS implementation compliant server receives one-way SSL/TLS communication with two-way SSL/TLS identity. Communication is conducted via both one-way and two-way SSL/TLS. The server uses a data store to take trusted identity and gather the roles for processing on behalf of the identity. Auditing maximizes the appropriate artifacts for continued verification and validation of the trusted access. The assurance occurs through the simplification and documentation of the multilevel access control mechanism. The BOPS implementation requires an administration console (hereafter "admin console"), which is available after the enrollment process which allows dynamic modification of users, groups, and roles (Figure 16.2).

BOPS shall be implemented with an active IDS that provides prevention of any form of brute-forcing or denial-of-service (distributed or single denial of service) attacks. The standard contains a custom rule that identifies and tracks the attempts to forge two-way SSL/TLS certificate impersonation, a session replay, forged packets, and a variety of other attempts to circumvent the BOPS server. See Figure 7.

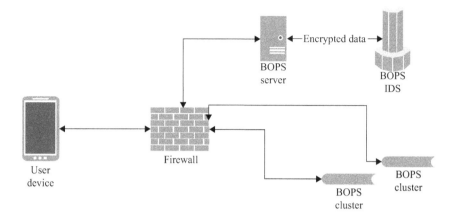

Figure 16.2   Client/server application architecture: variation of client session

16.6.2   Registration

16.6.2.1   Background

The registration process initiates the BOPS adoption within an organization. BOPS may appear as a cluster but is considered a business component. Before the BOPS administrator (hereafter "BOPS admin") sets up an environment, the organization shall register for an API key at the BOPS server. The individual developers may apply for the API key as well (Figure 16.3).

At enrollment completion, the original site administrator (hereafter "original site admin") may create additional site administrators (hereafter "site admins"). In the future, the enrollment information shall be associated with the API key of the organization. The API registration pertains to two domains: the enrolled original site admin and the issued API key, which is based on the enrollment information, the organization, and the use case. The registration process is complete when the application commencement is agreed. After the BOPS admin creates an original site admin for an organization, the original site admin may create a site admin (Figure 8). The steps after the registration are described in Clause 5 of this standard.

16.6.2.2   Developers and the BOPS service

Prior to the development process that utilizes the BOPS service, a developer needs to be registered in the BOPS admin console. By providing the application name and using a question oriented identification mechanism to identify the developer, the BOPS establishes a new account and creates the API key, which would be identified with the application name and associated with the application.

16.6.2.3   The end user and the BOPS service

The communication between the application and the BOPS server is established on top of the two-way SSL/TLS. Genesis is a mechanism that establishes this connection.

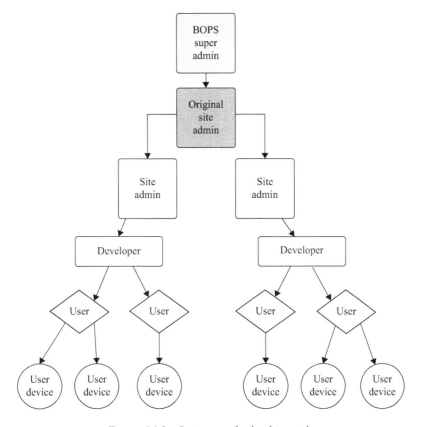

Figure 16.3   Instance of roles hierarchy

Genesis specifies how the users identify themselves to the BOPS server, so the server will generate a private key to set up the two-way SSL/TLS communication between the application and BOPS server. Axiom is one of the mechanisms that BOPS uses to identify users.

16.6.2.4   The end user device and the BOPS service

The application is responsible for providing a unique identifier (ID) that identifies the device of the end user. The application uses the device association's API to notify the BOPS server about the link between the user and the user's device.

16.6.3   Prevention of replay

BOPS specifies a description of the RESTFul calls and behavior necessary for an system to defeat attacks and attack vectors. In addition, BOPS specifies the format of requests necessary to protect data in real time from known and unknown attacks.

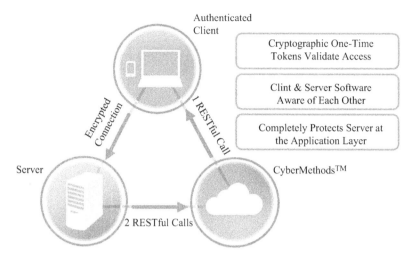

Figure 16.4   Protocol handshake

This functionality is present in the IDS. Replay mitigation uses a cryptographic one-time token to validate access. In this model, the IDS is a third tier ensuring the client and server are always aware of each other, thus ensuring the server is completely protected at the application layer.

The combination of the token and the IDS is to detect International Standards Organization (ISO) Layer 7 cyber-attacks, including replay, Distributed Denial of Service (DDoS), and similar attacks. The token is valid for one use and is usually passed from the client, to the server, and then returned to BOPS using RESTful calls (Figure 16.4).

The basic algorithm is described in BOPS. This standard recognizes that, for DDoS detection, every token must be distinct, there must be an algorithm between the client and server that takes into account that time may vary, and that the values must differ from client to client as well as access to access.

The token's algorithm works as follows. First, in the Genesis step, the Web, mobile or embedded device (client) issues a RESTful call to request a token. The token is then received and embedded in the client's encrypted message to the server. The server receives the token and checks the validity of the message by passing the token to the IDS which then verifies that the token is valid and ensures the difference between the creation time and the current time falls within a specified 60-s time period.

To initiate the Genesis step, the client may choose to establish a five tuple by specifying any or all of the values. The IDS will determine any of the five values that are not set by the client and will return a token to the client in a RESTful format.

The client and the server share the same five tuple which is then used to compute a timestamp which is, in turn, SHA512 encoded and compared by the IDS or BOPS. The computed timestamp moves backwards to a time based on the 5 tuple and is unique for each call.

The token never contains the timestamp itself, as all values in the token are converted into a SHA512 sum for comparison. This allows the values to change on each minute value interval to prevent Blind replay. And, finally, the token's minute range is 3 (and not 60) to allow a sufficiently large entropy (48,771,072) to prevent trial and error attacks.

An example of the entropy of the five tuple is below:

Value	Entropy
Year 0 to current year (2015)	2016
Month 0–11	12
Day 0–27	28
Hour 0–23	24
Minute 0–2 (the minute entropy is 3 so that the value will only be the same for 3 min which limits the number of concurrent attacks)	3
Total entropy $= 2016 \times 12 \times 28 \times 24 \times 3 = 48,771,072$	

The algorithm works as follows. First, all values rotate backwards. If the month is less than or equal to the current month then the year may be equal. If the month is greater than the current month then the year must rotate back. These two cases illustrate the algorithm.

GENESIS EXAMPLES

	Genesis Example 1			Genesis Example 2	
	GMT = 2015-06-10 15:30			GMT = 2015-06-10 15:30	
	Genesis	Value		Genesis	Value
Year	5	2010	Year	5	2015
Month	11	11	Month	4	4
Day	4	8	Day	4	8
Hour	6	12	Hour	6	12
Minute	2	28	Minute	2	28

Since Example 1's current month is 6 (June) and our Genesis value for month is 11, and $11 > 6$, we then scope the year down on an interval of 5 and the year becomes 2010. The remaining values are multiples of the Genesis that are less than the actual date value.

We next describe the second example using the same current date and time. Since Example 2's current month is 6 (June) and our Genesis value for month is 4 and $4 \leq 6$ we scope the year down to an interval of 5 which is equivalent to 2015. Thus, the year becomes 2015 and the remaining values are multiples of the Genesis that are less than the actual date value.

16.7   Summary

The intent of the chapter is not to reiterate IEEE 2410-2015 or BOPS. Rather, it is to give some background to the what and why of the specification and hopefully aid the reader in understanding the specification.

For the developer, hopefully there is meat and examples to aid in BOPS implementation.

Further Reading

Bonneau, J., C. Herley, P. C. van Oorschot, and F. Stajano, "The quest to replace passwords: A framework for comparative evaluation of Web authentication schemes," *Proceedings 2012 IEEE Symposium on Security and Privacy, S&P 2012*, San Francisco, CA, pp. 553–567, May 2012.

Chan, S. W., and R. Mordani, *Java™ Servlet Specification, Version 3.1*. Redwood Shores, CA: Oracle America, Inc., 2013.

Handley, M., *JAX-RS: Java™ API for RESTful Web Services, Version 1.0*. Santa Clara: CA: Sun Microsystems, Inc., 2008.

Nahamoo, D., "IBM Research," *IBM 5 in 5: Biometric data will be the key to personal security*, 2011.

PayPal Stories, "PayPal and the National Cyber Security Alliance Unveil Results of New Mobile Security Survey," 2015. Available at https://stories.paypal-corp.com/archive/paypal-and-the-national-cyber- security-alliance-unveil-results-of-new-mobile-security-survey.

Slideshare.net, "The 10 Hot Consumer Trends of 2014 Report," 2013. Available at http://www.slideshare.net/Ericsson/the-10-hot-consumer-trends-of-2014.

U.S. Department of Defense, DoD 5200.28-STD, "Department of Defense Trusted Computer System Evaluation Criteria," December 1985.

Chapter 17

Big data and cloud identity service for mobile authentication

Charles Y. Li[1] and Nalini K. Ratha[2]

17.1 Introduction

Increased individual mobility has pushed the modern society needs for a reliable individual identity verification system as a critical component in many transactions in commercial industries, public sectors and government domains. The requirement for an ideal human identity verification is critical to security and prevention of identity fraud. Thus, trusted identity management has become an essential part of contemporary system infrastructure. It is now well accepted that biometrics—the science of identifying a person (or verifying their identity) based on their physiological or behavioral characteristics—can provide significant value when building such systems. Three key cornerstones in a trusted biometrics-based identity system include the following:

(a) A trusted identity enrollment process
(b) A trusted identity verification process
(c) An identity credential management mechanism.

In this chapter, we present several emerging developments in mobile biometrics technologies with particular focus on futuristic cognitive authentication systems for enabling large-scale trusted identity management systems based on biometrics and also biometrics identity services in the cloud. Biometrics is excellent mechanisms for the authentication or identification of individuals because of the credential's uniqueness and persistence almost over the lifetime of the person. Biometric identity services in the cloud enable mobile biometrics wide adoption economically and also take advantages of adjacent technology advancements.

17.1.1 Identity establishment and management

To understand mobile biometrics technologies including identity services in the cloud, we need to visit how identity is established and managed within the identity management ecosystem. **To establish an identity**, one can combine many piece of

[1]IBM GBS Chief Technology Officer for Cybersecurity & Biometrics, USA
[2]IBM Thomas J. Watson Research Center, USA

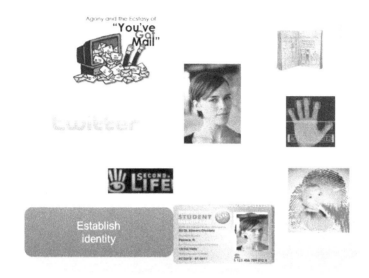

Figure 17.1   Identity establishment: many information including biometrics contributes to the identity establishment

information including a person's photo, social media behavior, biometrics, travel history, and social connections as shown in Figure 17.1. Biometrics is one enabler of the individual's identity and it doesn't equate to one's identity.

To manage the identities, there are several key attributes associated with the assigned electronic identification number as shown in Figure 17.2:

(a)   **Entitlement** (privileges)—your given privileges to what you can do—rights to board, rights to run certain applications, rights to access certain data.
(b)   **Reputations** (history)—You've crossed border many times, made reservations several times, ran an application many times
(c)   **Trust**—I trust you to run an application because I know your identity and you have the right reputations
(d)   **Status/Environment**—It is the context of which the identity has been established for and the status of all the attributes.

Within the identity management ecosystem, several players assume various key roles in order for service providers to provide the authorized services to the proper identity. As shown in Figure 17.3, the individual is the subject who requests for services; the service provider as a replying party to verify identity/entitlement/reputation/trust before providing services. The identity is issued to the "subject" by issuing "Identity Provider," and the subject is provided entitlements by authorities and service providers. Biometrics is a technology enabler and provides identity assertion information; it enables Identity Providers to provide accurate identity. As the discussion progresses in this section, this Identity Management Ecosystem fits well with Identity Cloud Services which will be elaborated in the later subsections.

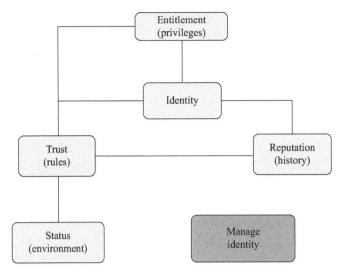

Figure 17.2   Identity management: identity attributes includes entitlement, reputation, and trust

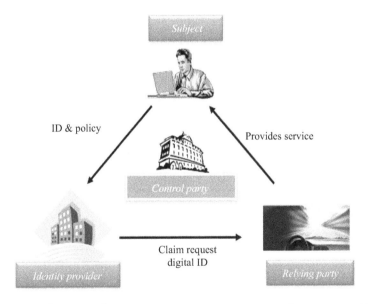

Figure 17.3   Players in identity management ecosystem: service requester, identity provider, and service provider

17.1.2   Mega trend impacts

The general users demand for mobility has been changing how we live and how we do business. As of 2015, there were about seven billion mobile phones of which close to two billion are smart mobile devices. Meanwhile, major mobile platforms including iOS, Android, Windows 10 mobile OS became available and mature; the platform developers have made their platforms available for third-party partners to develop mobile apps in an unprecedented manner that shows that we are in an API and mobile App economy—people build their business by selling Apps and APIs. In the past decade, the emergence and maturation of mobile apps and platforms have created a profound impact on people's lives whether they are an executive, a merchant, a teacher, a student, an IT developer, a reporter, or a politician. We use mobile apps for sending text messages, writing blogs, sharing locations and photos, exchanging money, banking, taking classes, and making reservations. A whole new social phenomenon has been emerging as a result of adapting mobile apps, enhancing people's participation, and enriching the number of activities occurring in the cyber society.

There are several major trends that have made a major impact on mobile biometrics. First, biometrics data collection has proliferated due to the convenience of mobile platforms, and mobile apps being used by more users. Big data techniques development has contributed to how we can leverage the big biometric data collected. Second, biometrics standard development and adoption have spread since 9/11 2001; some known standards forums/workgroups include ISO SC37, INCITS M1, NIST ITL. Third, biometrics commoditization has happened in the areas of capture devices, matching capabilities, analysis tools, and end-to-end solutions. Furthermore, another trend that has made a large impact is cloud technology advancement, with providers including Amazon AWS, Microsoft Azure, IBM Softlayer, cloud standards including OpenStack, Cloud foundry development has a great impact on mobile biometrics and its associated solutions.

17.1.3   Large-scale biometric applications and big data

In the mobile world, biometrics identity establishment generates a lot of data to help further assert a more accurate identity; at the same time, this creates many challenges due to the sheer scale of the data. With the proliferation of large amounts of data generated on the mobile platform, the computation demands are moving to the edge where the data is created. Mobile biometrics has emerged as one of the most active biometrics applications in the last few years due to demand for security and privacy needs. In order to maximize the value of big data, users are demanding for near real time responses, secure authentication, and transactions. Mobile biometrics techniques, solution, and architecture must embrace the change and meet demand.

To illustrate the data scale that is relevant to biometrics applications, we estimated the data scale by comparing among a few large biometrics systems and potential biometrics data collected via mobile devices shown in Figure 17.4 by using image size of a face 200 kB, a pair of irises of 10 KB, and a set of ten flat fingerprints of 1 MB.

Most biometrics systems can operate in two basic modes: 1:1, also called authentication or verification; and 1:N, called matching or identification. Authentication systems are simpler because they only process a single claimed identity and then

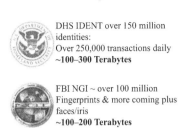 DHS IDENT over 150 million identities:
Over 250,000 transactions daily
~100–300 Terabytes

 ID Cards/Border Crossing/Benefits/ Multiple Instances 7,000,000,000x
(10 Print 0.5–1MB + Face 200KB + IRIS KB)
7 Exabytes

 FBI NGI ~ over 100 million Fingerprints & more coming plus faces/iris
~100–200 Terabytes

 Prolific usage of mobile phones
7 billion mobile phones
7 Exabytes of behavior data

Us DoS has in the range of 100 million faces & Others
~ at least 10–50 Terabytes

 1 Billion Arrivals 2012 worldwide
United States – 100–200 million international arrivals 2012
1 Exabytes traveler data

 EU VIS Biometrics Matching System (BMS) at 70 million individuals and 100K daily enrollment
~ 100 Terabytes

 Unique Identification Authority of India (UIDAI) plans to enroll 1.2 billion citizens. (UID Program)
(enroll million/day: half billion by 2014)
3–4 Exabytes Biometrics & Biographic Data

Figure 17.4   An illustrative data scale of biometrics applications

compare it against one or more new biometrics captures. On the other hand, 1:*N* identification systems are harder because of many reasons including the large amount of data, rapid processing requirements and high accuracy demands to prevent identity fraud. Biometric authentication systems on user facing devices such as laptops and mobile platforms have grown in popularity in such a way it is now common to encounter databases of millions of users on enterprise level and often for governmental applications it can easily reach hundreds of millions (or even more). Not only are there a large number of records, but biometrics systems also have to deal with a variety of multimodal data types (text, 1D signals, images, and video). Moreover, as with all Big Data systems, there are issues of managing the ever-changing database, rapid access to information, system security, and record integrity. Both forms of trusted identity system (1:1 and 1:*N*) are classic examples of Big Data systems. Generally speaking, such Big Data systems have four critical dimensions associated with them, namely: Volume, Velocity, Variety, and Veracity.

(a) **Volume:** There is an extremely large quantity of enrollment and verification data in some modern biometrics systems, such as FBI Next Generation Identification [1] (with plans to reach 100 million individuals) or India's Unique Identity Authority of India [2] (with >1.2 billion individuals). Clearly, large volumes are an issue for to both authentication and identification, but they pose a greater challenge for the 1:*N* matching required by identification than for the 1:1 matching involved in authentication

(b) **Velocity:** This term refers to the speed at which data arrives. It becomes an important concern when large civilian and enterprise systems, such as online commerce and banks, use biometrics for authentication for all of their daily transactions. Again, the velocity challenge is considerably more difficult for identification than it is for authentication, since the basic identification process is more time consuming. This challenge is further exacerbated in systems with a large number of concurrent users, or when attempting to apply "data in motion analytics" (e.g., Stream Computing) to provide timely alerts based on biometrics.

(c) **Variety:** Biometrics systems have been constructed to operate on a wide variety of data types, such as 2D gray-scale images (fingerprint), colored and nonvisual images (face and iris), 3D data (face), video data (face and gait), and 1D temporal signals (voice and signature). In addition, there are different types of enrolled database templates for matching each type of biometric, sometimes several different formats for the same biometric. Moreover, there are a variety of techniques used for matching—no one approach covers all the biometrics and representations—and sometimes even multiple matchers for the same representation.

(d) **Veracity:** The veracity aspect is largely about managing the intrinsic error rates in a biometrics system, particularly the tradeoff between false positives and false negatives. There are several parts to this. First, ensuring the biometrics signal came from a live person (detecting spoofing) is a key requirement in every biometric transaction. Second, good quality biometrics data (full coverage and low noise) is needed for acceptable results both in enrollment and verification systems. Third, it is essential that the enrollment database not contain duplicates records, especially with different associated identities.

17.1.4   Cloud computing

Large and complex biometric systems discussed in the previous sections require a significant effort from many angles: data collection, data curation, data enrollment, deduplication, issuance of credentials, and finally providing service to end users in a trusted and secure fashion. The computing, storage, and communication needs are extremely high and expensive to maintain with diverse set of regulations and compliance requirements in many parts of the world. That's where the cloud computing technology can be utilized efficiently. Cloud computing is a model of service delivery of on-demand network access to a shared pool of configurable computing resources including networks, network bandwidth, servers, processing, memory, storage, applications, virtual machines, and services that can be rapidly provisioned and released with minimal management effort or interaction with a provider of the service. The cloud computing infrastructure consists of a network of nodes. A cloud service has three components to decide: deployment model; cloud service and deployment model. Let's look at the choices and also discuss the options that may be well suited for hosting a cognitive authentication service.

(a) Deployment Models have four options:
   1. **Private cloud**: The cloud infrastructure is operated solely for an organization. It may be managed by the organization or a third party and may exist on-premises or off-premises.
   2. **Community cloud**: The cloud infrastructure is shared by several organizations and supports a specific community that has shared concerns (e.g., mission, security requirements, policy, and compliance considerations).

It may be managed by the organizations or a third party and may exist on-premises or off-premises.

3. **Public cloud**: The cloud infrastructure is made available to the general public or a large industry group and is owned by an organization selling cloud services.

4. **Hybrid cloud**: The cloud infrastructure is a composition of two or more clouds (private, community, or public) that remain unique entities but are bound together by standardized or proprietary technology that enables data and application portability (e.g., cloud bursting for load-balancing between clouds).

Very often authentication services need to be hosted in a private cloud or hybrid cloud.

(b) A cloud service can be availed in three ways:

1. **Software as a Service (SaaS):** The capability provided to the consumer is to use the provider's applications running on a cloud infrastructure. The applications are accessible from various client devices through a thin client interface such as a web browser (e.g., web-based e-mail). The consumer does not manage or control the underlying cloud infrastructure including network, servers, operating systems, storage, or even individual application capabilities, with the possible exception of limited user-specific application configuration settings.

2. **Platform as a Service (PaaS):** The capability provided to the consumer is to deploy onto the cloud infrastructure consumer-created or acquired applications created using programing languages and tools supported by the provider. The consumer does not manage or control the underlying cloud infrastructure including networks, servers, operating systems, or storage, but has control over the deployed applications and possibly application hosting environment configurations.

3. **Infrastructure as a Service (IaaS):** The capability provided to the consumer is to provision processing, storage, networks, and other fundamental computing resources where the consumer is able to deploy and run arbitrary software, which can include operating systems and applications. The consumer does not manage or control the underlying cloud infrastructure but has control over operating systems, storage, deployed applications, and possibly limited control of select networking components (e.g., host firewalls).

As can be seen from the above description, a biometric authentication system is more likely to be a SaaS.

(c) Characteristics includes the following five choices:

1. On-demand self-service: A cloud consumer can unilaterally provision computing capabilities, such as server time and network storage, as needed automatically without requiring human interaction with the service's provider.

2. Broad network access: Capabilities are available over a network and accessed through standard mechanisms that promote use by heterogeneous

thin or thick client platforms (e.g., mobile phones, laptops, and desktops).

3. Resource pooling: the provider's computing resources are pooled to serve multiple consumers using a multitenant model, with different physical and virtual resources dynamically assigned and reassigned according to demand. There is a sense of location independence in that the consumer generally has no control or knowledge over the exact location of the provided resources but may be able to specify location at a higher level of abstraction (e.g., country, state, or datacenter).

4. Rapid elasticity: capabilities can be rapidly and elastically provisioned, in some cases automatically, to quickly scale out and rapidly released to quickly scale in. To the consumer, the capabilities available for provisioning often appear to be unlimited and can be purchased in any quantity at any time.

5. Measured service: cloud systems automatically control and optimize resource use by leveraging a metering capability at some level of abstraction appropriate to the type of service (e.g., storage, processing, bandwidth, and active user accounts). Resource usage can be monitored, controlled, and reported providing transparency for both the provider and consumer of the utilized service.

In this category, the choice is not restricted to one of the options. Rather, a biometric authentication system would need all the five choices supported. Even with the big data and cloud-computing technology, we believe the modern authentication system is likely to be not responsive to the needs and diversity of the users and service providers.

The chapter is organized as follows. In Sections 17.2 and 17.3, we will describe the evolving era of mobile devices with an unbelievably large number of sensors, as well as the mobile characteristics and performance metrics pertinent to mobile biometrics. In Section 17.4, we will briefly summarize the applicability of different modalities on mobile platforms. The ecosystem involving a mobile transaction and an application architecture will be presented in Section 17.5. The identity services in the cloud and the cognitive authentication modules will be shown in Sections 17.6 and 17.7. Finally, we will provide our conclusions in terms where mobile biometric can impact this space in Section 17.8.

It is very fascinating to visualize the interplay between the smart mobile devices and the user needs supported through both big data and cloud infrastructure with a cognitive layer.

17.2 Characteristics of mobile biometrics

Mobile applications and their usage has become so widespread that it has created new opportunities, giving rise to many new technologies but also bringing about challenges at the same time. Among them are demands for easier and more secure access to users' blogs, chatrooms, and websites through mobile platforms. As a result of mobility proliferation, large amounts of data about users and their behavior activities has been generated. Commercial companies, special interest groups, and the government have started data mining mobile data in order to identify users in the wild world of cyber society. Biometric technologies, the primary identity assertion technique, can be used for both user authentication and identification.

Biometric development accelerated as a result of the tragedies of September 11, 2001. Prior to that, biometrics was primarily used in government and regulated industries including criminal justice, banking, border protection, critical infrastructure protection, and in the vetting process of critical infrastructure employees. According to many market research reports and industry projections from various market research groups [3], biometrics development and application will be a primary growth sector in mobile applications in the next decade.

In the last 10 years, biometric application on mobile platforms has been accelerated because of the following factors:

(a)   Commoditized capture sensors and matching technologies
(b)   Expiration of major patents
(c)   Development and adoption of biometric standards.

The following sections will discuss some key concepts and characteristics of biometrics pertinent to mobile biometrics.

17.2.1   Mobile biometric concepts

In order to discuss biometrics applications on mobile platforms, this section describes some basic concepts that are pertinent throughout this chapter and expands from identity discussion in Section 17.1.

Biometrics is the science of identifying, or verifying the identity of, a person based on physical or behavioral characteristics. The biometric modality needs to be measurable on the basis of physical and/or behavioral characteristics and to be used to verify the identity of an individual.

Person identity and biometrics—Typically, there are three traditional ways to verify the identity of a person:

(a)   What you have—physical tokens including keys, passports, and smartcards;
(b)   What you know—personal knowledge including secret (passwords, pass phrases, . . .) and nonsecret knowledge (user id, mother's maiden name, favorite color, . . .);
(c)   Who you are—personal biometrics including physical (fingerprints, face, iris, . . .) and behavioral (walk, keystroke pattern, vocal characteristics, . . .).

Physical tokens such as smartcards can be lost, stolen, duplicated, or forgotten, and passwords can be forgotten, shared, observed, or broken (or guessed). Generally speaking, biometric identification characteristics are unique to an individual, are measurable, and remain fixed over the individual's or specific application's lifetime (a property called "permanence"). Biometrics is used to establish or verify a person's physical identity. As shown in Figure 17.5, some modalities are more popular than

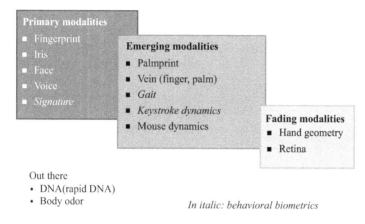

Out there
- DNA(rapid DNA)
- Body odor

In italic: behavioral biometrics

Figure 17.5   Biometric technology types

others due to different usage needs; however, mobile platforms further limit some modality's mobile adoption, some examples including palm print, signature, and gait.

In the context of mobile applications, users constantly face a dilemma: demanding privacy protection and requesting easy access to apps and/or websites through mobile platforms.

As you will discover throughout this chapter, we describe how users can authenticate themselves through various means on mobile platform. Initially, solutions were influenced by mainstream biometric techniques: physical biometrics such as faces, fingerprints, and irises, which all require quality sensors on mobile devices. As sensors have become higher quality and more miniaturized, they have been integrated into mobile or handheld devices. Behavioral biometrics, including voice, mouse dynamics, and keystroke dynamics, are further investigated and found to be viable authentication modalities.

17.2.2   Mobile biometric data

There are some intrinsic challenges that accompany mobile data collection for both verification and identification purposes due to the unconstrained capture environment. These characteristics include the following:

(a)   The data collected is frequently low quality due to unhabituated users and unconstrained collection environment. As a result, verification is more feasible; however, large-scale identification is a challenge.
(b)   Much behavioral data can be collected; however, technology is not yet mature enough to use the behavioral data for authentication and identification in real operations.
(c)   Biometric data collection requires quality sensor capabilities and extra computing resources. As a result, it creates latency and usability challenges for users.

Due to the large number of mobile users and the amount of data collected, there is a fundamental accuracy challenge when identification is performed using data collected from mobile platforms. We can call this a challenge—biometric performance at gig scale. For an individual to be enrolled into a biometric system, a 1:N search is performed initially to ensure no duplicate identity is enrolled, which is called the deduplication process. If we perform a "brute force" deduplication, the challenge looks like the following:

(a)   If the system performs a cumulative deduplication, this is a combination problem and the total number of biometric sample comparisons is $N(N-1)/2$.
(b)   If the system deduplicates a population of 100 million, then the enrollment results in 4,999,999,950,000,000 checks required.
(c)   The task above needs 15 years to complete if the matching speed is ten million matches per second—an expensive and challenging endeavor.

Another challenge is the biometric accuracy.

(a)   If the False Match Rate (FMR) was one identification false match per million, there would be 500,000 False Matches with 1 million enrollment population (deduplicate).
(b)   If the system needed to enroll 100 million subjects, there would be 500 million false matches.
(c)   Many social media websites have hundreds of millions users and users access them regularly via mobile platforms.

The question is posed as what you do with the 500 million false matches; this results in a mission impossible.

17.2.3   Biometric processes and performance metrics

Mobile platform as a system, which consists of users, system components, and data collected, has some measurable performance metrics when using biometrics to verify or identify users. Mobile application data collected are mainly for authentication and/or identification purposes. The three fundamental identity-related processes are enrollment, verification, and identification:

(a)   Enrollment is, for a known individual, when a high-quality biometric sample is obtained for each modality (e.g., face or voice) of interest and stored in a database of biometric samples.
(b)   Verification is commonly described as a 1-to-1 matching process and is typically used for access control. An individual's newly captured biometric sample is matched against the biometric sample on file for claimed identity. Classically, Receiver Operating Characteristic (ROC) depicts the accuracy for verification purpose.
(c)   Identification is commonly described as a 1-to-N matching process and is typically applied to determining an unknown identity, determining if someone is on a security watch list, and for deduplication of identities when enrolling biometric samples in a database. An individual's newly captured biometric sample

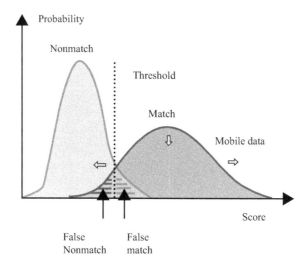

*Figure 17.6   Probability graph: false nonmatch is also referred to as type I error
and false match is also referred to as type II error in conventional
statistics*

(referred to as the probe) is searched against a database of biometric samples,
each of which is associated with a unique identity.

- For user driven searches and forensic analysis, the accuracy is based on
the number of top ranked candidates' retrievals and what percentage of
probe searches return the probe's gallery mate within the top rank-ordered
results. The Cumulative Match Characteristics (CMC) depicts this performance measurement because it reports the percentage of probes identified
within a given rank.
- For large-scale systems, human experts examine search results that produce a matcher score exceeding a preset match threshold. There is a
tradeoff between false alarms—FMR, and misses—False NonMatch Rate
(FNMR). These measurements are generally compared using a Detection
Error Tradeoff (DET) plot.

To understand how to implement and understand biometrics with mobile platforms,
we need to examine the metrics and graphs specifically used for the biometrics data
collected from social media and networking.

Score Probability Graph—The probability graph in Figure 17.6 illustrates
distributions of genuine biometrics matching scores versus imposters matching
score frequency distribution. The threshold line indicates false matches and false
nonmatches matches. With low quality data (both verification and identification) collected with mobile platforms, there would be larger number of false match and false
nonmatch.

Receiver Operating Characteristic (ROC) Curve—For verification system,
this graph illustrates the verification accuracy—(1 – FNMR) versus the operating

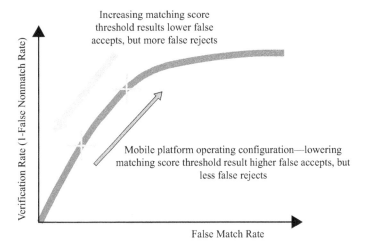

Figure 17.7   Receiver operating characteristics (ROC) curve

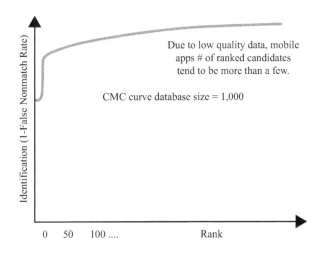

Figure 17.8   Cumulative match characteristic (CMC) curve

point of FMR. This is a good measurement for authentication type of application. For social media users, the false acceptance rate needs to be raised higher in order for "friendlier" access to the device or platform (Figure 17.7).

Cumulative Match Characteristic (CMC) Curve—For identification system, this is a measurement of accuracy to identify an individual in the top ranked matches. The graph shows the identification accuracy—(1 – FNMR)—with predefined gallery size and number of matches returned. In case of data mining in social media collection based on mobile platforms, this measures the chance an individual can be identified from the social media data. For examples, it answers the question of what the chances are that someone's face will be found in the sea of data collected in Facebook—actually, significant amount of data is collected via mobile platforms (Figure 17.8).

Table 17.1   Sensors in a smart mobile device

Apple iPhone 6S plus	Samsung Galaxy S7
Fingerprint scanner embedded in the home button	Fingerprint scanner
5/12 Megapixels	5/12 Megapixels
3-D touch screen	
For screen-based signatures	
Accelerometer	Accelerometer
GPS	GPS
Barometer	Barometer
NFC Proximity detection	NFC
Three-axis gyro	Gyroscope
Wi-Fi	Wi-Fi
Bluetooth	Bluetooth
Microphone	Microphone
External iris scanner	External iris scanner
External retina scanner	External retina scanner
Ambient light sensor	Heart rate monitor

A good source of biometrics glossary can be found at http://biometrics.gov/Documents/Glossary.pdf.

17.3   Smart mobile devices

17.3.1   Many mobile sensors available

With the advent of smart mobile devices and their ability to support network access, new forms of authentication are necessitated to enable the use of these devices for many meaningful applications. On these devices, typing passwords is also difficult because of small screens. However, these devices have a rich variety of sensors such as accelerometers, GPS, multitouch screen, camera, microphone, fingerprint scanner, and proximity sensor. This creates an interesting platform for multibiometrics [4]. The scanners themselves may not be of the forensic grade to provide recommended signals, but the plurality of the sensor can potentially overcome inherent noise and usability restrictions. Table 17.1 shows comparison between the sensors available on two top end smart phones; and Table 17.2 shows sensors available on some top smart wearable devices.

17.3.2   Multibiometrics fusion

Generally, the strength of a fusion system is best when the sources are largely independent and uncorrelated. The scores from multiple classifiers applied to a person

Table 17.2 Sensors in a smart wearable device

Microsoft band	Nymi band	Apple watch
Optical heart rate sensor	6-Axis accelerometer	"force touch" screen
3-Axis accelerometer	Gyroscope	Gyro
Gyro	ECG monitor "Heart ID"	Accelerometer
GPS		Infrared
Ambient light sensor		Photodiode
Skin temperature sensor		Visible-light LED
UV sensor		GPS + NFC for wireless
Capacitive sensor		payments
Galvanic skin response		
Microphone		

are said to the conditionally independent if

$$p(x_{f_1} x_{f_2} \ldots x_{f_P} | I_k) = p(x_{f_1} | I_k) p(x_{f_2} | I_k) \cdots p(x_{f_P} | I_k)$$

where x_{f_i} is the score from classifier i ($i = 1, 2, \ldots, P$), I_k represents the kth person, and $p(.)$ represents the probability density function. If we focus on the two-class verification problem, k can take value of either 1 (authentic) or 0 (impostor). If we have sufficient samples of each type of input, we can estimate the marginal and joint density functions to check if a set of classifiers are statistically independent. This would be a strict evaluation of statistical independence. A more nuanced approach is to look at the mutual information between them, which is the observed joint conditional probability (left side), divided by the product of the individual conditional probabilities (right side).

For example, assume we have separate matcher for each finger of a person (index, middle, ring, etc.). Let there be K persons, P fingers per person, and N fingerprint samples from each finger. N scores for each of the P fingers can be found by matching against corresponding fingers. For identification, marginal density functions can be estimated using N scores for each person, and the joint density function can be estimated using all possible N^P correspondences of these scores. For verification, marginal density functions can be estimated using these NK authentic scores and $N(K-1)K$ impostor scores. There would then be $N^P K$ joint authentic scores and $N^P(K-1)K$ joint impostor scores. In such a case, the P fingers are said to be conditionally independent if the equation above holds.

The nonbiometrics sensors, such as GPS and accelerometer, can additionally make the system aware of the environment and location in which a transaction occurs. This allows the device to effectively support a multilayer security model as is often needed in remote access applications. As in conventional Big Data systems, the past transaction history and the context of the application can also add value, and thus aid in providing the user a secure nonrepudiable authentication method. In a broader sense as shown in Figure 17.9, fusion now encompasses multiple issues related to the identity of the person based on biometrics, environment, and behavioral patterns; in [4], the authors discussed extensively on this topic.

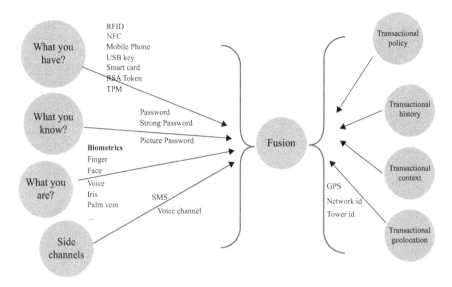

Figure 17.9   Fusion of multiple sources: the identity of a person based on biometrics, environment, and behavioral patterns

17.4   Emerging mobile biometrics techniques

17.4.1   Traditional biometrics—fingerprint, face, and iris

Fingerprint technology was the first adopted biometric modality for computer authentication; since then, from separate single-finger sensors to embedded swipe sensors, fingerprint capture sensors became widely available on mobile platforms. In recent years, fingerprint technology has become further commoditized on phones, tablets, and computers. Even though fingerprint technology has been used in criminal justice applications historically, making the technology more publicly acceptable and removing the "criminal stigma" of fingerprints has helped further adoption of fingerprint for both mobile and computer platforms authentication purposes. Mobile smart phone users can leverage the fingerprint-technology-enabled platforms for initial authentication and have access to mobile apps including social media and financial transactions. However, fingerprint technology usage in mobile platforms has vulnerabilities similar to those of other biometrics modalities. Especially when implemented in an unsupervised fashion, measures must be taken to mitigate the risk of spoofing. Furthermore, some known challenges related to fingerprint image capture remain to be addressed, including contaminated fingertips, ghost images, oil residue left on the sensor surface, improper finger pressure on the sensor surface. In fact, the study shows that fingerprint unlock screen for iPhone suffers from wet finger and usability remains a challenge. Therefore, fingerprint technology for mobile authentication still need improvement; and also, it would be interesting to see more studies on the performance of the system/sensor false reject and false accept rates.

Face recognition has been the most researched and most useful biometric technology on mobile platforms. Face recognition is noninvasive and supported by many commercial products; users can conveniently use face recognition for authentication by leveraging the on-board camera sensor of the mobile device. Historically, cameras on mobile platforms can capture the facial images that can be used either with on-board facial verification or with cloud-based services. At this present state of technology, most mobile device authentication requires significant computing resources for image processing. This poses a challenge to mobile device and many other sensors that require processing at the point of biometrics capture. Even if the authentication is supported via a cloud identity service, the face detection stage is processed at the point of capture. Mobile authentication applications generate better quality face images compared with faces that are collected simply for mobile app sharing because users concerned with authentication are cooperative and motivated to present images that are of sufficient quality to match previously stored images. Image quality characteristics that impact the success and accuracy of face recognition include subject lighting, pose, and facial expression; facial occlusions include eyeglasses, hats, headscarf, or even background. Over the years, researchers and technology developers have developed many algorithms and techniques to support each step of the face recognition process.

Actually, iris recognition is one of the most promising authentication modalities for mobile applications; however, various challenges of iris technology remain for mobile applications.

Some inherent challenges still remain in deploying iris applications because the iris is a small anatomical target of 1 cm diameter located behind a curved, wet, reflecting surface (the cornea). Humans regularly blink their eyelids for protection and lubrication, and the eyes are constantly in motion. In addition, the iris can be obscured by eyelashes, lenses, and reflections, and partially occluded by eyelids, which can droop.

In mobile platforms, the primary application of iris recognition is to authenticate users for reidentification. The challenges are how to deploy an economical sensor to the mobile phone platform and how to capture a user's iris conveniently and with sufficient clarity. Iris recognition utilizes infrared light to illuminate the eye to maximize the visibility of iris features, particularly for dark-colored irises. The illumination poses some challenges from a usability perspective because it creates a perception of causing health effects. Recently, miniaturized commercial iris capture sensors have been installed on smart phones as an additional camera. Ultimately, the front high-resolution camera is needed in order to capture the iris; however, the approach is a challenge because most on-board cameras have an infrared filter built-in, which is detrimental to iris capture.

Iris scanners stand out as the next best biometric solution for mobile devices. They promise a high reliability and can be designed in a very compact manner. The contrast of recorded images is very important for the quality of the scanner as the iris pattern can only then be securely identified. An advanced sensor can compensate various needs; in the very near future, we may have an iris scanning capability with the front camera of mobile phones. In the last couple of years, smart-phone makers started building iris recognition technology into their smartphones either using a separate iris sensor or using the higher resolution front-facing camera. It is desirable to have one single camera to capture both iris scans and normal images for better usability. With increasing acceptance and usability improvement, users can use mobile apps by staring at your home screen: The Android or iPhone iris scanner uses infrared light to unlock the phone and accounts

17.4.2   Behavior biometrics

In mobile applications, there are quite a few user behaviors induced by biometrics that can be used for authentication. Some modalities include keystroke dynamics, mouse dynamics, and signatures; the primary usage is for authentication and continuous identity monitoring. Because of the data quality and the size of population, behavioral biometrics application in identification problem could be a challenge.

Like any other biometric recognition modalities, **keystroke dynamics** include stages of acquisition, feature selection, matching, and decision. Keystroke dynamics biometrics is unlikely to replace existing knowledge-based authentication entirely and is also not robust enough to be a sole biometric authenticator. However, the advantages of keystroke dynamics are indisputable, such as the ability to operate in stealth mode, low implementation cost, high user acceptance, and ease of integration into existing security systems. These create the basis of a potentially effective way of enhancing overall security rating by playing a significant role in part of a larger multifactor authentication mechanism.

Recent research suggests that **mouse dynamics** is a viable solution for identification on a small scale; Yampolskiy and Govindaraju provide a good review of the subject [5]. This modality could be a part of the multifactor authentication process. Commercial products have not yet matured as a number of challenges exist, including:

(a)   identification of stable features for each user application;
(b)   developing adaptive algorithms as the user's mouse dynamics change over time;
(c)   scalability of mouse dynamics to large populations.

In addition, some external environmental and user emotional variables can cause subject behavior to change over time. Also, the time to authenticate a user based on mouse dynamics depends on their frequency of interaction with the mouse. Some research shows that it may require a long time to reliably authenticate a user; however, some recent studies indicate authentication times as fast as less than 10 seconds with some closed datasets. Reducing this time period is imperative to making this approach a viable option. Even though some work has achieved a low False Alarm Rate and a high True Positive Rate, most of the work has been on small population sizes. Much work remains to be done. Meanwhile, performance requirements will be attained through information fusion among keystroke analysis, mouse dynamics, and user behavioral patterns. In mobile applications, this technique is used for verification purposes, not for identifying an individual. Some pilot projects have been conducted but have not yet resulted in real deployment applications.

Most traditional **signature recognition** applications require a signature pad, commonly used for financial transactions and document signing. Verification is usually done by visual inspection. Unlike the iris, fingerprint, and face, which use physical features, signature biometrics is behavior based and it is already widely accepted by general public. There are some obvious advantages:

(a) achieving high accuracy academically;
(b) eliminating fraud;
(c) implementing economically;
(d) substituting PIN or password;
(e) attracting increasing interest, academic, and commercial.

However, signature recognition can be used for authentication on the mobile platform; this modality using mobile platforms has gained much attention in mobile applications last a few years [6]. Potentially, users could use a stylus to sign up their signatures on the standard smart phone screen or touchable workstations. The feasibility of dynamic signature verification as a user-centric validation service on mobile devices has been studied previously. While signature verification is still a challenging task, the performance of the systems is being continually improved with new approaches and algorithms; some algorithms can achieve EER rage in the range of low single digit. Moreover, the combination of signature with other biometric traits can lead to very low error rates. This is feasible in camera- or fingerprint-sensor-enabled mobile devices, which can perform face or fingerprint verification.

In recent years, the academic community has researched user social media behavior and their patterns particularly on mobile platforms. In fact, majority of the social media apps or website are accessed via mobile platforms. Some example topics include behavioral biometrics with audit logs, command line lexicon, e-mail behavior, human machine user interface interaction, keystroke dynamics, mouse dynamics, network traffic, programing style, registry access, storage activity, and any other user actions that are potentially unique to the user. Though most of the research is at its relative infancy, researchers still continue to look for discriminatory features, various combinations, and classification techniques to address the continuous authentication problem, since most modalities are not suitable for identification. In previous sections, several forms of conventional biometrics have been discussed pertinent to mobile social media applications, and additional areas of interest that researchers discovered include [7–9]:

(a) Application switches: this captures application behavior habits that a user possesses. For example, frequent switching vs staying with one application for a long time.
(b) Temporal patterns: this captures how long the user typically spends on a specific resource or application. For example, a user might usually spend 1 hour on program development and then 10 min on reading news.
(c) Resource Accesses: this captures the way the user accesses resources. For example, a user might usually go to a folder, open a file and start editing. If we observe

that the user then goes to one specific folder, keeps opening and closing different files, this could be a different user.

(d)   Resource used this captures the way the user creates and edits documents.
(e)   Content Signatures: this captures snippets of the resources the user is accessing. We can then build a topic model with those snippets to characterize various users.
(f)   Temporal activity fingerprints: this captures what the user is usually doing at a specific time of day. For example, a user might usually log out at 8 p.m.

However, there are some fundamental challenges to deploy this type of operations because:

(a)   searching for behavior patterns is computing expensive;
(b)   it is intrusive to analyze the user's interactions with applications/resources;
(c)   limited data and time are allowed to use behavioral biometrics for authentication.

In the very near future, we should see revolutionary advances in chip design, sensor design, and analytics technologies; subsequently, social media users can take advantage of more secure solutions based on various kinds of behavior biometrics in a real situation.

17.4.3   Risk-based continuous authentication and trust management

With increasing advanced sensors and various biometric modalities available, as discussed above, usability remains the most important and challenging issue for biometrics adoption in social media. As shown in Figure 17.10, a likely approach would be a combination of the traditional biometric modalities for initial authentication and continuous authentication based on behavior. The continuous authentication process will be a risk-based authentication approach, a dynamic authentication approach that takes into account the profile of the user or system requesting authentication.

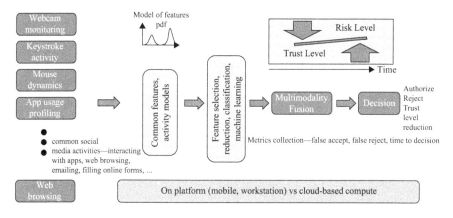

Figure 17.10   Risk-based continuous authentication and trust management

Risk-based implementation allows the application to build a trust level of users and manage the trust level based on risk. In this case, mobile platforms need to capture user activities, timing features of mouse and keyboard, and application profiles; therefore, the system can formulate a decision that incorporates trust levels continuously. As increasing computing power becomes available and analytics capability enhances, this approach will become economical and feasible. Some systems are developed to achieve some of the discussed capabilities through government or private funding (see http://www.darpa.mil/program/active-authentication).

In the near future, a social media user will be able to use a mobile smart phone with a built-in fingerprint sensor and a face and iris unified camera to access social media website. In addition to the traditional biometrics data captured, build-in software can also capture users' human machine interaction and applications. A typical use case can be described as in the following:

(a) User brings the mobile phone in front of the face
(b) User can also present fingerprint using the touch screen
(c) Mobile phone captures face and/or iris and/or fingerprint with different quality
(d) Phone unlocks its screen based on the biometrics captured and authenticates the user to enable myriad of application such as Instagram, Google, or Facebook
(e) The system detects some unusual activities because the owner usually doesn't use weather.com at the initial sessions
(f) The system still allows browsing on Facebook, but locks the features of chatting, editing, adding functions to avoid impersonation
(g) The system locks all apps accessing online banking apps
(h) The system locks all edits to user profile
(i) The system prompts and asks user to present fingerprint/face/iris to reauthenticate

The series of events above depicts a possible future scenario in which mobile app users potentially experience with enhanced initial biometrics-based authentication and continuous authentication based on behavior biometrics.

17.5 Conceptual mobile application architecture

In mobile application context, there are two types of biometric applications as discussed in previous sections:

(a) Verification and authentication authorize the user to access the website or apps remotely. This includes two steps: (a) initial authentication using sensors on the mobile platform and (b) continuous authentication based on behavior biometrics;
(b) Identification for forensic analysis or large-scale recognition applications by consuming user data collected from mobile apps including social media activities. Examples include videos, photos, blogs from Facebook, Google +, and Flicker.

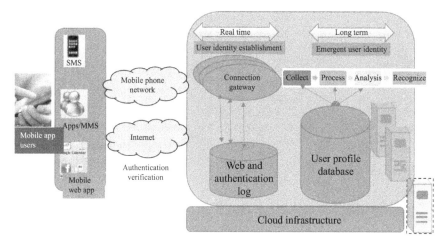

*Figure 17.11   Mobile application conceptual architecture from an identity
          perspective*

In Figure 17.11 of a mobile platform conceptual architecture for biometrics, users use apps on the mobile device or web browser to access remote websites; in both cases, the user can be authenticated using biometrics. All the data collected on user activities are stored in the data store(s); companies or government can apply analytics on the data through stages "Collect," "Process," "Analyze," and "Recognize" for identity analysis.

Several trends in commercial IT industry has been shaping how users behave and essentially making people latched on mobile apps. Cloud technology, as an example, has become tangible since its inception and major cloud companies include Microsoft, Amazon, IBM, and Cisco which all offer automatic IaaS. Recently, most users have observed that third-party apps or website depend on Google or Facebook's Single Sign On (SSO) service. This is a major step forward for identity management becoming a service. In addition, the FIDO (Fast IDentity Online) Alliance formed in 2012 [10] is to address the lack of interoperability among strong authentication devices as well as the problems users face with creating and remembering multiple usernames and passwords. The FIDO Alliance plans to change the nature of authentication by developing specifications that define an open, scalable, interoperable set of mechanisms that supplant reliance on passwords to securely authenticate users of online services. This new standard for security devices and browser plugins will allow any website or cloud application to interface with a broad variety of existing and future FIDO-enabled devices that the user has for online security. In the next section, we will discuss Identity Services in Cloud that are becoming a trend and will improve mobile identity authentication.

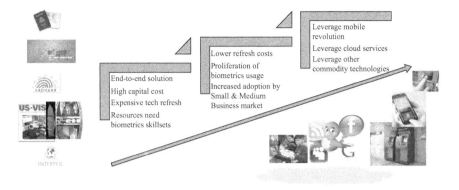

Figure 17.12   Biometrics application paradigm shifts

17.6   Biometric identity services in the cloud

17.6.1   Biometrics-enabled identity services

Current biometrics applications are mostly end-to-end solution with high capital cost; they are expensive for any technology refresh, need engineering resources who need biometrics skillsets (Figure 17.12).

In the last decade, there are a trend of lowering technology refresh costs because of proliferation of biometrics usage, increased adoption by small and medium business market. At present, mobility has become the most explosive platform; cloud services have become widely available and used in commercial industry with mature IaaS and emerging PaaS and SaaS. Furthermore, the biometrics capture and matching are also gradually becoming commodity due to popular demands, more vendors available, more standards adoption.

As a natural progression, biometrics-enabled identity services in the cloud are becoming a reality; in the market place, a lot of commercial companies are starting to offer biometric identity services in the cloud. To compare the traditional biometric application with a cloud-based biometrics architecture, a typical biometrics application architecture and a cloud-based solution are shown in in Figure 17.13. Typically, there are several tiers of capabilities in a multitiered architecture paradigm

(a)   Capture workstation including mobile, kiosk, and static stations
(b)   A workflow management and biometrics-based services including enrollment, identification, and verification services
(c)   A standard interface for interfacing with matching capabilities or subsystems
(d)   Matching subsystem including fingerprint, face, iris, and other modalities
(e)   Data storage and infrastructure

In the previous subsections, we have discussed cloud technologies and its key characteristics: on-demand self-service, ubiquitous network access, location independent

Biometrics architecture   A cloud architecture

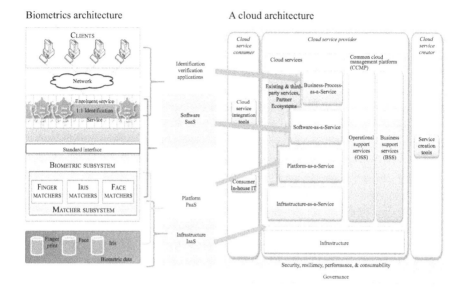

Figure 17.13   Traditional biometric application vs a cloud-based biometric architecture

resource pooling, rapid elasticity, and measured service; to appreciate the benefits, we look at the architecture from the cloud biometrics application perspective. As shown in Figure 17.13, the data storage and infrastructure lays in the IaaS layer; the matcher layer and the matcher interface can be part of PaaS; the workflow management and biometrics services including some matching service all can be part of SaaS; identification, verification, and other domain specific application services can be part of Business as a Service (BaaS) layer. From the deployment model perspective, some public section applications including school and health service providers would prefer to use public or hybrid cloud; defense and intelligence communities would require private and/or on-premises data center.

17.6.2   Biometric identity service cloud model

In this section, we attempt to define a model of Biometrics Identity Services in the Cloud. Figure 17.14 shows an example of a development and test cloud, an operational cloud to support field operations where the users and mobile clients consume cloud services. Within the development and test cloud environment, technology providers can be developing and testing matching algorithms; system integrators can deploy those technologies with operational data sources based on resources need. As a system owner and users of all the identity services, they will only need to define Service Level Agreement (SLA) based on accuracy, performance, and speed.

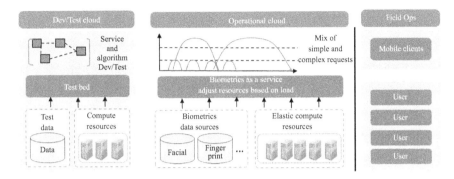

Figure 17.14   A biometric identity service cloud model

Some basic challenges including privacy, security, and anonymization are not different from traditional service oriented-architecture (SOA)-based design; cloud-based solution does not make the challenges easier or harder. Based on needs, one-premises, off premises and hybrid cloud are key decisions of which the system owner need to decide. With the current cloud technology and vendor offerings, all options are open. To compare Service Oriented-Architecture technology, cloud technology significantly differs in five major areas as we discussed previously.

There, industry groups, national and international bodies promote and standardize cloud technologies including NIST, Cloudfoundry, OpenStack, etc. Looking at Identity Service from the cloud perspective, we need to avoid vendor-lock in while maintaining competition and innovation to benefit the community and also the users. Several key challenges and questions that we need to answer

(a)   What type of standards do we need to have and how do we comply with cloud standards?
(b)   What will it take for biometric technology developers to work on or above hypervisor level?
(c)   How we can keep innovation from commercial companies while preventing monopoly?

As show in Figure 17.15, there are several touch points of the architecture laid out in the figure that standards are needed in order to promote biometric identity service in the cloud adoption. One or two standards are listed in the gray bubble are potential candidates

(a)   Mobile standards—Apache Cordova is the only standard here (and the one used in IBM's Worklight). There are other technologies that promise cross-platform, mobile app development capability but they are proprietary.
(b)   IaaS Standards—Listed OpenStack (the one IBM committed to). Other IaaS "standards" include Amazon EC2, technically a closed standard but a standard by market saturation; and Apache CloudStack, a waning standard.

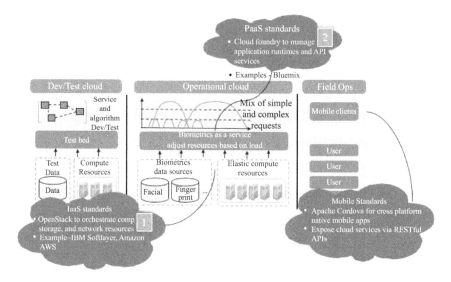

Figure 17.15   Biometric identity service standardization: PaaS standards, IaaS standards, and mobile standards

(c)   PaaS Standards—Listed Cloudfoundry (the one IBM committed to). Other PaaS "standards" include RedHat's OpenShift, which has a chance of picking up in the market; Heroku; Google's App Engine, technically closed but a market force; Microsoft's Azure, another technically closed but a market force.

As a community, we need to explore standards of biometric identity services at SaaS level in order to promote real adaption and promote competition.

ANSI INCITS 442-2010 Biometric Identity Assurance Services (BIAS) standard development is an excellent exercise that we should learn while developing SaaS for biometric identity services in the cloud. SaaS for biometric identity cloud services should provide basic biometric identity assurance functionality as modular and independent operations which can be assembled in many different ways to perform and support a variety of business processes. The services are intended to offer a consistent and common interface to various system resources, which may include:

(a)   A 1:1 fingerprint verification matching server
(b)   A 1:N iris search/match engine
(c)   A facial biometric watch list
(d)   A criminal or civil AFIS system
(e)   A name-based biographic identity database
(f)   An archive of biometric identifiers
(g)   A gallery/population of subjects

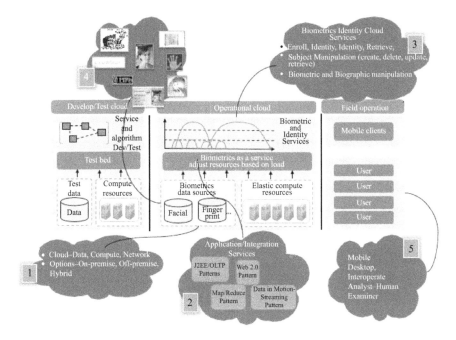

Figure 17.16   Implementation options of a biometric identity cloud services

In fact, there could be some primitive cloud services at subject level and composite cloud services to combine one or many primitive cloud services; multimodality fusion and all source fusion can also be at composite service level.

17.6.3   How to develop a biometrics-identity-service-cloud model?

In order to implement a biometrics-identity-service-cloud model in application, we need take into account the interests of all the IT infrastructure, matching technologies, middleware/workflow management, and capture technologies and solution providers. Ultimately, the users and system owners will spend less effort to define all the requirements of technologies and systems; they can spend more of their resources focusing on business service level agreement including performance, accuracy, and speed. When Service Oriented-Architecture products and technologies emerged more than a decade ago, the industry ends up with a lot of a lot of standard compliant products/solutions—this promotes competition and encourages innovation.

As shown in Figure 17.16, we take the following steps to implement a biometric identity cloud service application with a simple cloud model of one development and test cloud, and operational cloud

(1)   Decide on-premise and off-premise cloud options and plan data, compute and network cloud resources

(2)   Pick up matching software vendors based on different technologies

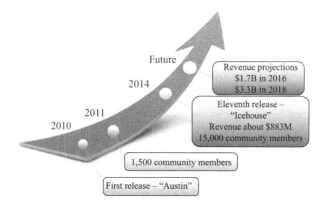

Figure 17.17   OpenStack growth in four years

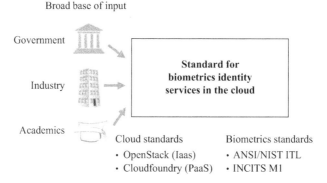

Figure 17.18   We need the whole biometrics identity community to promote standards to promote biometric identity cloud services

(3)   Define cloud services based on composite and primitive cloud SaaS
(4)   Taking into account all sources for Identity assertion
(5)   Client, mobile users, interoperable systems, and human in the loop (Figure 17.17)

OpenStack is an example of a Cloud Standard and Reference Architecture based on, openness, modularity, and well-defined interfaces. In short amount of time, OpenStack has ignited a market that promoted active community, reduced vendor lock-in, introduced agility, and innovation.

However, cloud is still an emergent market and cloud vendors face a tremendous competition among each other.

We need standardization in order to promote the biometric identity services in the cloud. As shown in Figure 17.18, government sponsors, industry, and academics need to collaborate to develop new standards and leverage the current cloud standards and biometric standards.

17.7   Cognitive authentication system: a point of view

The present method of authentication systems can be compared with the older era of computing. In contrast, let's look at the definition of cognitive computing: "Cognitive computing refers to systems that learn at scale, reason with purpose and interact with humans naturally. Rather than being explicitly programed, they learn and reason from their interactions with us and from their experiences with their environment. They are made possible by advances in a number of scientific fields over the past half-century, and are different in important ways from the information systems that preceded them." [11].

In order to achieve this new level of computing, cognitive systems must be: [Wikipedia]

(a)   Adaptive: They must learn as information changes, and as goals and require-ments evolve. They must resolve ambiguity and tolerate unpredictability. They must be engineered to feed on dynamic data in real time, or near real time.
(b)   Interactive: They must interact easily with users so that those users can define their needs comfortably. They may also interact with other processors, devices, and Cloud services, as well as with people.
(c)   Iterative and stateful: They must aid in defining a problem by asking ques-tions or finding additional source input if a problem statement is ambiguous or incomplete. They must "remember" previous interactions in a process and return information that is suitable for the specific application at that point in time.
(d)   Contextual: They must understand, identify, and extract contextual elements such as meaning, syntax, time, location, appropriate domain, regulations, user's profile, process, task, and goal. They may draw on multiple sources of informa-tion, including both structured and unstructured digital information, as well as sensory inputs (visual, gestural, auditory, or sensor provided).

The key goal of our approach is to define the next generation of human authentication system that is universally available in terms of geography and supports a multitude of devices and languages heavily inspired by the cognitive computing characteristics.

Cognitive-computing technologies have taken off lately because of significant research advances including machine learning application and deep learning neural network.

As we have observed recently, the next generation authentication requirements will not be a single shot authentication in a programmable way. Authentication will be a cognitive process with the ability to being contextual, iterative, interactive, and adaptive. Beyond the programmable techniques today applied in existing authenti-cation systems, employing cognitive computing techniques can result in a huge leap forward for authentication systems. This new era of cognitive computing will bring with it fundamental differences in how authentication systems are built and interact with humans. In the programmable systems era, humans do most of the directing. Hence, their goals serve at best a few designer's thinking. Traditional programmable systems are fed data and their results are based on processing that is preprogramed by humans. In contrast, the cognitive era is about thinking itself—how we gather information, access it and make decisions. Cognitive-based systems build knowledge and learn, understand natural language, and reason and interact more naturally with human beings than traditional programmable systems. The term "reasoning" refers to how cognitive systems demonstrate insights that are very similar to those of humans. Cognitive systems are able to put content into context, providing confidence-weighted responses, with supporting evidence. They are also able to quickly find the proverbial needle in a haystack, identifying new patterns and insights. Over time, cognitive systems will simulate even more closely how the brain actually works.

By design, cognitive systems are probabilistic as they are designed to adapt and make sense of the complexity and unpredictability of huge amount of unstructured information such as text, images, and natural speech. The cognitive systems analyze and interpret that information, organize it and offer explanations for their conclusions. Being probabilistic, they do not offer definitive answers but guide their decisions by weighing in information and ideas from multiple sources, to reason, and then offer hypotheses for consideration. These characteristics of the cognitive systems match perfectly with next generation authentication systems.

Key capabilities of a cognitive inspired authentication system would include [11]:

(a) Deeper human engagement: Authentication of a human requires a fuller engagement with the human. Cognitive systems create more fully human interactions with people—based on the mode, form, and quality each person prefers. The cognitive authentication system takes advantage of the data available today to create a fine-grained picture of individuals including geolocation data, web interactions, transaction Image, history, loyalty program patterns, and data from wearables including signals like heart rate and other sensory data from a mobile device. In addition, adding to that, a cognitive authentication system will have access to tone, sentiment, emotional state, environmental conditions, and the strength and nature of a person's relationships. The fusion of all these measurements and use of them in an authentication scenario would be a new engagement model with the person. By continuously learning, these engagements deliver greater and greater value, and become more natural, anticipatory, and emotionally appropriate.

(b) Scalability and elevate expertise: Many parts of the cognitive trusted identity system relies on human observation for fraud in enrollment and transaction. This information needs to be retained within the system for better overall experience. In addition, as the local geographic regulation change, today there in no quick way to accommodate these requirements. Cognitive authentication systems by nature would help organizations keep pace, serving as a companion for professionals to enhance their performance. This reduces the time required for a professional to become an expert. In addition, because these systems can be taught by leading practitioners the transfer of the learnt concept is possible to broad populations.

(c) New levels of authentication services with cognition: Cognitive algorithms enable new classes of services to sense, reason, and learn about their users and the world around them. This allows for continuous improvement and adaptation, and for augmentation of their capabilities to deliver uses not previously imagined. For example, in authentication systems, the cognitive model will learn about adapting to user nuances to provide sharper performance.

(d) Improving the overall process of authentication with efficient processes: Authentication requirements in business processes infused with cognitive capabilities can exploit the phenomenon of data, from internal and external sources resulting in heightened awareness of workflows, context, and environment, leading to continuous learning, better forecasting, and increased operational effectiveness.

(e) They enhance exploration and discovery: By applying cognitive authentication technologies to vast amounts of data, the systems can uncover patterns, opportunities and actionable hypotheses that would be virtually impossible to discover using traditional research or programmable systems alone.

17.8   Conclusions

In the world of mobile biometrics, we must embrace the emergence of data explosion with ever growing mobile sensors, proliferation of data collection, maturing cloud technologies, commoditized computing resources with big data appliances and broadband network, readily available analytics tools, popular open source frameworks and software, widespread adoption of standards including NIST-ITL, INCITS.

With wide and quick adoption of big data and cloud technologies, the establishment of identity services in the cloud gradually becomes a reality. Consequently, the technology brings paradigm shifts to new solutions compared to conventional vertical, silo, end-to-end expensive biometrics-based solutions. The biometrics identity community as an industry should revisit the initial motive and objective of the biometrics adoption—focusing on identity assertion because making asserted identity as the goal. As an enabling technology, biometrics solution should not mean a race for accuracy, unlimited computing resources for better sensors and more data; we should continue improving authentication and identification performances with the newly available data. Driven by economics and technology trends, future identity solutions will take advantage of the mobility, proliferation of data, advancement of big data infrastructure platform technologies, IT commoditization and all data sources available. However, many challenges remain including security, privacy, and protection of biometric data for cloud-based identity information. With mobile platforms, low quality data become a fact of "life" for system developers and users because we face all challenging factors including low resolution data, out of focus images, unconstrained capture, insufficient contract with color or gray scale.

References

[1]   USEN.PDF N. K. Ratha, J. H. Connell and S. Pankanti, "Big Data approach to biometric-based identity analytics," in IBM Journal of Research

and Development, vol. 59, no. March–May (2/3), pp. 4:1–4:11, 2015. http://public.dhe.ibm.com/common/ssi/ecm/gb/en/gbe03641usen/GBE03641

[2] The Unique Identification Authority of India (UIDAI), https://uidai.gov.in/about-uidai/about-uidai.html

[3] There are quite a few market research reports by industry groups yearly; some are based expert opinions, economic data, oil export values, very few based on actual planning data. Without providing a biased view on the market, I have intentionally not provided any specific report names.

[4] A. A. Ross, K. Nandakumar, and A. Jain. Handbook of Multibiometrics, Springer 2007.

[5] R. V. Yampolskiy and V. Govindaraju, "Behavioral biometric: A survey and classification," Int. J. Biometrics, vol. 1, no. Jun. (1), pp. 81–113, 2008.

[6] C. Avila, J. Casanova, F. Ballesteros, *et al.*, PCAS – State of the art of mobile biometrics, liveness and no-coercion detection (2014).

[7] O. Brdiczka. From documents to tasks: deriving user tasks from document usage patterns. In Proceedings of the 15th international conference on Intelligent user interfaces, pages 285–288, 2010.

[8] O. Brdiczka, M. Langet, J. Maisonnasse, and J. L. Crowley. Detecting human behavior models from multimodal observation in a smart home. IEEE Trams. Autom. Sci. Eng., 6(4):588–597, 2009.

[9] O. Brdiczka, N. M. Su, and J. B. Begole. Temporal task footprinting: identifying routine tasks by their temporal patterns. In Proceedings of the 15th international conference on intelligent user interfaces, pages 281–284, 2010.

[10] The FIDO (Fast IDentity Online) Alliance – https://fidoalliance.org/.

[11] John E Kelly III and Steve Hamm. Smart Machine: IBM's Watson and the Era of Cognitive Computing, Columbia Business School Publishing 2013.

Chapter 18
Outlook for mobile biometrics
Guodong Guo[1] and Harry Wechsler[2]

Mobile biometrics is an emerging research area with some unique capabilities compared to the traditional biometrics. For instance, as discussed in Chapters 6–9 in Part 2 of this book, some unique sensors embedded in smart phones or smart watches can be utilized for mobile identification or verification. Even with the traditional modalities, such as face, iris, and fingerprint, mobile biometrics need to address some special issues, such as power consumption, algorithm complexity, device memory limitations, frequent changes in operational environment, security, durability, reliability, connectivity, and so on.

One aim of this book is to provide some unique views and experience from industrial partners. Their experience in developing mobile sensors or devices, algorithms, and systems could be useful and valuable to academic researchers. The angle from industry to view mobile biometrics may be different from academia, but the rich information and unique experience from industrial experts could inspire academic researchers greatly with some fresh ideas. In this book, there are five chapters from the industry experts, about one-third of the whole book. Progress can also be made faster in industrial mobile biometrics product development, given that important challenges and relevant issues are addressed by the academic researchers in a timely manner.

As an emerging topic, more and more new research work will be seen in the near future on mobile biometrics. We hope this book can serve as a platform to motivate more research efforts from both academia and industry and to promote the research and development on novel mobile biometrics. We would like to see new ideas and products on mobile biometrics, inspired by this book. That would be the greatest reward to us for editing this book.

In addition to the chapters already presented in this book, we would like to raise some further issues for both academic researchers and industrial experts in order to advance the study on mobile biometrics:

1. What is the ideal range of power consumption for mobile biometrics in a typical mobile device? The mobile devices can be smart phones, smart watches, or other portable devices. In consumer mobile devices, power consumption is

[1] Department of Computer Science and Electrical Engineering, West Virginia University, USA
[2] Department of Computer Science, George Mason University, USA

a critical factor to consider for everyday use. Although the battery technology keeps improving, users may run more apps or do more things with their smart phones. Consuming less power can lead to more frequent use of mobile biometrics. In some special applications, such as a continuous user authentication on a mobile device, the power consumption can be a main concern for a day long use. So, a quantitative guidance of power consumption for mobile biometrics can help the research and development of mobile biometrics in various portable devices.

2. How to keep the computational complexity and memory usage low for mobile biometrics? Compared to desktop computers, mobile devices usually have less computational power and memory spaces. Even with significant progresses in hardware development, it is always a concern about the computational complexity and memory usage in developing practical mobile biometric applications. From the algorithmic development viewpoint, the methods developed for mobile biometrics often need to consider the percentage of CPU and memory usage. Sometimes, the developers even have to seek a tradeoff between the algorithm's (recognition) accuracy and complexity.

3. What are the typical and/or special operational environments for mobile biometrics? Different from traditional biometrics that usually operate from simple scenes, such as airports or banks, mobile biometrics may be adopted and used in different scenarios because of their small sizes and convenient moves. In outdoor applications, such as in a snowy day or in the desert, the operating temperatures, illumination conditions, etc. can vary dramatically. The mobile sensors and hardware may need special care for reliable use, and the algorithms need to account for different environment factors for robustness.

4. What security levels can be provided by various mobile biometric traits or modalities? As presented in the chapters, there are many different traits for mobile biometrics, some of which have never been used in traditional biometrics. How to characterize the security levels of these various mobile traits? Some modalities may be appropriate for high security-level applications, whereas some others may not. Further, some modalities are appropriate for long term, continuous authentication, even without the user's awareness, but not all of them are appropriate. It would be valuable to categorize the security levels of various mobile biometric traits for practical adoption and proper application.

5. How reliable is each mobile biometric modality? Reliability is critical for practical use of mobile biometrics. When many different traits are available in a mobile device, are they all reliable? How to evaluate, rank, and characterize the reliability for each mobile biometric modality including apparent image quality? How long it will take to extract a reliable feature for each trait? Can one combine different traits and how to do it in an adaptive fashion is yet another question. For continuous user authentication, what is the optimal time interval to monitor the user continuously?

6. Is there any connectivity issue for each mobile biometric trait or modality? In mobile recognition, especially operating in an identification mode, the captured query data may be matched to a gallery of subjects stored in a remote server

or cloud. In this case, the mobile device needs to have a network connection, such as through the WiFi or satellite signal, to access the remote server or cloud. The network connection might be not reliable or with varying signal strength. If this were to happen, can the mobile biometric system be adapted to different connection situations? Or use different mobile modalities depending on the network connection?

7. How universal are the available mobile biometric modalities? As discussed in this book, many different mobile traits have been available for recognition. Are these mobile traits universal, meaning usable for all subjects? Are there any mobile traits that are not universal, but more appropriate for some populations?

8. Is it more difficult or easier for mobile antispoofing compared to traditional biometrics? Defense against attacks and spoofing is a key component in developing real biometric systems. The mobile biometric systems may be used in various scenarios. For the consumer market, different users may operate mobile biometrics. Because of the mobility and variety of users, can mobile antispoofing be harder to address than that encountered for traditional biometrics? Or could it be feasible to adopt more mobile biometric traits to address the antispoofing problem?

9. Will mobile biometrics be more challenging than the traditional when the same trait is considered? There are some traits that are common to both mobile and traditional biometrics, such as face, iris, and fingerprint. However, the application scenarios might be quite different. For example, a face recognition system in an airport can have static cameras and controlled illumination conditions, whereas in mobile face recognition, the users can orient the mobile sensors in various ways, and the illumination can vary significantly. As a result, the mobile face recognition could be more challenging. With this in mind, the research on mobile face recognition can utilize the progress in traditional face recognition, but also focus on the special challenges of mobile development.

10. How to develop a common basis and database to evaluate various mobile biometrics? For consumer mobile devices, there are a number of platforms and operating systems. The mobile sensors can be very different as well. How to develop common benchmarks to compare and evaluate the mobile biometric systems? The comparisons can involve the same trait, e.g. fingerprint, but on different devices, or different traits, such as voice and gait. Is it possible to capture a database to evaluate different mobile biometric systems? How to measure both the mobile sensors and algorithms? Evaluation can be critical for large scale adoption of mobile biometrics.

11. How to develop multibiometric systems in a mobile device? In a smart phone or smart watch, different sensors are integrated and embedded, which provides an opportunity for developing multibiometric systems. For instance, a regular smart phone can have visual cameras, voice sensor, accelerometers, gyroscopes, fingerprint sensor, etc. All these can naturally equip a smart phone with a multibiometric recognition system. So the research problems are as follows: what traits can be used together, how to use them effectively, and how to make it still convenient for users. It might not be a good idea to use all available traits for

one application, because the users do not like to take a long time and provide too many biometric traits. Depending on the security levels and application scenario, the mobile multibiometric system can be flexible rather than using fixed modalities or setup.

12. In a mobile multibiometric system, suppose different sensors, modalities, and algorithms are available, can it perform a kind of self-management? The self-management can include self-configuration (determining what sensors, modalities, and algorithms to integrate together to form multibiometric authentication or identification; the configuration may be user-adaptive, environment-adaptive, and power-adaptive), self-optimization (based on the power concern, security level concern, and efficacy concern), self-protection (with the vulnerability of some modalities detected during some time period, and void the use of those modalities in creating the multibiometric system; or combining with multi-biometric antispoofing during recognition). The self-management of mobile multibiometric systems is similar to Ganek and Corbi's Autonomic Computing, but with a focus on mobile biometrics.

The issues raised above are some thoughts that both researchers and developers need to address to further the development and use of mobile biometrics. R&D of consequence has to consider throughout the interplay between theory and practice in order to benefit the ever growing reach and scope for real applications. This mobile biometrics book would hopefully catalyze for years to come critical thinking and technology development for mobile and portable biometric systems and products.

Index